BIRKHÄUSER

Progress in Mathematics
Volume 279

For other titles in this series, go to
http://www.springer.com/series/4848

Arithmetic and Geometry Around Quantization

Özgür Ceyhan
Yuri I. Manin
Matilde Marcolli
Editors

Birkhäuser
Boston • Basel • Berlin

Editors
Özgür Ceyhan
MPI für Mathematik
Vivatsgasse 7
53111 Bonn
Germany
ceyhan@mpim-bonn.mpg.de

Yuri I. Manin
MPI für Mathematik
Vivatsgasse 7
53111 Bonn
Germany
manin@mpim-bonn.mpg.de

Matilde Marcolli
California Institute of Technology
Department of Mathematics
1200 E. California Blvd.
Pasadena, CA 91125
USA
matilde@caltech.edu

ISBN 978-0-8176-4830-5 e-ISBN 978-0-8176-4831-2
DOI 10.1007/978-0-8176-4831-2
Springer New York Dordrecht Heidelberg London

Library of Congress Control Number: 2009942422

Mathematics Subject Classification (2000): 53C38, 17B67, 11F27, 18D05, 18E30, 14N35

Printed on acid-free paper

Birkhäuser Boston is part of Springer Science+Business Media (www.birkhauser.com)

Preface

Quantization has been a potent source of interesting ideas and problems in various branches of mathematics. A European Mathematical Society activity, the Arithmetic and Geometry around Quantization (AGAQ) conference has been organized in order to present hot topics in and around quantization to younger mathematicians, and to highlight possible new research directions.

This volume comprises lecture notes, and survey and research articles originating from AGAQ. A wide range of topics related to quantization is covered, thus aiming to give a glimpse of a broad subject in very different perspectives

- Symplectic and algebraic geometry, in particular, mirror symmetry and related topics
  by *S. Akbulut, G. Ben Simon, Ö. Ceyhan, K. Fukaya, S. Salur.*
- Representation theory, in particular quantum groups, the geometric Langlands program and related topics
  by *S. Arkhipov, D. Gaitsgory, E. Frenkel, K. Kremnizer.*
- Quantum ergodicity and related topics
  by *S. Gurevich, R. Hadani.*
- Non-commutative geometry and related topics
  by *S. Mahanta, W. van Suijlekom.*

In their chapter, Akbulut and Salur introduce a new construction of certain 'mirror dual' Calabi–Yau submanifolds inside of a G_2 manifold. The question of constructing central extensions of (2-)groups using (2-)group actions on categories is addressed by Arkhipov and Kremnizer. In his chapter, Ceyhan introduces quantum cohomology and mirror symmetry to real algebraic geometry. Frenkel and Gaitsgory discuss the representation theory of affine Kac–Moody algebras at the critical level and the local geometric Langlands program of the authors. Motivated by constructions of Lagrangian Floer theory, Fukaya explains the fundamental structures such as A_∞-structure and operad structures in Lagrangian Floer theory using an abstract and unifying framework. Ben Simon's chapter provides the first results about the universal covering of the group of quantomorphisms of the prequantization space. In their two chapters, Hadani and Gurevich give a detailed account of self-reducibility of the Weil representation and quantization of symplectic vector spaces over finite fields. These two subjects are the main ingredients of their proof of the Kurlberg–Rudnick Rate Conjecture. The expository chapter of Mahanta is a survey on the construction of motivic rings associated to the category of differential graded categories. Finally, van Suijlekom introduces and discusses Connes–Kreimer type Hopf algebras of Feynman graphs that are relevant for quantum field theories with gauge symmetries.

Yuri Tschinkel
Yuri Zarhin

Contents

Mirror Duality via G_2 and $Spin(7)$ Manifolds

Selman Akbulut* and Sema Salur

Dept. of Mathematics, Michigan State University, East Lansing MI, 48824
akbulut@math.msu.edu
Dept. of Mathematics, University of Rochester, Rochester NY, 14627
salur@math.rochester.edu

Summary. The main purpose of this chapter is to give a construction of certain "mirror dual" Calabi–Yau submanifolds inside of a G_2 manifold. More specifically, we explain how to assign to a G_2 manifold (M, φ, Λ), with the calibration 3-form φ and an oriented 2-plane field Λ, a pair of parametrized tangent bundle valued 2- and 3-forms of M. These forms can then be used to define different complex and symplectic structures on certain 6-dimensional subbundles of $T(M)$. When these bundles are integrated they give mirror CY manifolds. In a similar way, one can define mirror dual G_2 manifolds inside of a $Spin(7)$ manifold (N^8, Ψ). In case N^8 admits an oriented 3-plane field, by iterating this process we obtain Calabi–Yau submanifold pairs in N whose complex and symplectic structures determine each other via the calibration form of the ambient G_2 (or $Spin(7)$) manifold.

AMS Classification Codes (2000). 53C38, 53C29

1 Introduction

Let (M^7, φ) be a G_2 manifold with a calibration 3-form φ. If φ restricts to be the volume form of an oriented 3-dimensional submanifold Y^3, then Y is called an associative submanifold of M. Associative submanifolds are very interesting objects as they behave very similarly to holomorphic curves of Calabi–Yau manifolds.

In [AS], we studied the deformations of associative submanifolds of (M, φ) in order to construct Gromov-Witten-like invariants. One of our main observations was that oriented 2-plane fields on M always exist by a theorem of Thomas [T], and by using them one can split the tangent bundle $T(M) = \mathbf{E} \oplus \mathbf{V}$ as an orthogonal direct sum of an associative 3-plane bundle $\mathbf{E}$ and a complex 4-plane bundle $\mathbf{V}$. This allows us to define "complex associative submanifolds" of M, whose deformation equations may be reduced to the

* partially supported by NSF grant 0805858 DMS 0505638

Ö. Ceyhan et al. (eds.), *Arithmetic and Geometry Around Quantization*,
Progress in Mathematics 279, DOI: 10.1007/978-0-8176-4831-2_1,

Seiberg–Witten equations, and hence we can assign local invariants to them, and assign various invariants to (M, φ, Λ), where Λ is an oriented 2-plane field on M. It turns out that these Seiberg–Witten equations on the submanifolds are restrictions of global equations on M.

In this chapter, we explain how the geometric structures on G_2 manifolds with oriented 2-plane fields (M, φ, Λ) provide complex and symplectic structures to certain 6-dimensional subbundles of $T(M)$. When these bundles are integrated we obtain a pair of Calabi–Yau manifolds whose complex and symplectic structures are remarkably related into each other. We also study examples of Calabi–Yau manifolds which fit nicely in our mirror set-up. Later, we do similar constructions for $Spin(7)$ manifolds with oriented 3-plane fields. We then explain how these structures lead to the definition of "dual G_2 manifolds" in a $Spin(7)$ manifold, with their own dual Calabi–Yau submanifolds.

In the main part of this chapter we give a geometric approach to the "mirror duality" problem. One aspect of this problem is about finding a pair of Calabi–Yau 3-folds whose complex and symplectic structures are related to each other, more specifically their Hodge numbers are related by $h^{p,q} \leftrightarrow h^{3-p,q}$. We propose a method of finding such pairs in 7-dimensional G_2-manifolds. The basic examples are the six-tori $T^6 = T^3 \times T^3 \leftrightarrow T^2 \times T^4$ in the G_2 manifold $T^6 \times S^1$. Also, by this method we pair the non-compact manifolds $\mathcal{O}(-1) \oplus \mathcal{O}(-1)$ and $T^*(S^3)$ as duals to each other. Later in [AES] we show that by applying this method to one of Joyce's G_2 manifolds indeed produces a pair of mirror dual Borcea–Voisin's Calabi–Yau manifolds.

Acknowledgments: We thank R. Bryant and S. Gukov for their valuable comments.

2 Associative and complex distributions in G_2 manifolds

Let us review quickly the basic definitions concerning G_2 manifolds. The main references are the two foundational papers [HL] and [B1], as well as [S], [B2], [BS], and [J]. We also need some properties introduced in [AS]. Now let $\mathbb{O} = \mathbb{H} \oplus l\mathbb{H} = \mathbb{R}^8$ be the octonions, which is an 8-dimensional division algebra generated by $\langle 1, i, j, k, l, li, lj, lk \rangle$, and let $\mathrm{Im}\mathbb{O} = \mathbb{R}^7$ be the imaginary octonions with the cross product operation $\times : \mathbb{R}^7 \times \mathbb{R}^7 \to \mathbb{R}^7$, defined by $u \times v = \mathrm{Im}(\bar{v}.u)$. The exceptional Lie group G_2 is the linear automorphisms of $\mathrm{Im}\mathbb{O}$ preserving this cross product operation, it can also be defined in terms of the orthogonal 3-frames in $\mathbb{R}^7$:

$$G_2 = \{ (u_1, u_2, u_3) \in (\mathrm{Im}\mathbb{O})^3 \mid \langle u_i, u_j \rangle = \delta_{ij},\ \langle u_1 \times u_2, u_3 \rangle = 0 \}.$$

Another useful definition of G_2, which was popularized in [B1], is the subgroup of $GL(7,\mathbb{R})$ which fixes a particular 3-form $\varphi_0 \in \Omega^3(\mathbb{R}^7)$. Denote $e^{ijk} = dx^i \wedge dx^j \wedge dx^k \in \Omega^3(\mathbb{R}^7)$; then

$$G_2 = \{A \in GL(7,\mathbb{R}) \mid A^*\varphi_0 = \varphi_0\}.$$

$$\varphi_0 = e^{123} + e^{145} + e^{167} + e^{246} - e^{257} - e^{347} - e^{356}. \tag{1}$$

Definition 1 *A smooth 7-manifold M^7 has a G_2 structure if its tangent frame bundle reduces to a G_2 bundle. Equivalently, M^7 has a G_2 structure if there is a 3-form $\varphi \in \Omega^3(M)$ such that at each $x \in M$ the pair $(T_x(M), \varphi(x))$ is isomorphic to $(T_0(\mathbb{R}^7), \varphi_0)$ (pointwise condition). We call (M,φ) a manifold with G_2 structure.*

A G_2 structure φ on M^7 gives an orientation $\mu \in \Omega^7(M)$ on M, and μ determines a metric $g = g_\varphi = \langle\ ,\ \rangle$ on M, and a cross product structure $\times$ on the tangent bundle of M as follows: Let $i_v = v\lrcorner$ be the interior product with a vector v, then

$$\langle u, v\rangle = [i_u(\varphi) \wedge i_v(\varphi) \wedge \varphi]/6\mu. \tag{2}$$

$$\varphi(u,v,w) = \langle u \times v, w\rangle. \tag{3}$$

Definition 2 *A manifold with G_2 structure (M,φ) is called a G_2 manifold if the holonomy group of the Levi–Civita connection (of the metric g_φ) lies inside of G_2. Equivalently (M,φ) is a G_2 manifold if φ is parallel with respect to the metric g_φ, that is $\nabla_{g_\varphi}(\varphi) = 0$; which is equivalent to $d\varphi = 0$, $d(*_{g_\varphi}\varphi) = 0$. Also equivalently, at each point $x_0 \in M$ there is a chart $(U, x_0) \to (\mathbb{R}^7, 0)$ on which φ equals to φ_0 up to second order terms, i.e., on the image of U $\varphi(x) = \varphi_0 + O(|x|^2)$.*

Remark 1 *One important class of G_2 manifolds are the ones obtained from Calabi–Yau manifolds. Let (X, ω, Ω) be a complex 3-dimensional Calabi–Yau manifold with Kähler form ω and a nowhere vanishing holomorphic 3-form Ω, then $X^6 \times S^1$ has holonomy group $SU(3) \subset G_2$, hence is a G_2 manifold. In this case $\varphi = \mathrm{Re}\Omega + \omega \wedge dt$. Similarly, $X^6 \times \mathbb{R}$ gives a noncompact G_2 manifold.*

Definition 3 *Let (M,φ) be a G_2 manifold. A 4-dimensional submanifold $X \subset M$ is called* coassociative *if $\varphi|_X = 0$. A 3-dimensional submanifold $Y \subset M$ is called* associative *if $\varphi|_Y \equiv \mathrm{vol}(Y)$; this condition is equivalent to the condition $\chi|_Y \equiv 0$, where $\chi \in \Omega^3(M, TM)$ is the tangent bundle valued 3-form defined by the identity:*

$$\langle \chi(u,v,w), z\rangle = *\varphi(u,v,w,z) \tag{4}$$

The equivalence of these conditions follows from the 'associator equality' [HL]

$$\varphi(u,v,w)^2 + |\chi(u,v,w)|^2/4 = |u \wedge v \wedge w|^2$$

Similar to the definition of χ one can define a tangent bundle 2-form, which is just the cross product of M (nevertheless, viewing it as a 2-form has its advantages).

Definition 4 *Let (M,φ) be a G_2 manifold. Then $\psi \in \Omega^2(M,TM)$ is the tangent bundle valued 2-form defined by the identity:*

$$\langle \psi(u,v), w \rangle = \varphi(u,v,w) = \langle u \times v, w \rangle \tag{5}$$

Now we have two useful properties from [AS]; the first property basically follows from definitions, the second property fortunately applies when the first property fails to give anything useful.

Lemma 1 *([AS]) To any 3-dimensional submanifold $Y^3 \subset (M,\varphi)$, χ assigns a normal vector field, which vanishes when Y is associative.*

Lemma 2 *([AS]) For any associative manifold $Y^3 \subset (M,\varphi)$ with a nonvanishing oriented 2-plane field, χ defines a complex structure on its normal bundle (notice in particular that any coassociative submanifold $X \subset M$ has an almost complex structure if its normal bundle has a nonvanishing section).*

Proof. Let $L \subset \mathbb{R}^7$ be an associative 3-plane, that is $\varphi_0|_L \equiv \mathrm{vol}(L)$. Then for every pair of orthonormal vectors $\{u,v\} \subset L$, the form χ defines a complex structure on the orthogonal 4-plane $L^\perp$, as follows: Define $j : L^\perp \to L^\perp$ by

$$j(X) = \chi(u,v,X) \tag{6}$$

This is well defined, i.e., $j(X) \in L^\perp$, because when $w \in L$ we have:

$$\langle \chi(u,v,X), w \rangle = *\varphi_0(u,v,X,w) = - * \varphi_0(u,v,w,X) = \langle \chi(u,v,w), X \rangle = 0$$

Also $j^2(X) = j(\chi(u,v,X)) = \chi(u,v,\chi(u,v,X)) = -X$. We can check the last equality by taking an orthonormal basis $\{X_j\} \subset L^\perp$ and calculating

$$\begin{aligned}\langle \chi(u,v,\chi(u,v,X_i)), X_j \rangle &= *\varphi_0(u,v,\chi(u,v,X_i), X_j) \\ &= - * \varphi_0(u,v,X_j,\chi(u,v,X_i)) \\ &= -\langle \chi(u,v,X_j), \chi(u,v,X_i) \rangle = -\delta_{ij}\end{aligned}$$

The last equality holds since the map j is orthogonal, and the orthogonality can be seen by polarizing the associator equality, and by noticing $\varphi_0(u,v,X_i) = 0$. Observe that the map j only depends on the oriented 2-plane $\Lambda = \langle u,v \rangle$ generated by $\{u,v\}$ (i.e., it only depends on the complex structure on Λ).

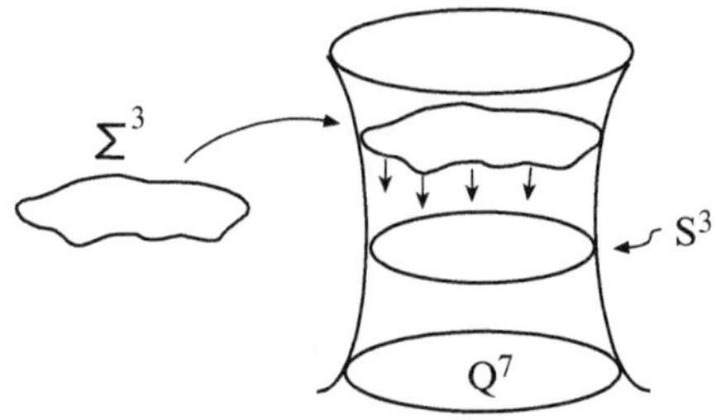

Fig. 1.

Remark 2 *Notice that Lemma 1 gives an interesting flow on the 3-dimensional submanifolds of G_2 manifolds $f : Y \hookrightarrow (M, \varphi)$ (call χ-flow), described by*

$$\frac{\partial}{\partial t} f = \chi(f_* \mathrm{vol}\,(Y))$$

For example, by [BS] the total space of the spinor bundle $Q^7 \to S^3$ (with $\mathbb{C}^2$ fibers) is a G_2 manifold, and the zero section $S^3 \subset Q$ is an associative submanifold. We can imbed any homotopy 3-sphere Σ^3 into Q (homotopic to the zero-section). We conjecture that the χ-flow on $\Sigma \subset Q$, takes Σ diffeomorphically onto the zero section S^3. Note that, since any S^3 smoothly unknots in S^7 it is not possible to produce quick counterexamples by tying local knots.

Finally, we need some identities from [B2] (also see [K]) for (M^7, φ), which follow from local calculations by using definition (1). For $\beta \in \Omega^1(M)$ we have:

$$|\, \varphi \wedge \beta \,|^2 = 4|\beta|^2, \quad \text{and} \quad |\, * \varphi \wedge \beta \,|^2 = 3|\beta|^2, \tag{7}$$

$$(\xi \lrcorner\, \varphi) \wedge \varphi = 2 * (\xi \lrcorner\, \varphi)\ , \quad \text{and} \quad * \; [\, *(\beta \wedge *\varphi) \wedge *\varphi \,] = 3\beta, \tag{8}$$

$$\beta^{\#} \times (\beta^{\#} \times u) = -|\beta|^2 u + \langle \beta^{\#}, u \rangle \beta^{\#}, \tag{9}$$

where $*$ is the star operator, and $\beta^{\#}$ is the vector field dual of β. Let ξ be a vector field on any Riemannian manifold (M, g), and $\xi^{\#} \in \Omega^1(M)$ be its dual 1-form, i.e.,$\xi^{\#}(v) = \langle \xi, v \rangle$. Then for $\alpha \in \Omega^k(M)$:

$$* \, (\xi \lrcorner\, \alpha) = (-1)^{k+1} (\xi^{\#} \wedge *\alpha). \tag{10}$$

3 Mirror duality in G_2 manifolds

On a local chart of a G_2 manifold (M, φ), the form φ coincides with the form $\varphi_0 \in \Omega^3(\mathbb{R}^7)$ up to quadratic terms, we can express the corresponding tangent valued forms χ and ψ in terms of φ_0 in local coordinates. More generally, if $e_1, \ldots, e_7$ is any local orthonormal frame and $e^1, \ldots, e^7$ is the dual frame, from the definitions we get:

$$
\begin{aligned}
\chi = \ & (e^{256} + e^{247} + e^{346} - e^{357})e_1 \\
& + (-e^{156} - e^{147} - e^{345} - e^{367})e_2 \\
& + (e^{157} - e^{146} + e^{245} + e^{267})e_3 \\
& + (e^{127} + e^{136} - e^{235} - e^{567})e_4 \\
& + (e^{126} - e^{137} + e^{234} + e^{467})e_5 \\
& + (-e^{125} - e^{134} - e^{237} - e^{457})e_6 \\
& + (-e^{124} + e^{135} + e^{236} + e^{456})e_7.
\end{aligned}
$$

$$
\begin{aligned}
\psi = \ & (e^{23} + e^{45} + e^{67})e_1 \\
& + (e^{46} - e^{57} - e^{13})e_2 \\
& + (e^{12} - e^{47} - e^{56})e_3 \\
& + (e^{37} - e^{15} - e^{26})e_4 \\
& + (e^{14} + e^{27} + e^{36})e_5 \\
& + (e^{24} - e^{17} - e^{35})e_6 \\
& + (e^{16} - e^{25} - e^{34})e_7.
\end{aligned}
$$

The forms χ and ψ induce complex and symplectic structures on certain subbundles of $T(M)$ as follows: Let ξ be a nonvanishing vector field of M. We can define a symplectic ω_ξ and a complex structure J_ξ on the 6-plane bundle $V_\xi := \xi^\perp$ by

$$\omega_\xi = \langle \psi, \xi \rangle \quad \text{and} \quad J_\xi(X) = X \times \xi. \tag{11}$$

Now we can define

$$\text{Re } \Omega_\xi = \varphi|_{V_\xi} \quad \text{and} \quad \text{Im } \Omega_\xi = \langle \chi, \xi \rangle. \tag{12}$$

In particular $\omega_\xi = \xi \lrcorner\, \varphi$, and Im $\Omega_\xi = \xi \lrcorner\, *\varphi$. Call $\Omega_\xi = \text{Re } \Omega_\xi + i \text{ Im } \Omega_\xi$. The reason for defining these is to pin down a Calabi–Yau like structure on any G_2 manifold. In case $(M, \varphi) = CY \times S^1$ these quantities are related to the ones in Remark 1. Notice that when $\xi \in \mathbf{E}$ then J_ξ is an extension of J of Lemma 2 from the 4-dimensional bundle $\mathbf{V}$ to the 6-dimensional bundle V_ξ.

By choosing different directions, i.e., different ξ, one can find the corresponding complex and symplectic structures. In particular we will get two different complex structures if we choose ξ in the associative subbundle $\mathbf{E}$ (where φ restricts to be 1), or if we choose ξ in the complementary subbundle $\mathbf{V}$, which we will call the coassociative subbundle. Note that φ restricts to zero on the coassociative subbundle.

In local coordinates, it is a straightforward calculation that by choosing $\xi = e_i$ for any i, from equations (11) and (12), we can easily obtain the corresponding structures ω_ξ, J_ξ, Ω_ξ. For example, let us assume that $\{e_1, e_2, e_3\}$ is the local orthonormal basis for the associative bundle $\mathbf{E}$, and $\{e_4, e_5, e_6, e_7\}$ is the local orthonormal basis for the coassociative bundle $\mathbf{V}$. Then if we choose $\xi = e_3 = e_1 \times e_2$ we get $\omega_\xi = e^{12} - e^{47} - e^{56}$ and $\mathrm{Im}\Omega_\xi = e^{157} - e^{146} + e^{245} + e^{267}$. On the other hand, if we choose $\xi = e_7$ then $\omega_\xi = e^{16} - e^{25} - e^{34}$ and $\mathrm{Im}\Omega_\xi = -e^{124} + e^{135} + e^{236} + e^{456}$ which will give various symplectic and complex structures on the bundle V_ξ.

3.1 A useful example

Let us take a Calabi–Yau 6-torus $\mathbb{T}^6 = \mathbb{T}^3 \times \mathbb{T}^3$, where $\{e_1, e_2, e_3\}$ is the basis for one $\mathbb{T}^3$ and $\{e_4, e_5, e_6\}$ is the basis for the other (terms expressed with a slight abuse of notation). We can take the product $M = \mathbb{T}^6 \times S^1$ as the corresponding G_2 manifold with the calibration 3-form $\varphi = e^{123} + e^{145} + e^{167} + e^{246} - e^{257} - e^{347} - e^{356}$, and with the decomposition $T(M) = \mathbf{E} \oplus \mathbf{V}$, where $\mathbf{E} = \{e_1, e_2, e_3\}$ and $\mathbf{V} = \{e_4, e_5, e_6, e_7\}$. Now, if we choose $\xi = e_7$, then $V_\xi = \langle e_1, \ldots, e_6 \rangle$ and the symplectic form is $\omega_\xi = e^{16} - e^{25} - e^{34}$, and the complex structure is

$$J_\xi = \begin{pmatrix} \mathbf{e_1} \mapsto -e_6 \\ \mathbf{e_2} \mapsto \quad e_5 \\ \mathbf{e_3} \mapsto \quad e_4 \end{pmatrix}$$

and the complex valued (3, 0)-form is $\Omega_\xi = (e^1 + ie^6) \wedge (e^2 - ie^5) \wedge (e^3 - ie^4)$; note that this is just $\Omega_\xi = (e^1 - iJ_\xi(e^1)) \wedge (e^2 - iJ_\xi(e^2)) \wedge (e^3 - iJ_\xi(e^3))$.

On the other hand, if we choose $\xi' = e_3$ then $V_{\xi'} = \langle e_1, \ldots, \hat{e}_3, \ldots, e_7 \rangle$ and the symplectic form is $\omega_{\xi'} = e^{12} - e^{47} - e^{56}$ and the complex structure is

$$J_{\xi'} = \begin{pmatrix} \mathbf{e_1} \mapsto -\mathbf{e_2} \\ e_4 \mapsto \quad e_7 \\ e_5 \mapsto \quad e_6 \end{pmatrix}$$

Also, $\Omega_{\xi'} = (e^1 + ie^2) \wedge (e^4 - ie^7) \wedge (e^5 - ie^6)$; as above this can be expressed more tidily as $\Omega_{\xi'} = (e^1 - iJ_{\xi'}(e^1)) \wedge (e^4 - iJ_{\xi'}(e^4)) \wedge (e^5 - iJ_{\xi'}(e^5))$. In the expressions of J's the basis of the associative bundle $\mathbf{E}$ is indicated by bold face letters to indicate the differing complex structures on $\mathbb{T}^6$. To sum up: If we choose ξ from the coassociative bundle $\mathbf{V}$ we get the complex structure which decomposes the 6-torus as $\mathbb{T}^3 \times \mathbb{T}^3$. On the other hand if we choose ξ from the associative bundle $\mathbf{E}$ then the induced complex structure on the 6-torus corresponds to the decomposition as $\mathbb{T}^2 \times \mathbb{T}^4$. This is the phenomenon known as "mirror duality." Here these two $SU(3)$ and $SU(2)$ structures are different but they come from the same φ hence they are dual. These examples suggest the following definition of "mirror duality" in G_2 manifolds:

Definition 5 *Two Calabi–Yau manifolds are mirror pairs of each other, if their complex structures are induced from the same calibration 3-form in a G_2 manifold. Furthermore we call them strong mirror pairs if their normal vector fields ξ and ξ' are homotopic to each other through nonvanishing vector fields.*

Remark 3 *In the above example of $CY \times S^1$, where $CY = \mathbb{T}^6$, the calibration form $\varphi = \mathrm{Re}\Omega + \omega \wedge dt$ gives Lagrangian torus fibration in X_ξ and complex torus fibration in $X_{\xi'}$. They are different manifestations of φ residing on one higher dimensional G_2 manifold M^7. In the next section this correspondence will be made precise.*

In Section 4.2 we will discuss a more general notion of mirror Calabi–Yau manifold pairs, when they sit in different G_2 manifolds, which are themselves mirror duals of each other in a $Spin(7)$ manifold.

3.2 General setting

Let (M^7, φ, Λ) be a manifold with a G_2 structure and a nonvanishing oriented 2-plane field. As suggested in [AS] we can view (M^7, φ) as an analog of a symplectic manifold, and the 2-plane field Λ as an analog of a complex structure taming φ. This is because Λ along with φ gives the associative/complex bundle splitting $T(M) = \mathbf{E}_{\varphi,\Lambda} \oplus \mathbf{V}_{\varphi,\Lambda}$. Now, the next object is a choice of a nonvanishing unit vector field $\xi \in \Omega^0(M, TM)$, which gives a codimension one distribution $V_\xi := \xi^\perp$ on M, which is equipped with the structures $(V_\xi, \omega_\xi, \Omega_\xi, J_\xi)$ as given by (11) and (12).

Let $\xi^\#$ be the dual 1-form of ξ. Let $e_{\xi^\#}$ and $i_\xi = \xi \lrcorner$ denote the exterior and interior product operations on differential forms. Clearly $e_{\xi^\#} \circ i_\xi + i_\xi \circ e_{\xi^\#} = \mathrm{id}$.

$$\varphi = e_{\xi^\#} \circ i_\xi(\varphi) + i_\xi \circ e_{\xi^\#}(\varphi) = \omega_\xi \wedge \xi^\# + Re\ \Omega_\xi. \tag{13}$$

This is just the decomposition of the form φ with respect to $\xi \oplus \xi^\perp$. Recall that the condition that the distribution V_ξ be integrable (the involutive condition which implies $\xi^\perp$ comes from a foliation) is given by

$$d\xi^\# \wedge \xi^\# = 0. \tag{14}$$

Even when V_ξ is not integrable, by [Th] it is homotopic to a foliation. Assume X_ξ is a page of this foliation, and for simplicity assume this 6-dimensional manifold is smooth.

It is clear from the definitions that J_ξ is an almost complex structure on X_ξ. Also, the 2-form ω_ξ is non-degenerate on X_ξ, because from (2) we can write

$$\omega_\xi^3 = (\xi \lrcorner\, \varphi)^3 = \xi \lrcorner\, [\,(\xi \lrcorner\, \varphi) \wedge (\xi \lrcorner\, \varphi) \wedge \varphi\,] = \xi \lrcorner\, (6|\xi|^2 \mu) = 6\mu_\xi \tag{15}$$

where $\mu_\xi = \mu|_{V_\xi}$ is the induced orientation form on V_ξ.

Lemma 3 *J_ξ is compatible with ω_ξ, and it is metric invariant.*

Proof. Let $u, v \in V_\xi$

$$\begin{aligned}\omega_\xi(J_\xi(u), v) &= \omega_\xi(u \times \xi, v) = \langle \psi(u \times \xi, v), \xi \rangle = \varphi(u \times \xi, v, \xi) && \text{by (5)}\\ &= -\varphi(\xi,\ \xi \times u,\ v) = -\langle\ \xi \times (\xi \times u), v\ \rangle && \text{by (3)}\\ &= -\langle\ -|\xi|^2 u + \langle \xi, u \rangle \xi, v\ \rangle = |\xi|^2 \langle u, v \rangle - \langle \xi, u \rangle \langle \xi, v \rangle && \text{by (9)}\\ &= \langle u, v \rangle.\end{aligned}$$

By plugging in $J_\xi(u)$, $J_\xi(v)$ for u, v: We get $\langle J_\xi(u), J_\xi(v) \rangle = -\omega_\xi(u, J_\xi(v)) = \langle u, v \rangle$. □

Lemma 4 *Ω_ξ is a nonvanishing $(3,0)$-form.*

Proof. By a local calculation as in Section 3.1 we see that Ω_ξ is a $(3,0)$ form, and is nonvanishing because $\Omega_\xi \wedge \overline{\Omega}_\xi = 8i \operatorname{vol}(X_\xi)$, i.e.,

$$\begin{aligned}\frac{1}{2i}\, \Omega_\xi \wedge \overline{\Omega}_\xi = Im\ \Omega_\xi \wedge Re\ \Omega_\xi &= (\xi \lrcorner * \varphi) \wedge [\ \xi \lrcorner\, (\xi^\# \wedge \varphi)\ ]\\ &= -\xi \lrcorner\, [\ (\xi \lrcorner * \varphi) \wedge (\xi^\# \wedge \varphi)\ ]\\ &= \xi \lrcorner\, [*(\xi^\# \wedge \varphi) \wedge (\xi^\# \wedge \varphi)\ ] && \text{by (10)}\\ &= |\xi^\# \wedge \varphi|^2\ \xi \lrcorner \operatorname{vol}(M)\\ &= 4|\xi^\#|^2\ (*\xi^\#) = 4 \operatorname{vol}(X_\xi). && \text{by (7)}\end{aligned}$$

□

We can easily calculate $*\mathrm{Re}\ \Omega_\xi = -\mathrm{Im}\ \Omega_\xi \wedge \xi^\#$ and $*\mathrm{Im}\ \Omega_\xi = \mathrm{Re}\ \Omega_\xi \wedge \xi^\#$. In particular if $\star$ is the star operator of X_ξ (so by (15) $\star\omega_\xi = \omega_\xi^2/2$), then

$$\star\, \mathrm{Re}\ \Omega_\xi = \mathrm{Im}\ \Omega_\xi. \tag{16}$$

Notice that ω_ξ is a symplectic structure on X_ξ whenever $d\varphi = 0$ and $\mathcal{L}_\xi(\varphi)|_{V_\xi} = 0$, where $\mathcal{L}_\xi$ denotes the Lie derivative along ξ. This is because $\omega_\xi = \xi \lrcorner\, \varphi$ and

$$d\omega_\xi = \mathcal{L}_\xi(\varphi) - \xi \lrcorner\, d\varphi = \mathcal{L}_\xi(\varphi).$$

Also $d^*\varphi = 0 \implies d^*\omega_\xi = 0$, without any condition on the vector field ξ, since

$$* \varphi = \star\ \omega_\xi - \mathrm{Im}\ \Omega_\xi \wedge \xi^\#, \tag{17}$$

and hence $d(\star\omega_\xi) = d(*\varphi|_{X_\xi}) = 0$. Also $d\varphi = 0 \implies d(\mathrm{Re}\ \Omega_\xi) = d(\varphi|_{X_\xi}) = 0$.

Furthermore, $d^*\varphi = 0$ and $\mathcal{L}_\xi(*\varphi)|_{V_\xi} = 0 \implies d(\mathrm{Im}\ \Omega_\xi) = 0$; this is because $\mathrm{Im}\ \Omega_\xi = \xi \lrcorner\, (*\varphi)$, where $*$ is the star operator on (M, φ). Also, J_ξ is integrable when $d\Omega = 0$ (e.g., [Hi1]). By using the following definition, we can sum up all the conclusions of the above discussion as Theorem 5 below.

Definition 6 *(X^6,ω,Ω,J) is called an almost Calabi–Yau manifold, if X is a Riemannian manifold with a non-degenerate 2-form ω (i.e.,$\omega^3 = 6\mathrm{vol}(X)$) which is co-closed, and J is a metric invariant almost complex structure which is compatible with ω, and Ω is a nonvanishing $(3,0)$ form with $\mathrm{Re}\ \Omega$ closed. Furthermore, when ω and $\mathrm{Im}\ \Omega$ are closed, we call this a Calabi–Yau manifold.*

Theorem 5 *Let (M,φ) be a G_2 manifold, and ξ be a unit vector field which comes from a codimension one foliation on M; then $(X_\xi,\omega_\xi,\Omega_\xi,J_\xi)$ is an almost Calabi–Yau manifold with $\varphi|_{X_\xi} = \mathrm{Re}\ \Omega_\xi$ and $*\varphi|_{X_\xi} = \star\omega_\xi$. Furthermore, if $\mathcal{L}_\xi(\varphi)|_{X_\xi} = 0$ then $d\omega_\xi = 0$, and if $\mathcal{L}_\xi(*\varphi)|_{X_\xi} = 0$ then J_ξ is integrable; when both of these conditions are satisfied then $(X_\xi,\omega_\xi,\Omega_\xi,J_\xi)$ is a Calabi–Yau manifold.*

Remark 4 *If ξ and ξ' are sections of $\mathbf{V}$ and $\mathbf{E}$, respectively, then from [M] the condition $\mathcal{L}_\xi(*\varphi)|_{X_\xi} = 0$ (complex geometry of X_ξ) implies that deforming associative submanifolds of X_ξ along ξ in M keeps them associative; and $\mathcal{L}_{\xi'}(\varphi)|_{X_{\xi'}} = 0$ (symplectic geometry of $X_{\xi'}$) implies that deforming coassociative submanifolds of $X_{\xi'}$ along ξ' in M keeps them coassociative (for an example, see Example 1).*

The idea of inducing an $SU(3)$ structure on a hypersurface of a G_2 manifold goes back to Calabi [Ca] and Gray [G]. Also versions of Theorem 5 appear in [CS] and [C1], where they refer *almost Calabi–Yau's* above as "*half-flat $SU(3)$ manifolds*" (see also the related discussion in [GM] and [Hi2]). Also in [C1] it was shown that, if the hypersurface X_ξ is totally geodesic then it is Calabi–Yau.

Notice that both the complex and symplectic structures of the CY-manifold X_ξ in Theorem 5 are determined by φ when they exist. Recall that (c.f., [V]) elements $\Omega \in H^{3,0}(X_\xi,\mathbb{C})$ along with the topology of X_ξ (i.e., the intersection form of $H^3(X_\xi,\mathbb{Z})$) parametrize complex structures on X_ξ as follows: We compute the third Betti number $b_3(M) = 2h^{2,1} + 2$ since

$$H^3(X_\xi,\mathbb{C}) = H^{3,0} \oplus H^{2,1} \oplus H^{1,2} \oplus H^{0,3} = 2(\mathbb{C} \oplus H^{2,1}).$$

Let $\{A^i, B_j\}$ be a symplectic basis of $H_3(X,\mathbb{Z})$, $i = 1,..,h^{2,1}+1$, then

$$X_i = \int_{A^i} \Omega \tag{18}$$

give complex numbers which are local homogeneous coordinates of the moduli space of complex structures on X_ξ, which is an $h^{2,1}$-dimensional space (there is an extra parameter here since Ω is defined up to scale multiplication).

As we have seen in the example of Section 3.1, the choice of ξ can give rise to quite different complex structures on X_ξ (e.g., $SU(2)$ and $SU(3)$ structures). For example, assume $\xi \in \Omega^0(M,\mathbf{V})$ and $\xi' \in \Omega^0(M,\mathbf{E})$ be unit

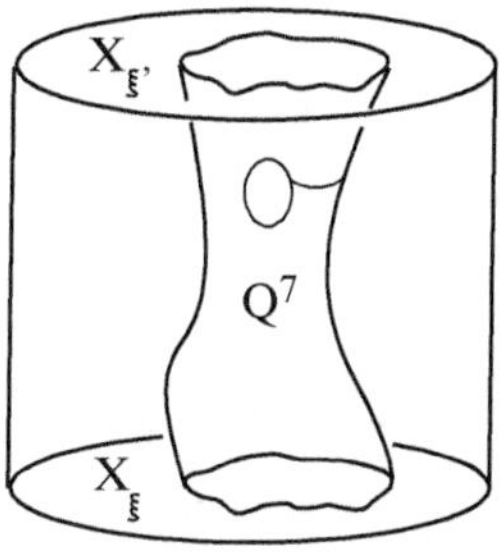

Fig. 2.

vector fields, such that the codimension one plane fields $\xi^{\perp}$ and $\xi'^{\perp}$ come from foliations. Let X_ξ and $X_{\xi'}$ be pages of the corresponding foliations. By our definition X_ξ and $X_{\xi'}$ are mirror duals of each other. Decomposing $T(M) = \mathbf{E}\oplus\mathbf{V}$ gives rise to splittings $TX_\xi = \mathbf{E}\oplus\bar{\mathbf{E}}$, and $TX_{\xi'} = \mathbf{C}\oplus\mathbf{V}$, where $\bar{\mathbf{E}} = \xi^{\perp}(\mathbf{V}) \subset \mathbf{V}$ is a 3-dimensional subbundle, and $\mathbf{C} = (\xi')^{\perp}(\mathbf{E}) \subset \mathbf{E}$ is a 2-dimensional subbundle. Furthermore, $\mathbf{E}$ is Lagrangian in TX_ξ i.e. $J_\xi(\mathbf{E}) = \bar{\mathbf{E}}$, and $\mathbf{C}$, $\mathbf{V}$ are complex in $TX_{\xi'}$ i.e., $J_{\xi'}(\mathbf{C}) = \mathbf{C}$ and $J_{\xi'}(\mathbf{V}) = \mathbf{V}$. Also notice that Re Ω_ξ is a calibration form of $\mathbf{E}$, and ω_ξ is a calibration form of $\mathbf{C}$. In particular, $\langle \Omega_\xi, \mathbf{E}\rangle = 1$ and $\langle \omega_\xi \wedge \xi^{\#}, \mathbf{E}\rangle = 0$; and $\langle \Omega_{\xi'}, \mathbf{E}\rangle = 0$ and $\langle \omega_{\xi'} \wedge (\xi')^{\#}, \mathbf{E}\rangle = 1$.

If X_ξ and $X_{\xi'}$ are strong duals of each other, we can find a homotopy of nonvanishing unit vector fields ξ_t $(0 \leq t \leq 1)$ starting with $\xi \in \mathbf{V}$ and ending with $\xi' \in \mathbf{E}$. This gives a 7-plane distribution $\Xi = \xi_t^{\perp} \oplus \frac{\partial}{\partial t}$ on $M \times [0,1]$ with integral submanifolds $X_\xi \times [0,\epsilon)$ and $X_{\xi'} \times (1-\epsilon, 1]$ on a neighborhood of the boundary. Then by [Th] and [Th1] we can homotop Ξ to a foliation extending the foliation on the boundary (possibly by taking ϵ smaller). Let $Q^7 \subset M \times [0,1]$ be the smooth manifold given by this foliation, with $\partial Q = X_\xi \cup X_{\xi'}$, where $X_\xi \subset M \times \{0\}$ and $X_{\xi'} \subset M \times \{1\}$.

We can define $\Phi \in \Omega^3(M \times [0,1])$ with $\Phi|_{X_\xi} = \Omega_\xi$ and $\Phi|_{X_{\xi'}} = \xi \lrcorner \star \omega_{\xi'}$

$$\Phi = \Phi(\varphi, \Lambda, t) = \langle \omega_{\xi_t} \wedge \xi_t^{\#}, \mathbf{E}\rangle\, \xi_t'' \lrcorner \star \omega_{\xi_t} + \langle \text{Re}\, \Omega_{\xi_t}, \mathbf{E}\rangle\, \Omega_{\xi_t}$$

where $\xi_t'' = J_{\xi\times\xi'}(\xi_t) = \xi_t \times (\xi \times \xi')$ (hence $\xi_0'' = -\xi'$ and $\xi_1'' = \xi$). This can be viewed as a correspondence between the complex structure of X_ξ and the symplectic structure of $X_{\xi'}$. In general, the manifold pairs X_α and X_β (as constructed in Theorem 5) determine each other's almost Calabi -Yau structures via φ provided they are defined.

Proposition 6 *Let $\{\alpha, \beta\}$ be orthonormal vector fields on (M, φ). Then on X_α the following holds* *(i)* Re $\Omega_\alpha = \omega_\beta \wedge \beta^{\#} + \text{Re } \Omega_\beta$

(ii) Im $\Omega_\alpha = \alpha \lrcorner (\star\omega_\beta) - (\alpha \lrcorner \text{Im } \Omega_\beta) \wedge \beta^{\#}$

(iii) $\omega_\alpha = \alpha \lrcorner \text{Re } \Omega_\beta + (\alpha \lrcorner \omega_\beta) \wedge \beta^{\#}$

Proof. Re $\Omega_\alpha = \varphi|_{X_\alpha}$ gives (i). Since Im $\Omega_\alpha = \alpha \lrcorner * \varphi$, the following gives (ii)

$$\begin{aligned}
\alpha \lrcorner (\star\omega_\beta) &= \alpha \lrcorner [\, \beta \lrcorner * (\beta \lrcorner \varphi) \,] \\
&= \alpha \lrcorner \beta \lrcorner (\beta^{\#} \wedge *\varphi) \\
&= \alpha \lrcorner * \varphi + \beta^{\#} \wedge (\alpha \lrcorner \beta \lrcorner * \varphi) \\
&= \alpha \lrcorner * \varphi + (\alpha \lrcorner \text{Im } \Omega_\beta) \wedge \beta^{\#}
\end{aligned}$$

(iii) follows from the following computation:

$$\alpha \lrcorner \text{Re } \Omega_\beta = \alpha \lrcorner \beta \lrcorner (\beta^{\#} \wedge \varphi) = \alpha \lrcorner \varphi + \beta^{\#} \wedge (\alpha \lrcorner \beta \lrcorner \varphi) = \alpha \lrcorner \varphi - (\alpha \lrcorner \omega_\beta) \wedge \beta^{\#}$$

□

Notice that even though the identities of Proposition 6 hold only after restricting the right hand side to X_α, all the individual terms are defined everywhere on (M, φ). Also, from the construction, X_α and X_β inherit vector fields β and α, respectively.

Corollary 7 *Let $\{\alpha, \beta\}$ be orthonormal vector fields on (M, φ). Then there are $A_{\alpha\beta} \in \Omega^3(M)$, and $W_{\alpha\beta} \in \Omega^2(M)$ satisfying*

$$\begin{array}{lllll}
(a) & \varphi|_{X_\alpha} = \text{Re } \Omega_\alpha & \textit{and} & \varphi|_{X_\beta} = \text{Re } \Omega_\beta \\
(b) & A_{\alpha\beta}|_{X_\alpha} = \text{Im } \Omega_\alpha & \textit{and} & A_{\alpha\beta}|_{X_\beta} = \alpha \lrcorner (\star\omega_\beta) \\
(c) & W_{\alpha\beta}|_{X_\alpha} = \omega_\alpha & \textit{and} & W_{\alpha\beta}|_{X_\beta} = \alpha \lrcorner \text{Re } \Omega_\beta
\end{array}$$

For example, when φ varies through metric preserving G_2 structures [B2], (hence fixing the orthogonal frame $\{\xi, \xi'\}$), it induces variations of ω on one side and Ω on the other side.

Remark 5 *By using Proposition 6, in the previous torus example of 3.1 we can show a natural correspondence between the groups $H^{2,1}(X_\xi)$ and $H^{1,1}(X_{\xi'})$. Even though $\mathbb{T}^7$ is a trivial example of a G_2 manifold, it is an important special case, because the G_2 manifolds of Joyce are obtained by smoothing quotients of $\mathbb{T}^7$ by finite group actions. In fact, this process turns the subtori X_ξ's into a pair of Borcea–Voisin manifolds with a similar correspondence of their cohomology groups [AES].*

For the discussion of the previous paragraph to work, we need a nonvanishing vector field ξ in $T(M) = \mathbf{E} \oplus \mathbf{V}$, moving from $\mathbf{V}$ to $\mathbf{E}$. The bundle $\mathbf{E}$ always has a non-zero section, in fact it has a nonvanishing orthonormal3-frame field;

but $\mathbf{V}$ may not have a non-zero section. Nevertheless the bundle $\mathbf{V} \to M$ does have a nonvanishing section in the complement of a 3-manifold $Y \subset M$, which is a transverse self intersection of the zero section. In [AS], Seiberg–Witten equations of such 3-manifolds were related to associative deformations. So we can use these partial sections and, as a consequence, X_ξ and $X_{\xi'}$ may not be closed manifolds. The following is a useful example:

Example 1 *Let X_1, X_2 be two Calabi–Yau manifolds, where X_1 is the cotangent bundle of S^3 and X_2 is the $\mathcal{O}(-1) \oplus \mathcal{O}(-1)$ bundle of S^2. They are proposed to be the mirror duals of each other by physicists (c.f., [Ma]). By using the approach of this paper, we identify them as 6-dimensional submanifolds of a G_2 manifold. Let's choose $M = \Lambda^2_+(S^4)$; this is a G_2 manifold by Bryant–Salamon [BS].*

Let $\pi : \Lambda^2_+(S^4) \to S^4$ be the bundle projection. The sphere bundle of π (which is also $\overline{\mathbb{CP}^3}$) is the so-called twistor bundle, let us denote it by $\pi_1 : Z(S^4) \to S^4$. It is known that the normal bundle of each fiber $\pi_1^{-1}(p) \cong S^2$ in $Z(S^4)$ can be identified with $\mathcal{O}(-1) \oplus \mathcal{O}(-1)$ [S]. Now we take $\mathbf{E}$ to be the bundle of vertical tangent vectors of π, and $\mathbf{V} = \pi^*(TS^4)$, lifted by connection distribution. Let ξ be the pullback of the vector field on S^4 with two zeros (flowing from north pole $\mathbf{n}$ to south pole $\mathbf{s}$), and let ξ' be the radial vector field of $\mathbf{E}$. Clearly $X_\xi = T^*(S^3)$ and $X_{\xi'} = \mathcal{O}(-1) \oplus \mathcal{O}(-1)$.

Note that ξ is nonvanishing in the complement of $\pi^{-1}\{\mathbf{n}, \mathbf{s}\}$, whereas ξ' is nonvanishing in the complement of the zero section of π. Clearly on the set where they are both defined, ξ and ξ' are homotopic through nonvanishing vector fields ξ_t. This would define a cobordism between the complements of the zero sections of the bundles $T^*(S^3)$ and $\mathcal{O}(-1) \oplus \mathcal{O}(-1)$, if the distributions $\xi_t^\perp$ were involutive.

Here the change of complex structures $X_{\xi'} \rightsquigarrow X_\xi$ happens as follows. Let $S^3_\lambda \to S^2$ be the Hopf map with fibers consisting of circles of radius λ; clearly $S^3_\infty = S^2 \times \mathbb{R}$

$$(\mathbb{C}^2 - 0) \times S^2 \to (S^3_\lambda \times \mathbb{R}) \times S^2 \overset{\lambda \to \infty}{\longrightarrow} S^3_\infty \times S^3_\infty$$

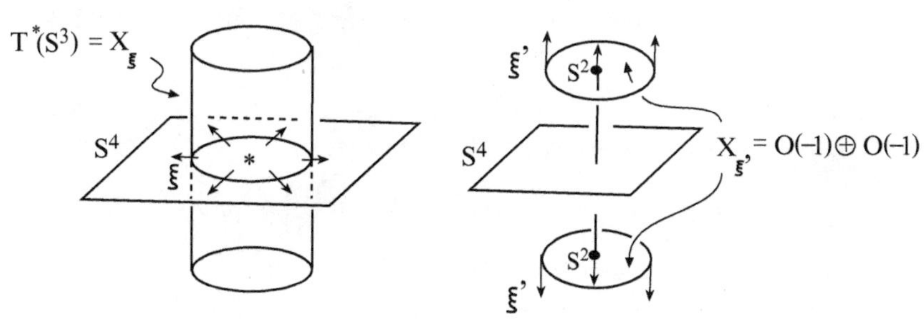

Fig. 3.

where the complex structure on $S^3_\infty \times S^3_\infty$ is the obvious one, induced from exchanging the factors. In general, if we allow the vector fields ξ and ξ' to be homotopic through vector fields ξ_t possibly with zeros, or the family $\xi_t^\perp$ do not remain involutive, the cobordism between X_ξ and $X_{\xi'}$ will have singularities.

Remark 6 *If we apply the construction of Example 1 to the total space of the spinor bundle $Q \to S^3$ (see Remark 2), the two dual 6-manifolds we get are $S^2 \times \mathbb{R}^4$ and $S^3 \times \mathbb{R}^3$.*

There is also a concept of mirror-dual G_2 manifolds in a $Spin(7)$ manifold, hence we can talk about mirror dual CY manifolds coming from two different mirror dual G_2 submanifolds of a $Spin(7)$ manifold. This is the subject of the next section.

4 Mirror duality in *Spin*(7) manifolds

Similar to the Calabi–Yau case there is a notion of mirror duality between G_2 manifolds [Ac], [AV], [GYZ], [SV]. In this section we will give a definition of mirror G_2 pairs, and an example which shows that associative and coassociative geometries in mirror G_2 pairs are induced from the same calibration 4-form in a $Spin(7)$ manifold, and hence these geometries are dual to each other. Let us first recall the basic definitions and properties of $Spin(7)$ geometries. The main references in this subject are [HL] and [Ti].

Definition 7 *An 8-dimensional Riemannian manifold (N, Ψ) is called a* $Spin(7)$ manifold *if the holonomy group of its metric connection lies in $Spin(7) \subset GL(8)$.*

Equivalently, a $Spin(7)$ manifold is an 8-dimensional Riemannian manifold with a triple cross product $\times$ on its tangent bundle, and a closed 4-form $\Psi \in \Omega^4(N)$ with

$$\Psi(u, v, w, z) = \langle u \times v \times w, z \rangle.$$

Definition 8 *A 4-dimensional submanifold X of a $Spin(7)$ manifold (N, Ψ) is called* Cayley *if $\Psi|_X \equiv \mathrm{vol}(X)$.*

Analogous to the G_2 case, we introduce a tangent bundle valued 3-form, which is just the triple cross product of N.

Definition 9 *Let (N, Ψ) be a $Spin(7)$ manifold. Then $\Upsilon \in \Omega^3(N, TN)$ is the tangent bundle valued 3-form defined by the identity:*

$$\langle \Upsilon(u, v, w), z \rangle = \Psi(u, v, w, z) = \langle u \times v \times w, z \rangle.$$

$Spin(7)$ manifolds can be constructed from G_2 manifolds. Let (M, φ) be a G_2 manifold with a 3-form φ, then $M \times S^1$ (or $M \times \mathbb{R}$) has holonomy group $G_2 \subset Spin(7)$, hence is a $Spin(7)$ manifold. In this case $\Psi = \varphi \wedge dt + *_7\varphi$, where $*_7$ is the star operator of M^7.

Now we will repeat a similar construction for a $Spin(7)$ manifold (N, Ψ), which we did for G_2 manifolds. Here we make an assumption that $T(M)$ admits a nonvanishing 3-frame field $\Lambda = \langle u, v, w \rangle$; then we decompose $T(M) = \mathbf{K} \oplus \mathbf{D}$, where $\mathbf{K} = \langle u, v, w, u \times v \times w \rangle$ is the bundle of Cayley 4-planes (where Ψ restricts to be 1) and $\mathbf{D}$ is the complementary subbundle (note that this is also a bundle of Cayley 4-planes since the form Ψ is self dual). In the G_2 case, existence of an analogous decomposition of the tangent bundle followed from [T] (in this case we can just restrict to a submanifold which a 3-frame field exists). On a chart in N let $e_1, \ldots, e_8$ be an orthonormal frame and $e^1, \ldots, e^8$ be the dual coframe, then the calibration 4-form is given as (c.f., [HL])

$$\begin{aligned} \Psi = \; & e^{1234} + (e^{12} - e^{34}) \wedge (e^{56} - e^{78}) \\ & + (e^{13} + e^{24}) \wedge (e^{57} + e^{68}) \\ & + (e^{14} - e^{23}) \wedge (e^{58} - e^{67}) + e^{5678} \end{aligned} \tag{19}$$

which is a self-dual 4-form, and the corresponding tangent bundle valued 3-form is

$$\begin{aligned} \Upsilon = \; & (e^{234} + e^{256} - e^{278} + e^{357} + e^{368} + e^{458} - e^{467})e_1 \\ & + (-e^{134} - e^{156} + e^{178} + e^{457} + e^{468} - e^{358} + e^{367})e_2 \\ & + (e^{124} - e^{456} + e^{478} - e^{157} - e^{168} + e^{258} - e^{267})e_3 \\ & + (-e^{123} + e^{356} - e^{378} - e^{257} - e^{268} - e^{158} + e^{167})e_4 \\ & + (e^{126} - e^{346} + e^{137} + e^{247} + e^{148} - e^{238} + e^{678})e_5 \\ & + (-e^{125} + e^{345} + e^{138} + e^{248} - e^{147} + e^{237} - e^{578})e_6 \\ & + (-e^{128} + e^{348} - e^{135} - e^{245} + e^{146} - e^{236} + e^{568})e_7 \\ & + (e^{127} - e^{347} - e^{136} - e^{246} - e^{145} + e^{235} - e^{567})e_8. \end{aligned}$$

This time we show that the form Υ induces G_2 structures on certain subbundles of $T(N)$. Let γ be a nowhere vanishing vector field of N. We define a G_2 structure φ_γ on the 7-plane bundle $V_\gamma := \gamma^\perp$ by (where $*_8$ is the star operator on N^8)

$$\varphi_\gamma := \langle \Upsilon, \gamma \rangle = \gamma \lrcorner \Psi = *_8(\Psi \wedge \gamma^\#). \tag{20}$$

Assuming that V_γ comes from a foliation, we let M_γ be an integral submanifold of V_γ. We have $d\varphi_\gamma = 0$, provided $\mathcal{L}_\gamma(\Psi)|_{V_\gamma} = 0$. On the other hand, we always have $d(\star\varphi_\gamma) = 0$ on M_γ. To see this, we use

$$\Psi = \varphi_\gamma \wedge \gamma^\# + *_7\varphi_\gamma$$

where $*_7$ is the star operator on $M\gamma$, and use $d\Psi = 0$ and the foliation condition $d\gamma^\# \wedge \gamma^\# = 0$, and the identity $\theta|_{M_\gamma} = \gamma \lrcorner \, [\, \theta \wedge \gamma^\# \,]$ for forms θ. In order to state the next theorem, we need a definition:

Definition 10 *A manifold with G_2 structure (M, φ) is called an almost G_2-manifold if φ is co-closed.*

Theorem 8 *Let (N^8, Ψ) be a Spin(7) manifold, and γ be a unit vector field which comes from a foliation, then $(M_\gamma, \varphi_\gamma)$ is an almost G_2 manifold. Furthermore, if $\mathcal{L}_\gamma(\Psi)|_{M_\gamma} = 0$ then $(M_\gamma, \varphi_\gamma)$ is a G_2 manifold.*

Proof. Follows by checking Definition 1 and by the discussion above. □

Theorem 8 was previously proven in [C2], where it was also shown that if the hypersurface M_γ is totally geodesic then it is a G_2 manifold (also see [G] and [IC] for related discussion). The following theorem says that the induced G_2 structures on M_α, M_β determine each other via Ψ; more specifically φ_α and φ_β are restrictions of a global 3-form of N.

Proposition 9 *Let (N, Ψ) be a Spin(7) manifold, and $\{\alpha, \beta\}$ be an orthonormal vector fields on N. Then the following holds on M_α*

$$\varphi_\alpha = -\alpha \lrcorner \, (\varphi_\beta \wedge \beta^\# + *_7\, \varphi_\beta)$$

Proof. The proof follows from the definitions, and by expressing φ_α and φ_β in terms of $\beta^\#$ and $\alpha^\#$ by the formula (13). □

As in the G_2 case, by choosing different γs, one can find various different G_2 manifolds M_γ with interesting structures. Most interestingly, we will get certain "dual" M_γs by choosing γ in $\mathbf{K}$ or in $\mathbf{D}$. This will shed light on more general version of mirror symmetry of Calabi–Yau manifolds. First we will discuss an example.

4.1 An example

Let $\mathbb{T}^8 = \mathbb{T}^4 \times \mathbb{T}^4$ be the $Spin(7)$ 8-torus, where $\{e_1, e_2, e_3, e_4\}$ is the basis for the Cayley $\mathbb{T}^4$ and $\{e_5, e_6, e_7, e_8\}$ is the basis for the complementary $\mathbb{T}^4$. We can take the corresponding calibration 4-form (20) above, and take the decomposition $T(N) = \mathbf{K} \oplus \mathbf{D}$, where $\{e_1, e_2, e_3, e_4\}$ is the orthonormal basis for the Cayley bundle $\mathbf{K}$, and $\{e_5, e_6, e_7, e_8\}$ is the local orthonormal basis for the complementary bundle $\mathbf{D}$. Then if we choose $\gamma = e_4 = e_1 \times e_2 \times e_3$ we get

$$\varphi_\gamma = -e^{123} + e^{356} - e^{378} - e^{257} - e^{268} - e^{158} + e^{167}$$

On the other hand, if we choose $\gamma' = e_5$ then we get

$$\varphi_{\gamma'} = e^{126} - e^{346} + e^{137} + e^{247} + e^{148} - e^{238} + e^{678}$$

which give different G_2 structures on the 7 toris M_γ and $M_{\gamma'}$.

Note that if we choose γ from the Cayley bundle $\mathbf{K}$, we get the G_2 structure on the 7-torus M_γ which reduces the Cayley 4-torus $\mathbb{T}^4 = \mathbb{T}^3 \times S^1$ (where γ is tangent to S^1 direction) to an associative 3-torus $\mathbb{T}^3 \subset M_\gamma$ with respect to this G_2 structure. On the other hand if we choose γ' from the complementary bundle **D**, then the Cayley 4-torus $\mathbb{T}^4$ will be a coassociative submanifold of the 7-torus $M_{\gamma'}$with the corresponding G_2 structure. Hence associative and coassociative geometries are dual to each other as they are induced from the same calibration 4-form Ψ on a $Spin(7)$ manifold. This suggests the following definition of the "mirror duality" for G_2 manifolds.

Definition 11 *Two 7-manifolds with G_2 structures are mirror pairs, if their G_2-structures are induced from the same calibration 4-form in a $Spin(7)$ manifold. Furthermore, they are strong duals if their normal vector fields are homotopic.*

Remark 7 *For example, by [BS] the total space of an $\mathbb{R}^4$ bundle over S^4 has a $Spin(7)$ structure. By applying the process of Example 1, we obtain mirror pairs M_γ and $M_{\gamma'}$ to be $S^3 \times \mathbb{R}^4$ and $\mathbb{R}^4 \times S^3$ with dual G_2 structures.*

4.2 Dual Calabi–Yau's inside of $Spin(7)$

Let (N^8, Ψ) be a $Spin(7)$ manifold, and let $\{\alpha, \beta\}$ be an orthonormal 2-frame field in N, each coming from a foliation. Let $(M_\alpha, \varphi_\alpha)$ and (M_β, φ_β) be the G_2 manifolds given by Theorem 8. Similarly to Theorem 5, the vector fields β in M_α, and α in M_β give almost Calabi–Yau's $X_{\alpha\beta} \subset M_\alpha$ and $X_{\beta\alpha} \subset M_\beta$. Let us denote $X_{\alpha\beta} = (X_{\alpha\beta}, \omega_{\alpha\beta}, \Omega_{\alpha\beta}, J_{\alpha\beta})$ likewise $X_{\beta\alpha} = (X_{\beta\alpha}, \omega_{\beta\alpha}, \Omega_{\beta\alpha}, J_{\beta,\alpha})$. Then we have

Proposition 10 *The following relations hold:*

(i) $J_{\alpha\beta}(u) = u \times \beta \times \alpha$

(ii) $\omega_{\alpha\beta} = \beta \lrcorner\, \alpha \lrcorner\, \Psi$

(iii) $\mathrm{Re}\, \Omega_{\alpha\beta} = \alpha \lrcorner\, \Psi|_{X_{\alpha\beta}}$

(iv) $\mathrm{Im}\, \Omega_{\alpha\beta} = \beta \lrcorner\, \Psi|_{X_{\alpha\beta}}$

Proof. (i), (ii), (iii) follow from definitions, and from $X \times Y = (X \lrcorner\, Y \lrcorner\, \varphi)^{\#}$.

$$\begin{aligned}
\mathrm{Im}\, \Omega_{\alpha\beta} &= \beta \lrcorner *_7 \varphi_\alpha = \beta \lrcorner *_7 (\alpha \lrcorner\, \Psi) \\
&= \beta \lrcorner\, [\alpha \lrcorner\, *_8 (\alpha \lrcorner\, \Psi)] \\
&= \beta \lrcorner\, [\alpha \lrcorner\, (\alpha^{\#} \wedge \Psi)] \qquad \text{by (10)} \\
&= \beta \lrcorner\, [\Psi - \alpha^{\#} \wedge (\alpha \lrcorner\, \Psi)] \\
&= \beta \lrcorner\, \Psi - \alpha^{\#} \wedge (\beta \lrcorner\, \alpha \lrcorner\, \psi)].
\end{aligned}$$

The left-hand side of the equation is already defined on $X_{\alpha\beta}$, by restricting to $X_{\alpha\beta}$ we get (iii). □

Corollary 11 *When $X_{\alpha,\beta}$ and $X_{\beta,\alpha}$ coincide, they are oppositely oriented manifolds and $\omega_{\alpha,\beta} = -\omega_{\beta,\alpha}$, and* Re $\Omega_{\alpha\beta} = -$Im $\Omega_{\beta\alpha}$ *(as forms on $X_{\alpha\beta}$).*

Now let $\{\alpha, \beta, \gamma\}$ be an orthonormal 3-frame field in (N^8, Ψ), and $(M_\alpha, \varphi_\alpha)$, (M_β, φ_β), and $(M_\gamma, \varphi_\gamma)$ be the corresponding almost G_2 manifolds. As before, the orthonormal vector fields $\{\gamma, \beta\}$ in M_α and $\{\gamma, \alpha\}$ in M_β give rise to corresponding almost Calabi–Yau's $X_{\alpha,\gamma}$, $X_{\alpha,\beta}$ in M_α, and $X_{\beta,\gamma}$, $X_{\beta,\alpha}$ in M_β.

In this way (N^8, Ψ) give rise to 4 Calabi–Yau descendants. By Corollary 11, $X_{\alpha\beta}$ and $X_{\beta\alpha}$ are different geometrically; they may not even be the same as smooth manifolds, but for simplicity we may consider it to be the same smooth manifold obtained from the triple intersection of the three G_2 manifolds.

In case we have a decomposition $T(N) = \mathbf{K} \oplus \mathbf{D}$ of the tangent bundle of (N^8, Ψ) by Cayley plus its orthogonal bundles (Section 4); we can choose our frame to be special and obtain interesting CY-manifolds. For example, if we choose $\alpha \in \Omega^0(M, \mathbf{K})$ and $\beta, \gamma \in \Omega^0(N, \mathbf{D})$ we get one set of complex structures, whose types are indicated by the first row of the following diagram. On the other hand, if we choose all $\{\alpha, \beta, \gamma\}$ to lie entirely in $\mathbf{K}$ or $\mathbf{D}$ we get another set of complex structures, as indicated by the second row of the diagram.

$$\begin{array}{ccccccc}
 & & & (N^8, \Psi) & & & \\
 & & \swarrow & & \searrow & & \\
 & (M_\alpha, \varphi_\alpha) & & & & (M_\beta, \varphi_\beta) & \\
\swarrow & & \searrow & & \swarrow & & \searrow \\
\mathbf{X}_{\alpha\gamma} & & X_{\alpha\beta} & & \mathbf{X}_{\beta\gamma} & & X_{\beta\alpha} \\
SU(3) & & SU(3) & & SU(2) & & SU(3) \\
SU(2) & & SU(2) & & SU(2) & & SU(2)
\end{array}$$

Here all the corresponding symplectic and the holomorphic forms of the resulting Calabi–Yau's come from restriction of global forms induced by Ψ. The following gives relations between the complex/symplectic structures of these induced CY-manifolds; i.e.,the structures $X_{\alpha\gamma}$, $X_{\beta\gamma}$ and $X_{\alpha\beta}$ satisfy a certain triality relation.

Proposition 12 *We have the following relations;*

(i) Re $\Omega_{\alpha\gamma} = \alpha \lrcorner (*_6\, \omega_{\beta\gamma}) + \omega_{\alpha\beta} \wedge \beta^{\#}$

(ii) Im $\Omega_{\alpha\gamma} = \omega_{\beta\gamma} \wedge \beta^{\#} - \gamma \lrcorner *_6\, (\omega_{\alpha\beta})$

(iii) $\omega_{\alpha\gamma} = \alpha \lrcorner$ Im $\Omega_{\beta\gamma} + (\gamma \lrcorner\, \omega_{\alpha\beta}) \wedge \beta^{\#}$

First we need to prove a lemma;

Lemma 13 *The following relations hold;*

$$\begin{aligned}
\alpha\lrcorner *_6(\omega_{\beta\gamma}) &= \alpha\lrcorner\Psi + \gamma^\# \wedge (\alpha\lrcorner\gamma\lrcorner\Psi) + \beta^\# \wedge (\alpha\lrcorner\beta\lrcorner\Psi) \\
&\quad -\gamma^\# \wedge \beta^\# \wedge (\alpha\lrcorner\gamma\lrcorner\beta\lrcorner\Psi). \\
\mathrm{Im}\,\Omega_{\beta\gamma} &= -\gamma\lrcorner\Psi - \beta^\# \wedge (\gamma\lrcorner\beta\lrcorner\Psi). \\
\mathrm{Re}\,\Omega_{\alpha\gamma} &= \alpha\lrcorner\Psi - \gamma^\# \wedge (\gamma\lrcorner\alpha\lrcorner\Psi).
\end{aligned}$$

Proof.

$$\begin{aligned}
\alpha\lrcorner *_6(\omega_{\beta\gamma}) &= \alpha\lrcorner[\gamma\lrcorner\beta\lrcorner *_8 (\gamma\lrcorner\beta\lrcorner\Psi)] \\
&= -\alpha\lrcorner\gamma\lrcorner\beta\lrcorner(\gamma^\# \wedge \beta^\# \wedge \Psi) \\
&= -\alpha\lrcorner\gamma\lrcorner[-\gamma^\# \wedge \Psi + \gamma^\# \wedge \beta^\# \wedge (\beta\lrcorner\Psi)] \\
&= \alpha\lrcorner\Psi + \gamma^\# \wedge (\alpha\lrcorner\gamma\lrcorner\Psi) + \beta^\# \wedge (\alpha\lrcorner\beta\lrcorner\Psi) \\
&\quad -\gamma^\# \wedge \beta^\# \wedge (\alpha\lrcorner\gamma\lrcorner\beta\lrcorner\Psi).
\end{aligned}$$

$$\begin{aligned}
\mathrm{Im}\,\Omega_{\beta\gamma} &= \gamma\lrcorner *_7 (\beta\lrcorner\Psi) \\
&= \gamma\lrcorner[\beta\lrcorner *_8 (\beta\lrcorner\Psi)] \\
&= -\gamma\lrcorner\beta\lrcorner(\beta^\# \wedge \Psi) \\
&= -\gamma\lrcorner\Psi - \beta^\# \wedge (\gamma\lrcorner\beta\lrcorner\Psi).
\end{aligned}$$

$$\mathrm{Re}\,\Omega_{\alpha\gamma} = (\alpha\lrcorner\Psi)|_{X_{\alpha\gamma}} = \gamma\lrcorner[\gamma^\# \wedge (\alpha\lrcorner\Psi)] = \alpha\lrcorner\Psi - \gamma^\# \wedge (\gamma\lrcorner\alpha\lrcorner\Psi).$$

□

Proof of Proposition 12. We calculate the following by using Lemma 13:

$$\begin{aligned}
\alpha\lrcorner *_6(\omega_{\beta\gamma}) + \omega_{\alpha\beta} \wedge \beta^\# &= \alpha\lrcorner\Psi + \gamma^\# \wedge (\alpha\lrcorner\gamma\lrcorner\Psi) + \beta^\# \wedge (\alpha\lrcorner\beta\lrcorner\Psi) \\
&\quad -\gamma^\# \wedge \beta^\# \wedge (\alpha\lrcorner\gamma\lrcorner\beta\lrcorner\Psi) + (\beta\lrcorner\alpha\lrcorner\Psi) \wedge \beta^\# \\
&= \alpha\lrcorner\Psi - \gamma^\# \wedge (\gamma\lrcorner\alpha\lrcorner\Psi) - \gamma^\# \wedge \beta^\# \wedge (\alpha\lrcorner\gamma\lrcorner\beta\lrcorner\Psi).
\end{aligned}$$

Since we are restricting to $X_{\alpha\gamma}$ we can throw away terms containing $\gamma^\#$ and get (i). We prove (ii) similarly:

$$\begin{aligned}
\mathrm{Im}\,\Omega_{\alpha\gamma} &= (\gamma\lrcorner *_7 \varphi_\alpha) = \gamma\lrcorner[\,\alpha\lrcorner *_8 \varphi_\alpha\,] \\
&= \gamma\lrcorner[\alpha\lrcorner *_8 (\alpha\lrcorner\Psi)] = -\gamma\lrcorner[\alpha\lrcorner(\alpha^\# \wedge \Psi)) \\
&= -\gamma\lrcorner\Psi + \gamma\lrcorner[\alpha^\# \wedge (\alpha\lrcorner\Psi)] \\
&= -\gamma\lrcorner\Psi - \alpha^\# \wedge (\gamma\lrcorner\alpha\lrcorner\Psi].
\end{aligned}$$

$$\begin{aligned}
\omega_{\beta\gamma} \wedge \beta^\# - \gamma\lrcorner(*_6\,\omega_{\alpha\beta}) &= -\gamma\lrcorner(*_6\,\omega_{\alpha\beta}) + (\gamma\lrcorner\beta\lrcorner\Psi) \wedge \beta^\# \\
&= -\gamma\lrcorner\Psi - \beta^\# \wedge (\gamma\lrcorner\beta\lrcorner\Psi) - \alpha^\# \wedge (\gamma\lrcorner\alpha\lrcorner\Psi) \\
&\quad +\beta^\# \wedge \alpha^\# \wedge (\gamma\lrcorner\beta\lrcorner\alpha\lrcorner\Psi) + (\gamma\lrcorner\beta\lrcorner\Psi) \wedge \beta^\#.
\end{aligned}$$

Here, we used Lemma 13 with different indices $(\alpha, \beta, \gamma) \mapsto (\gamma, \alpha, \beta)$, and since we are restricting to $X_{\alpha\gamma}$ we threw away terms containing $\alpha^{\#}$. Finally, (iii) follows by plugging in Lemma 13 to definitions. □

The following says that the Calabi–Yau structures of $X_{\alpha\gamma}$ and $X_{\beta\gamma}$ determine each other via Ψ. Proposition 14 is basically a consequence of Proposition 6 and Corollary 11.

Proposition 14 *We have the following relations*

(i) $\text{Re } \Omega_{\alpha\gamma} = \alpha \lrcorner (*_6 \, \omega_{\beta\gamma}) - (\alpha \lrcorner \text{ Re } \Omega_{\beta\gamma}) \wedge \beta^{\#}$

(ii) $\text{Im } \Omega_{\alpha\gamma} = \omega_{\beta\gamma} \wedge \beta^{\#} + Im \ \Omega_{\beta\gamma}$

(iii) $\omega_{\alpha\gamma} = \alpha \lrcorner \text{ Im } \Omega_{\beta\gamma} + (\alpha \lrcorner \, \omega_{\beta\gamma}) \wedge \beta^{\#}$

Proof. All follow from the definitions and Lemma 11 (and by ignoring $\alpha^{\#}$ terms). □

References

[Ac] B. Acharya, *On Mirror Symmetry for Manifolds of Exceptional Holonomy*, Nucl.Phys. B524 (1998) 269–282, hep-th/9707186.

[AV] M. Aganagic and C. Vafa, *G_2 Manifolds, Mirror Symmetry and Geometric Engineering*, hep-th/0110171.

[AS] S. Akbulut and S. Salur, *Deformations in G_2 manifolds*, Adv. Math. 217 (2008), no. 5, 2130–2140.

[AES] S. Akbulut, B. Efe and S. Salur, *Mirror Duality in a Joyce Manifold*, Adv. Math, 9 September 2009, math.GT/0707.1512.

[ASa] V. Apostolov and S. Salamon, *Kahler Reduction of metrics with holonomy G_2*, math.DG/0303197.

[AW] M. Atiyah and E. Witten, *M-theory dynamics on a manifold of G_2 holonomy*, Adv. Theor. Math. Phys. 6 (2003), 1–106.

[B1] R.L. Bryant, *Metrics with exceptional holonomy*, Ann. of Math 126 (1987), 525–576.

[B2] ———, *Some remarks on G_2-structures*, Proceedings of Gökova Geometry-Topology Conference 2005, 75–109, Gökova Geometry/Topology Conference (GGT), Gökova, 2006.

[B3] ———, *Submanifolds and special structures on the Octonions*, J. Differential Geom. 17 (1982), no. 2, 185–232.

[BS] R.L. Bryant and M.S. Salamon, *On the construction of some complete metrics with exceptional holonomy*, Duke Math. Jour., vol. 58, no 3 (1989), 829–850.

[C1] F.M. Cabrera, *SU(3)-structures on hypersurfaces of manifolds with G_2-structures*, Monatsh. Math. 148 (2006), no. 1, 29–50.

[C2] ———, *Orientable hypersurfaces of Riemannian manifolds with Spin(7)-structure*, Acta Math. Hungar. 76 (1997), no. 3, 235–247.

[Ca] E. Calabi, *Construction and properties of some 6-dimensional almost complex manifolds*, Trans. Amer. Math. Soc. 87 (1958), 407–438.

[CS] S. Chiossi and S. Salamon, *The intrinsic torsion of $SU(3)$ and G_2-structures*, math.DG/0202282.

[G] A. Gray, *Vector cross product on manifolds*, Trans. Amer. Math. Soc. 141 (1969), 463–504, Correction 148 (1970), 625.

[GM] S. Gurrieri and A. Micu *Type IIB Theory on half-flat manifolds*, Classical Quantum Gravity 20 (2003), no. 11, 2181–2192.

[GYZ] S. Gukov, S.T. Yau and E. Zaslow, *Duality and Fibrations on G_2 Manifolds*, Turkish J. Math. 27 (2003), no. 1, 61–97.

[Hi1] N.J. Hitchin, *The moduli space of special Lagrangian submanifolds* , Ann. Scuola Norm. Sup. Pisa Cl. Sci. (4) 25 (1997), no. 3-4, 503–515 (1998).

[Hi2] ———, *Stable forms and special metrics*, Global differential geometry: the mathematical legacy of Alfred Gray (Bilbao, 2000), 70–89, Contemp. Math., 288, Amer. Math. Soc., Providence, RI, 2001.

[HL] F.R. Harvey and H.B. Lawson, *Calibrated geometries*, Acta. Math. 148 (1982), 47–157.

[IC] S. Ivanov and M. Cabrera *$SU(3)$-structures on submanifolds of a Spin(7)-manifold*, Differential Geom. Appl. 26 (2008), no. 2, 113–132.

[J] D.D. Joyce, *Compact Manifolds with special holonomy*, OUP, Oxford, 2000.

[K] S. Karigiannis, *Deformations of G_2 and $Spin(7)$ structures on manifolds*, Canad. J. Math. 57 (2005), no. 5, 1012–1055.

[M] R.C. McLean, *Deformations of calibrated submanifolds*, Comm. Anal. Geom. 6 (1998), 705–747.

[Ma] M. Marino, *Enumerative geometry and knot invariants*, Infinite dimensional groups and manifolds, 27–92, IRMA Lect. Math. Theor. Phys., 5, de Gruyter, Berlin, 2004. hep-th/0210145.

[S] S. Salamon, *Riemannian geometry and holonomy groups*, Pitman Res. Notes in Math. Series, no. 201.

[SV] S.L. Shatashvili and C. Vafa *Superstrings and Manifolds of Exceptional Holonomy*, Selecta Math. 1 (1995) 347–381.

[T] E. Thomas, *Postnikov invariants and higher order cohomology operations*, Ann. Math. (2) 85, (1967) 184–217.

[Th] W. Thurston, *Existence of codimension-one foliations*, Ann. Math. (2) 104, (1976) 249–268.

[Th1] W. Thurston, *Private communication.*

[Ti] G. Tian, *Gauge Theory and Calibrated Geometry I*, Ann. Math. 151, (2000), 193–268.

[V] C. Vafa, *Simmons Lectures (Harvard).*

2-Gerbes and 2-Tate Spaces

Sergey Arkhipov[1] and Kobi Kremnizer[2]

[1] Department of Mathematics, University of Toronto, Toronto, Ontario, Canada
hippie@math.toronto.edu
[2] Mathematical Institute, University of Oxford, Oxford, UK
kremnitzer@maths.ox.ac.uk

Summary. We construct a central extension of the group of automorphisms of a 2-Tate vector space viewed as a discrete 2-group. This is done using an action of this 2-group on a 2-gerbe of gerbal theories. This central extension is used to define central extensions of double loop groups.

AMS Subject Codes: 18D05, 22E67

1 Introduction

In this chapter we study the question of constructing central extensions of groups using group actions on categories.

Let G be a group. The basic observation is that the category of $\mathbb{G}_m$ central extensions of G is equivalent to the category of $\mathbb{G}_m$-gerbes over the classifying stack of G. This is in turn equivalent to the category of $\mathbb{G}_m$-gerbes over a point with an action of G. Thus by producing categories with a G action we get central extensions.

We then take this observation one category theoretic level higher. We want to study central extensions of 2-groups. Here a 2-group is a monoidal groupoid such that its set of connected components is a group with the induced product. We look at the case of a discrete 2-group, that is, we can think of any group G as a 2-group with objects the elements of the group, morphisms the identities, and monoidal structure the product.

We see that $\mathbb{G}_m$-central extensions of a discrete 2-group are the same as 2-gerbes over the classifying stack of the group. This also can be interpreted as a 2-gerbe with G-action. Thus to get extensions as a 2-group we should find 2-categories with G-action.

These observations are used to define central extensions of automorphism groups of 1-Tate spaces and discrete automorphism 2-groups of 2-Tate spaces.

The category of n-Tate spaces is defined inductively. 0-Tate spaces are finite dimensional vector spaces. $(n+1)$-Tate spaces are certain indpro objects of the category of n Tate spaces. To a 1-Tate space we can associate a 1-gerbe

Ö. Ceyhan et al. (eds.), *Arithmetic and Geometry Around Quantization*,
Progress in Mathematics 279, DOI: 10.1007/978-0-8176-4831-2_2,

of determinant theories. This 1-gerbe has a natural action of the automorphism group of the 1-Tate space. This gives the central extension of the group.

Similarly, to a 2-Tate space we can associate a 2-gerbe of gerbal theories with an action of the automorphism group of the 2-Tate space. This action gives a central extension of the discrete 2-group.

If G is a finite dimensional reductive group and V is a finite dimensional representation we get an embedding of the formal double loop group $G((s))((t))$ into the automorphism group of the 2-tate space $V((s))((t))$. Thus we can restrict the central extension to the double loop group. These central extensions of the double loop group as a 2-group will be used in the future to study the (2-)representation theory of these groups and relating it to the 2-dimensional Langlands program.

The idea of constructing the higher central extension in categorical terms belongs essentially to Michael Kapranov. S.A would like to thank him for sharing the idea in 2004.

After writing this chapter we found out that a similar result was obtained by Osipov in his unpublished Preprint. S.A. would like to thank Osipov for sharing the manuscript with him.

K.K. was supported in part by NSF grant DMS-0602007. S.A. was supported in part by NSERC.

2 Group actions on gerbes and central extensions

2.1 $\mathbb{G}_m$-gerbes and central extensions

Let's recall the notion of a group acting on a category.

Definition 1 *An action of a group G on a category C consists of a functor $F_g : C \to C$ for each $g \in G$ and a natural transformation $\tau_{g,h} : F_{gh} \to F_g F_h$ s.t.*

$$\begin{array}{ccc} F_{g_1 g_2 g_3} & \xrightarrow{\tau_{g_1, g_2 g_3}} & F_{g_1} F_{g_2 g_3} \\ \downarrow \tau_{g_1 g_2, g_3} & & \downarrow F_{g_1}(\tau_{g_2, g_3}) \\ F_{g_1 g_2} F_{g_3} & \xrightarrow{\tau_{g_1, g_2} F_{g_3}} & F_{g_1} F_{g_2} F_{g_3} \end{array} \qquad (1)$$

commutes for any $g_1, g_2, g_3 \in G$.

We also require that $F_1 = Id$ *and that* $\tau_{1,g} = Id$ *and* $\tau_{g,1} = Id$.

Suppose that C is a $\mathbb{G}_m$-gerbe (over a point). By this we mean that:

- C is a groupoid.
- C is connected (there exists an arrow between any two objects)
- For any object A of C, $Aut(A) \simeq \mathbb{G}_m$ in a coherent way.

Note that this implies that all the Hom spaces are $\mathbb{G}_m$-torsors.

Remark If C and D are 1-gerbes then their product $C \times D$ is also a 1-gerbe. This will be used below.

In this case we have the following theorem [5]:

Theorem 1 *Let G act on a $\mathbb{G}_m$-gerbe C. For each object A of C we get a $\mathbb{G}_m$-central extension $\widetilde{G}_A$. These central extensions depend functorially on A (hence are all isomorphic). If there exists an equivariant object this extension splits.*

Proof: Let $A \in obC$. Define

$$\widetilde{G}_A = \{(g, \phi) : g \in G, \phi \in Hom(F_g(A), A)\} \tag{2}$$

with product given by

$$(g_1, \phi_1)(g_2, \phi_2) = (g_1 g_2, \phi_1 \circ F_{g_1}(\phi_2)) \tag{3}$$

Associativity follows from Definition 1.

Another way of interpreting this theorem is as follows. An action of G on a gerbe C over a point is the same (by descent) as a gerbe over $\mathbb{BG}$. By taking the cover

$$\begin{array}{c} pt \\ \Big\downarrow \\ \mathbb{BG} \end{array} \tag{4}$$

we get that such a gerbe gives (again by descent) a $\mathbb{G}_m$-torsor L over G with an isomorphism

$$p_1^*(L) \otimes p_2^*(L) \to m^*(L) \tag{5}$$

Hence we get

Theorem 2 *The category of $\mathbb{G}_m$-central extensions of G is equivalent to the category of $\mathbb{G}_m$-gerbes over $\mathbb{BG}$.*

Remark 1 *In the above discussion we have used the notion of a gerbe over $\mathbb{BG}$. For this we could either use the theory of gerbes over stacks or treat $\mathbb{BG}$ as a simplicial object and pt as the universal fibration $\mathbb{EG}$. The same remark would apply later when we talk about 2-gerbes over $\mathbb{BG}$.*

2.2 Central extension of the automorphism group of a 1-Tate space

Let $\mathcal{V}$ be a 1-Tate space. Recall (or see section 4) that this is an ind-pro object in the category of finite dimensional vector spaces, thus equivalent to $\mathcal{V}$ having a locally linearly compact topology. Any such is isomorphic to $V((t))$ (formal loops into V) but non-canonically. Recall also the notion of a lattice $\mathcal{L} \subseteq \mathcal{V}$ (pro-subspace or linearly compact subspace) and that if $\mathcal{L}_1 \subseteq \mathcal{L}_2$ are two lattices then $\mathcal{L}_2/\mathcal{L}_1$ is finite dimensional.

Definition 2 *A determinant theory is a rule that assigns to each lattice $\mathcal{L}$ a one-dimensional vector space $\Delta_{\mathcal{L}}$ and to each pair $\mathcal{L}_1 \subset \mathcal{L}_2$ an isomorphism*

$$\Delta_{\mathcal{L}_1\mathcal{L}_2} : \Delta_{\mathcal{L}_1} \otimes Det(\mathcal{L}_2/\mathcal{L}_1) \to \Delta_{\mathcal{L}_2} \tag{6}$$

such that for each triple $\mathcal{L}_1 \subset \mathcal{L}_2 \subset \mathcal{L}_3$ the following diagram commutes

$$\begin{array}{ccc} \Delta_{\mathcal{L}_1} \otimes Det(\mathcal{L}_2/\mathcal{L}_1) \otimes Det(\mathcal{L}_3/\mathcal{L}_2) & \longrightarrow & \Delta_{\mathcal{L}_1} \otimes Det(\mathcal{L}_3/\mathcal{L}_1) \\ \downarrow & & \downarrow \\ \Delta_{\mathcal{L}_2} \otimes Det(\mathcal{L}_3/\mathcal{L}_2) & \longrightarrow & \Delta_{\mathcal{L}_3} \end{array} \tag{7}$$

We have the obvious notion of a morphism between two determinant theories and it is easy to see that the category of determinant theories is in fact a $\mathbb{G}_m$-gerbe.

Let $GL(\mathcal{V})$ be the group of continuous automorphisms of $\mathcal{V}$. This group acts on the gerbe of determinant theories and hence we get using Theorem 1 a central extension $\widetilde{GL(\mathcal{V})}_\Delta$ for each choice of determinant theory Δ. Unless $\mathcal{V}$ itself is a lattice, this central extension does not split.

3 Group actions on 2-gerbes and central extensions of 2-groups

3.1 2-Groups

Definition 3 *A 2-group is a monoidal groupoid C s.t. its set of connected components $\pi_0(C)$ with the induced multiplication is a group.*

The basic example is the discrete 2-group associated with any group G: the set of objects is G itself and the only morphisms are the identities. The monoidal structure comes from the group multiplication. We will denote this discrete 2-group by $\mathcal{G}$.

Note that 2-groups can be defined in any category with products (or better in any topos) so we have topological, differential, and algebraic 2-groups.

One can define a general notion of extensions of 2-groups but we are only interested in the following case:

Definition 4 *Let G be a group (in a topos) and A an abelian group (again in the topos). A central extension $\widetilde{\mathcal{G}}$ of the discrete 2-group associated to G by A is a 2-group s.t.:*

- $\pi_0(\widetilde{\mathcal{G}}) \simeq G$
- $\pi_1(\widetilde{\mathcal{G}}, I) \simeq A$

Here I is the identity object for the monoidal structure and π_1 means the automorphism group of the identity object.

3.2 Action of a group on a bicategory

Lets recall first the notion of a bicategory (one of the versions of a lax 2-category) [3].

Definition 5 *A bicategory $\mathcal{C}$ is given by:*

- *Objects $A, B, ..$*
- *Categories $\mathcal{C}(A,B)$ (whose objects are called 1-arrows and morphisms are called 2-arrows)*
- *Composition functors $\mathcal{C}(A,B) \times \mathcal{C}(B,C) \longrightarrow \mathcal{C}(A,C)$*
- *Natural transformations (associativity constraints)*

$$\begin{array}{ccc} \mathcal{C}(A,B)\times\mathcal{C}(B,C)\times\mathcal{C}(C,D) & \longrightarrow & \mathcal{C}(A,B)\times\mathcal{C}(B,D) \\ \downarrow & \swarrow & \downarrow \\ \mathcal{C}(A,C)\times\mathcal{C}(C,D) & \longrightarrow & \mathcal{C}(A,D) \end{array} \tag{8}$$

This data should satisfy coherence axioms of the MacLane pentagon form.

Remark As a bicategory with one object is the same as a monoidal category the coherence axioms should become clear (though lengthy to write).

Definition 6 *Let $\mathcal{C}$ and $\mathcal{D}$ be two bicategories. A functor $\mathcal{F} : \mathcal{C} \to \mathcal{D}$ consists of:*

- *For each object $A \in Ob(\mathcal{C})$ an object $\mathcal{F}(A) \in Ob(\mathcal{D})$*
- *A functor $\mathcal{F}_{AB} : \mathcal{C}(A,B) \to \mathcal{D}(\mathcal{F}(A), \mathcal{F}(B))$ for any two objects $A, B \in Ob\mathcal{C}$*
- *A natural transformation*

$$\begin{array}{ccc} \mathcal{C}(A,B)\times\mathcal{C}(B,C) & \longrightarrow & \mathcal{C}(A,C) \\ \Big\downarrow{\scriptstyle \mathcal{F}_{AB}\times\mathcal{F}_{BC}} & \swarrow\!\!\!\swarrow & \Big\downarrow{\scriptstyle \mathcal{F}_{AC}} \\ \mathcal{D}(\mathcal{F}(A),\mathcal{F}(B))\times\mathcal{D}(\mathcal{F}(B),\mathcal{F}(C)) & \longrightarrow & \mathcal{D}(\mathcal{F}(A),\mathcal{F}(C)) \end{array} \tag{9}$$

This natural transformation should be compatible with the associativity constraints.

Again the comparison with monoidal categories should make it clear what are the compatibilities.

Definition 7 *Let $\mathcal{F}$ and $\mathcal{G}$ be two functors between $\mathcal{C}$ and $\mathcal{D}$. A natural transformation (Ξ, ξ) is given by:*

- *A functor $\Xi_{AB} : \mathcal{D}(\mathcal{F}(A), \mathcal{F}(B)) \to \mathcal{D}(\mathcal{G}(A), \mathcal{G}(B))$ for each pair of objects*
- *A natural transformation*

$$\begin{array}{ccc} \mathcal{C}(A,B) & \xrightarrow{\mathcal{F}_{AB}} & \mathcal{D}(\mathcal{F}(A),\mathcal{F}(B)) \\ & {\scriptstyle \mathcal{G}_{AB}}\searrow \quad \overset{\xi_{AB}}{\swarrow\!\!\!\swarrow} & \Big\downarrow{\scriptstyle \Xi_{AB}} \\ & & \mathcal{D}(\mathcal{G}(A),\mathcal{G}(B)) \end{array} \tag{10}$$

These should be compatible with the structures.

Definition 8 *Given two natural transformations $(\Xi^1, \xi^1), (\Xi^2, \xi^2) : \mathcal{F} \to \mathcal{G}$ a modification is a natural transformation $\phi_{AB} : \Xi^1_{AB} \to \Xi^2_{AB}$ such that*

$$\begin{array}{ccc} \Xi^1_{AB}\mathcal{F}_{AB} & \overset{\xi_{AB}}{\Longrightarrow} & \mathcal{G}_{AB} \\ {\scriptstyle \phi_{AB}\mathcal{F}_{AB}}\Big\Downarrow & & \Big\Downarrow{\scriptstyle Id} \\ \Xi^2_{AB}\mathcal{F}_{AB} & \overset{\xi_{AB}}{\Longrightarrow} & \mathcal{G}_{AB} \end{array} \tag{11}$$

commutes for all A and B and is compatible with all the structures.

Now we can define an action of a group on a bicategory:

Definition 9 *Let G be a group and $\mathcal{C}$ a bicategory. An action of G on $\mathcal{C}$ is given by a functor $\mathcal{F}_g : \mathcal{C} \to \mathcal{C}$ for each $g \in G$ and a natural transformation $(\Xi, \xi)_{g,h} : \mathcal{F}_{gh} \to \mathcal{F}_g \mathcal{F}_h$ such that there exists a modification*

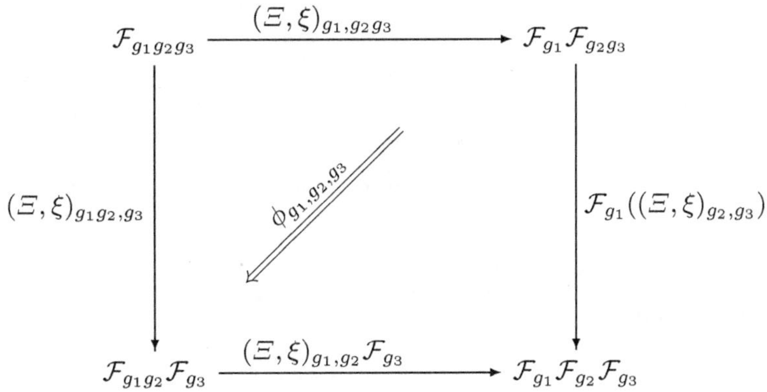

(12)

for any $g_1, g_2, g_3 \in G$ satisfying a cocycle condition.

3.3 2-gerbes and central extensions of 2-groups

Let A be an abelian group.

Definition 10 *A 2-gerbe (over a point) with band A is a bicategory $\mathcal{C}$ such that:*

- *It is a 2-groupoid: every 1-arrow is invertible up to a 2-arrow and all 2-arrows are invertible.*
- *It is connected: there exists a 1-arrow between any two objects and a 2-arrow between any two 1-arrows.*
- *The automorphism group of any 1-arrow is isomorphic to A.*

In other words all the categories $\mathcal{C}(A, B)$ are 1-gerbes with band A and the product maps are maps of 1-gerbes.

Theorem 3 *Suppose G acts on a 2-gerbe $\mathcal{C}$ with band A. To this we can associate a central extension $\widetilde{\mathcal{G}}$ of the discrete 2-group associated to G by A.*

The construction is the same as in 1 (with more diagrams to check). A better way of presenting the construction is using descent: a 2-gerbe with an action of G is the same as a 2-gerbe over $\mathbb{BG}$ (we haven't defined 2-gerbes in general. See [4]). Using the same cover as before $pt \to \mathbb{BG}$ we get a gerbe over G which is multiplicative. That means that we are given an isomorphism

$$p_1^*(\mathcal{F}) \otimes p_2^*(\mathcal{F}) \to m^*(\mathcal{F}) \tag{13}$$

satisfying a cocycle condition on the threefold product (here $m : G \times G \to G$ is the multiplication). This gerbe gives in turn an A-torsor over $G \times G$ giving the Hom-spaces of the 2-group and the multiplicative structure gives the monoidal structure.

This construction also works in the other direction. Suppose we have a central extension of the discrete 2-group $\mathcal{G}$ associated to the group G by the abelian group A. Then the Hom spaces define an A-torsor $\mathcal{HOM}$ over $G \times G$ and the existence of composition means that over $G \times G \times G$ we are given an isomorphism:

$$p_{12}^*(\mathcal{HOM}) \otimes p_{23}^*(\mathcal{HOM}) \to p_{13}^*(\mathcal{HOM}) \tag{14}$$

satisfying a cocycle condition over the fourfold product (associativity). Here p_{ij} are the projections. Thus we have a gerbe over G with band A. Let's denote this gerbe by $\mathcal{F}$.

The existence of the monoidal structure implies that we are given an isomorphism over $G \times G$

$$p_1^*(\mathcal{F}) \otimes p_2^*(\mathcal{F}) \to m^*(\mathcal{F}) \tag{15}$$

satisfying a cocycle condition on the threefold product. Hence the gerbe is multiplicative. In other words we have:

Lemma 1 *A central extension of the discrete 2-group associated to G by A is the the same as a 2-gerbe over $\mathbb{BG}$ with band A.*

Actually also here we have an equivalence of categories.

Remark 2 *Today's technology ([8])enables one to define n-gerbes with nice descent theory. So we can generalize the whole discussion to:*

Theorem 4 *The category of n-gerbes with band A and with action of G is equivalent to that of central extensions by A of the discrete n-group associated to G.*

This will be done in another paper.

4 2-Tate spaces and 2-groups

In this section we introduce the notion of a locally compact object introduced by Beilinson and Kato [2, 7].

4.1 Locally compact objects in a category

Definition 11 *Let C be a category. The category of locally compact objects of C is the full subcategory of $Ind(Pro(C))$ consisting of functors that are isomorphic to diagrams of the following sort: Let I, J be linearly directed orders.*

Let $F : I^{op} \times J \to C$ be a diagram such that for all $i, i' \in I$ and $j, j' \in J$ $i \leq i'$ and $j \leq j'$ the diagram:

$$\begin{array}{ccc} F(i',j) & \longrightarrow & F(i',j') \\ \downarrow & & \downarrow \\ F(i,j) & \longrightarrow & F(i',j) \end{array} \tag{16}$$

is both cartesian and cocartesian and vertical arrows are surjections and horizontal arrows are injections. A compact object is a locally compact object isomorphic to one which is constant in the Ind direction.

The following statement follows easily from set-theory and the Yoneda lemma:

Lemma 2 *If F is locally compact then the functors $\underset{\leftarrow}{lim}\,\underset{\rightarrow}{lim}\, F$ and $\underset{\rightarrow}{lim}\,\underset{\leftarrow}{lim}\, F$ are naturally isomorphic.*

From now on we will assume that the indexing sets I, J are countable.

Suppose C is an exact category. Say a sequence of locally compact objects is exact if it can be represented by a map of diagrams $F_1 \to F_2 \to F_3 : I^{op} \times J \to C$ where all the arrows are exact in C. A routine check shows :

Lemma 3 *The category of locally compact objects of C is exact.*

Remark 3 *Note that if C is Abelian (and nontrivial) the category of locally compact objects is not Abelian.*

Using the standard reindexing trick (Appendix of [1]) we also get

Lemma 4 *Let $F_1 \to F_2$ be an admissible injection (w.r.t. the exact structure) of compact objects then $coker(F_1 \to F_2)$ is also a compact object.*

Lemma 5 *Let F_1 and F_2 be two admissible compact subobjects of F, then $F_1 \times_F F_2$ is also compact.*

Now we can define inductively n-Tate spaces (we still assume that the indexing sets are countable):

Definition 12 *A 0-Tate space is a finite dimensional vector space. Suppose we have defined the category of n-Tate spaces. A $n+1$-Tate space is a locally compact object of n-Tate spaces. A lattice of an $(n+1)$-Tate space is an admissible compact subobject.*

Note that any 2-Tate space is of the form $\mathcal{V}((t))$ where $\mathcal{V}$ is a 1-Tate space. An example of a lattice in this case is $\mathcal{V}[[t]]$.

4.2 Some facts on 1-Tate spaces

We have from the previous section that:

Lemma 6 *The category of 1-Tate spaces is an exact category with injections set-theoretic injections and surjections dense morphisms.*

Recall also the notion of the determinant grebe associated to a Tate space $\mathcal{V}$. From now on we will denote it by $\mathcal{D}_{\mathcal{V}}$.

Lemma 7 *Let*

$$0 \to \mathcal{V}' \to \mathcal{V} \to \mathcal{V}'' \to 0 \tag{17}$$

be an admissible exact sequence of Tate spaces. Then we have an equivalence of $\mathbb{G}_m$-gerbes

$$\mathcal{D}_{\mathcal{V}'} \otimes \mathcal{D}_{\mathcal{V}''} \to \mathcal{D}_{\mathcal{V}} \tag{18}$$

such that if $\mathcal{V}_1 \subset \mathcal{V}_2 \subset \mathcal{V}_3$ then we have a natural transformation

$$\begin{array}{ccc} \mathcal{D}_{\mathcal{V}_1} \otimes \mathcal{D}_{\mathcal{V}_2/\mathcal{V}_1} \otimes \mathcal{D}_{\mathcal{V}_3/\mathcal{V}_2} & \longrightarrow & \mathcal{D}_{\mathcal{V}_1} \otimes \mathcal{D}_{\mathcal{V}_3/\mathcal{V}_1} \\ \downarrow & \swarrow\!\!\!\!\swarrow & \downarrow \\ \mathcal{D}_{\mathcal{V}_2} \otimes \mathcal{D}_{\mathcal{V}_3/\mathcal{V}_2} & \longrightarrow & \mathcal{D}_{\mathcal{V}_3} \end{array} \tag{19}$$

and if we have $\mathcal{V}_1 \subset \mathcal{V}_2 \subset V_3 \subset \mathcal{V}_4$ then the cubical diagram of natural transformations commutes.

4.3 2-Tate spaces and gerbal theories

It follows from the previous discussion that:

Lemma 8 *Let $\mathbb{V}$ be a 2-Tate space.*

1. *If $\mathbb{L}' \subset \mathbb{L}$ are two lattices then $\mathbb{L}/\mathbb{L}'$ is a 1-Tate space.*
2. *For any two lattices $\mathbb{L}$ and $\mathbb{L}'$ there exists a third lattice $\mathbb{L}'' \subset \mathbb{L} \cap \mathbb{L}'$.*

Now we can define a gerbal theory.

Definition 13 *Let $\mathbb{V}$ be a 2-Tate space. A gerbal theory $\mathbb{D}$ is*

- *For each lattice $\mathbb{L} \subset \mathbb{V}$ a $\mathbb{G}_m$-gerbe $\mathbb{D}_{\mathbb{L}}$*
- *If $\mathbb{L}' \subset \mathbb{L}$ are two lattices then we have an equivalence*

$$\mathbb{D}_{\mathbb{L}} \xrightarrow{\phi_{\mathbb{L}\mathbb{L}'}} \mathbb{D}_{\mathbb{L}'} \otimes \mathcal{D}_{\mathbb{L}/\mathbb{L}'} \tag{20}$$

- *For $\mathcal{V}_1 \subset \mathcal{V}_2 \subset \mathcal{V}_3$ we have a natural transformation*

$$\begin{array}{ccc} \mathbb{D}_{\mathcal{V}_1} \otimes \mathcal{D}_{\mathcal{V}_2/\mathcal{V}_1} \otimes \mathcal{D}_{\mathcal{V}_3/\mathcal{V}_2} & \longrightarrow & \mathbb{D}_{\mathcal{V}_1} \otimes \mathcal{D}_{\mathcal{V}_3/\mathcal{V}_1} \\ \downarrow & \swarrow\!\!\!\!\swarrow & \downarrow \\ \mathbb{D}_{\mathcal{V}_2} \otimes \mathcal{D}_{\mathcal{V}_3/\mathcal{V}_2} & \longrightarrow & \mathbb{D}_{\mathcal{V}_3} \end{array} \tag{21}$$

Given $\mathcal{V}_1 \subset \mathcal{V}_2 \subset \mathcal{V}_3$ these natural transformations should commute on a cubical diagram.

Now we have

Theorem 5 *gerbal theories on a given 2-Tate space $\mathbb{V}$ form a $\mathbb{G}_m$ 2-gerbe $\mathbb{GERB}_{\mathbb{V}}$.*

Let's denote $\mathbb{GL}(\mathbb{V})$ the group of continuous automorphisms of a 2-Tate space $\mathbb{V}$. This group acts naturally on the 2-gerbe $\mathbb{GERB}_{\mathbb{V}}$. *Remark* the action is actually a strict one. We get:

Theorem 6 *Let $\mathbb{V}$ be a 2-Tate space. Given a lattice $\mathbb{L} \subset \mathbb{V}$ we get a $\mathbb{G}_m$ central extension of the discrete 2-group associated to $\mathbb{GL}(\mathbb{V})$.*

Remark 4 *Using Theorem 4 we can go on and define central extensions of discrete n-groups of automorphism of n-Tate spaces.*

Application: central extension of a double loop group

Let G be a finite dimensional reductive group over a field. Let V be a finite dimensional representation of G. From this data we get a map

$$G((s))((t)) \to \mathbb{GL}(V((s))((t))) \tag{22}$$

where $G((s))((t))$ is the formal double loop group of G. From this embedding we get a central extension of the discrete 2-group $G((s))((t))$.

A variant

There is another way to think about $\mathbb{G}_m$-gerbes.

Definition 14 *Let Pic be the symmetric monoidal groupoid of 1-dimensional vector spaces. A $\mathbb{G}_m$-gerbe is a module category over this monoidal category equivalent to Pic as module categories (where Pic acts on itself by the monoidal structure).*

This definition is equivalent to the definition given before. Now, following Drinfeld [6] we define a graded version of a $\mathbb{G}_m$-gerbe.

Definition 15 *Let $Pic^{\mathbb{Z}}$ be the symmetric monoidal groupoid of $\mathbb{Z}$-graded 1-dimensional vector spaces with the super-commutativity constraint ($a \otimes b \to (-1)^{deg(a)deg(b)} b \otimes a$). A $\mathbb{Z}$-graded $\mathbb{G}_m$-gerbe is a module category over $Pic^{\mathbb{Z}}$ equivalent to it as module categories.*

We have a map from $Pic^{\mathbb{Z}}$ to the discrete 2-group $\mathbb{Z}$ which sends a 1-dimensional graded vector space to its degree. This map induces a functor between graded $\mathbb{G}_m$-gerbes and $\mathbb{Z}$-torsors. We can now repeat the entire story with $\mathbb{Z}$-graded gerbes. For instance, instead of a determinant theory we will get a graded determinant theory. The $\mathbb{Z}$-torsor corresponding to it will be the well known dimension torsor of dimension theories. A dimension theory for a 1-Tate space is a rule of associating an integer to each lattice satisfying conditions similar to those of a determinant theory.

In this way we will get for a 2-Tate space an action of $\mathbb{GL}(\mathbb{V})$ on the $\mathbb{G}_m$-gerbe of dimension torsors. This action will give us a central extension of the group $\mathbb{GL}(\mathbb{V})$ (not the 2-group!). And similarly we can get central extensions of groups of the form $G((s))((t))$. Thus we see that if we work with graded determinant theory we get a central extension of the discrete 2-group $\mathbb{GL}(\mathbb{V})$ which induces the central extension of the group $\mathbb{GL}(\mathbb{V})$ (For this central extension see [9]).

Remark 5 *Another reason to work with graded theories is that they behave much better for the direct sum of 1-Tate spaces. It is true that the determinant gerbe of the direct sum of 1-Tate spaces is equivalent to the tensor product of the gerbes but this equivalence depends on the ordering. If one works with graded determinant theories this equivalence will be canonical.*

References

1. M. Artin, B. Mazur, *Etale Homotopy,* Lecture Notes in Mathematics 100, Springer-Verlag 1969.
2. A. A. Beilinson, *How to glue perverse sheaves* in: *K-theory, Arithmatic and Geometry* Editor: Y. I. Manin, Lecture Notes in Mathematics 1289, 42–51, Springer-Verlag 1987.
3. Benabou, Jean *Introduction to bicategories.* 1967 Reports of the Midwest Category Seminar pp. 1–77 Springer, Berlin.
4. Breen, Lawrence *On the classification of 2-gerbes and 2-stacks.* Astrisque No. 225 (1994).
5. Brylinski, Jean-Luc *Central extensions and reciprocity laws.* Cahiers Topologie Gom. Diffrentielle Catg. 38 (1997), no. 3, 193–215.
6. Drinfeld, Vladimir *Infinite-dimensional vector bundles in algebraic geometry: an introduction.* The unity of mathematics, 263–304, Progr. Math., 244, Birkhäuser Boston, Boston, MA, 2006.

7. Kato, Kazuya *Existence theorem for higher local fields.* Invitation to higher local fields (Mnster, 1999), 165–195 (electronic), Geom. Topol. Monogr., 3, Geom. Topol. Publ., Coventry, 2000.
8. Lurie, Jacob *Higher topos theory.* Annals of Mathematics Studies, 170. Princeton University Press, Princeton, NJ, 2009.
9. Osipov, D. V. *Central extensions and reciprocity laws on algebraic surfaces.* (Russian) Mat. Sb. 196 (2005), no. 10, 111–136; translation in Sb. Math. 196 (2005), no. 9–10, 1503–1527.

The Geometry of Partial Order on Contact Transformations of Prequantization Manifolds

Gabi Ben Simon

Department of Mathematics, ETH, Zürich, Switzerland
gabi.ben.simon@math.ethz.ch

Summary. In this chapter we find a connection between the Hofer metric of the group of Hamiltonian diffeomorphisms of a closed symplectic manifold, with an integral symplectic form, and the geometry, defined in [12], of the quantomorphisms group of its prequantization manifold. This gives two main results: First, we calculate, partly, the geometry of the quantomorphism group of a prequantization manifold of an integral symplectic manifold which admits a certain Lagrangian foliation. Second, for every prequantization manifold we give a formula for the distance between a point and a distinguished curve in the metric space associated to its group of quantomorphisms. Moreover, our first result is a full computation of the geometry related to the symplectic linear group which can be considered as a subgroup of the contactomorphism group of suitable prequantization manifolds of complex projective space. In the course of the proof we use in an essential way the Maslov quasimorphism.
Ams codes: 53D50, 53D35.

1 Introduction and results

The motivating background of this chapter can be described as follows. Let (M, ω) be a closed symplectic manifold. Denote by $Ham(M, \omega)$ the group of all Hamiltonian symplectomorphisms of (M, ω). This important group carries a natural bi-invariant metric called the Hofer metric (a detailed description of the investigation of this metric structure can be found in [20]). An analogue of $Ham(M, \omega)$ in the case of contact geometry is the group of Hamiltonian contactomorphisms $Cont(P, \xi)$, where (P, ξ) is a contact manifold (see [2] appendix-4, [3] and [1] for basic information on contact manifolds). The groups Ham and $Cont$ are closely related. However, in contrast to the case of Ham, no interesting bi-invariant metrics on the group $Cont$ are known. In [12] Eliashberg and Polterovich noticed that for certain contact manifolds the universal cover of $Cont$ carries a bi-invariant partial order from which we get a natural metric space Z associated to $Cont$. The definition of Z is somewhat

Ö. Ceyhan et al. (eds.), *Arithmetic and Geometry Around Quantization*,
Progress in Mathematics 279, DOI: 10.1007/978-0-8176-4831-2_3,

indirect, and its geometry is far from being understood even in the simplest examples.
In this chapter we study the geometry of Z for certain subgroups of $Cont(P, \xi)$ where (P, ξ) is a prequantization space of a closed symplectic manifold (M, ω). As we remarked above an important tool is the establishment of a connection between the geometry of Z and the geometry of the universal cover of $Ham(M, \omega)$ endowed with the Hofer metric.
Finally let us mention that further developments on this subject appear in [11], whereas in this work we refer mostly to [12].

1.1 Preliminaries on partially ordered groups

A basic idea in this work is the implementation of basic notions from the theory of partially ordered groups to the universal cover of the groups of contact transformations of the relevant contact manifolds. So we start with the following basic definitions and constructions.

Definition 1. Let $\mathcal{D}$ be a group. A subset $\mathcal{K} \subset \mathcal{D}$ is called a *normal cone* if
- **a.** $f, g \in \mathcal{K} \Rightarrow fg \in \mathcal{K}$
- **b.** $f \in \mathcal{K}, h \in \mathcal{D} \Rightarrow hfh^{-1} \in \mathcal{K}$
- **c.** $\mathbf{1}_{\mathcal{D}} \in \mathcal{K}$

We define for $f, g \in \mathcal{D}$ that $f \geq g$ if $fg^{-1} \in \mathcal{K}$. It is not hard to check that this relation is reflexive and transitive.

Definition 2. If the above relation is also anti-symmetric, then we call it a *bi-invariant partial order* induced by $\mathcal{K}$. In this situation, an element $f \in \mathcal{K} \backslash \mathbf{1}$ is called a *dominant* if for every $g \in \mathcal{D}$ there exists $n \in \mathbb{N}$ such that $f^n \geq g$. If this partial order is not trivial we say that $\mathcal{D}$ is *orderable.*

Remark. Notice that the normality of the cone $\mathcal{K}$ implies that for every $f, g, d, e \in \mathcal{D}$

$$\text{if } f \geq g \text{ and } d \geq e \text{ then } fd \geq ge. \tag{1}$$

Definition 3. Let f be a dominant and $g \in \mathcal{D}$. Then the *relative growth* of f with respect to g is

$$\gamma(f, g) = \lim_{n \to \infty} \frac{\gamma_n(f, g)}{n}$$

where $\gamma_n(f, g)$ is defined by

$$\gamma_n(f, g) = inf\{p \in \mathbb{Z} |\ f^p \geq g^n\}.$$

The above limit exists, as the reader can check for himself (see also [12] section 1).

Now we want to relate a geometrical structure to the function γ defined above. Denote by $\mathcal{K}^+ \subset \mathcal{K}$ the set of all dominants. We define the metric space (Z, d) in the following way. First note that

$$f, g, h \in \mathcal{K}^+ \Rightarrow \gamma(f, h) \leq \gamma(f, g) \cdot \gamma(g, h).$$

We define the function

$$K : \mathcal{K}^+ \times \mathcal{K}^+ \to [0, \infty)$$

by

$$K(f, g) = \max\{\log \gamma(f, g), \log \gamma(g, f)\}.$$

It is straightforward to check that K is non-negative, symmetric, vanishes on the diagonal, and satisfies the triangle inequality. Thus K is a pseudo-distance. Define an equivalence relation on $\mathcal{K}^+$ by setting $f \sim g$ provided $K(f, g) = 0$. Put

$$Z = \mathcal{K}^+ / \sim .$$

The function K on $\mathcal{K}^+$ projects in a natural way to a genuine metric d on Z. Thus we get a metric space naturally associated to a partially ordered group.

We denote by $Z(G)$ the metric space associated to G, where G is a group which admits a nontrivial partial order (see also [12] for more information).

1.2 Geometry of the symplectic linear group

Let $Sp(2n, \mathbb{R})$ be the symplectic linear group. We denote by $\mathcal{S}$ its universal cover with base point $\mathbb{1}$, the identity matrix. We can think of $\mathcal{S}$ as the space of paths starting at $\mathbb{1}$ up to a homotopy relation between paths with the same end point. Throughout this subsection we consider the group $\mathcal{D}$ to be $\mathcal{S}$. Next, consider the equation:

$$\dot{X}(t)X^{-1}(t) = JH(t) \tag{2}$$

where $t \in S^1$, $X(t) \in \mathcal{S}$, H is a time-dependent symmetric matrix on $\mathbb{R}^{2n}$ and J is the matrix $\begin{pmatrix} 0 & -\mathbb{1} \\ \mathbb{1} & 0 \end{pmatrix}$ on $\mathbb{R}^{2n}$. The quadratic form represented by the matrix H will be called the *Hamiltonian* generating $X(t)$. It is easy to verify the following two facts:

A). The set of elements $X(t)$ in $\mathcal{S}$ generated by $H(t)$ which are non-negative as a quadratic form for each t form a normal cone.
B). In particular, those elements which are generated by a **strictly** positive $H(t)$ are dominants in $\mathcal{S}$. We shall denote this set by $\mathcal{S}^+$.
G Olshanskii has proved the following theorem (see [19] as a reference and [9] as a background source).

Theorem 1. *In the above setting the cone establishes a nontrivial partial order on* $\mathcal{S}$.

In view of theorem 1 we can explore the metric space Z. And indeed we have the following result.

Theorem 2. *The metric space (Z,d), derived from the partial order above, is isometric to $\mathbb{R}$ with the standard metric.*

The proof will be given in section 2.

Remark. One can prove theorem 1 using a criterion developed in [12]. Briefly this criterion says that in order to establish a nontrivial partial order it is enough to prove that there is no contractible loop in Sp which is generated by a strictly positive Hamiltonian. We should remark that the criterion appears in a larger context in [12]. In our case the criterion follows from the fact that the Maslov index of any positive loop is positive, and hence the loop cannot be contractible (see [17] and [23] for more information on the Maslov index).

1.3 The geometry of quantomorphisms and the Hofer metric

We start with some preliminary definitions and constructions before presenting the main results of this section (see [11] and [12]).

Let (P,ξ) be a prequantization space of a symplectic manifold (M,ω) where $[\omega] \in H^2(M,\mathbb{Z})$. Topologically, P is a principal S^1-bundle over M. We denote by $p : P \to M$ the bundle projection. P carries a distinguished S^1-invariant contact form α whose differential coincides with the lift of ω. The subgroup $Q \subset Cont(P,\xi)$ consisting of all contactomorphisms which preserve α is called *the group of quantomorphisms* of the prequantization space. We Denote by $\widetilde{Q}$ the universal cover of Q, and by $\widetilde{Ham}(M,\omega)$ the universal cover of $Ham(M,\omega)$ relative to the identity.

For the convenience of the reader we remind the following basic facts (we will give some more basic information at the beginning of Section 3).

The field of hyperplanes $\xi := \{\ker \alpha\}$ is called the *contact structure* of P. Every smooth function, *a contact Hamiltonian*, $H : P \times S^1 \to \mathbb{R}$ gives rise to a smooth isotopy of diffeomorphisms which preserve the contact structure (the set of time-1 maps of these flows is the the group $Cont(P,\xi)$). To see this, first define the Reeb vector field Y of α to be the unique vector field which satisfies the following equations:

$$\iota(Y)d\alpha = 0 \ , \alpha(Y) = 1.$$

Now given a smooth function H as above define X_H to be the unique vector field which satisfies the following two equations

$$(1)\ \iota(X_H)\alpha = H \ , (2)\ \iota(X_H)d\alpha = -dH + (\iota(Y)dH)\alpha \tag{3}$$

It is well known that the elements of the flow generated by X_H preserve the contact structure. Another important fact is that the vector field which corresponds to the constant Hamiltonian $H \equiv 1$ is the Reeb vector field, and the flow is the obvious S^1 action on P.

We remind the reader of two more basic facts.

Let $F : M \times S^1 \to \mathbb{R}$ be any Hamiltonian on M and let X_F be its Hamiltonian vector field. Let p^*F be its lift to P (by definition p^*F is constant along the fibers of P). Then it is well known (this is done by using formulas (3)) that

$$p_* X_{p^*F} = X_F.$$

Finally, denote by $\mathcal{F}$ the set of all time-1-periodic functions

$$F : M \times S^1 \to \mathbb{R} \text{ such that } \int_M F(x,t)\omega^n = 0 \text{ for every } t \in [0,1]. \tag{4}$$

Then it is not hard to check that for every representative, $\{f_t\}_{t\in[0,1]}$, of an element of $\widetilde{Ham}(M,\omega)$ there is a unique Hamiltonian $F \in \mathcal{F}$ which generates it.

The following lemma plays a fundamental role in section 3.

Lemma 1. *There is an inclusion*

$$\widetilde{Ham}(M,\omega) \hookrightarrow \widetilde{Q} \tag{5}$$

of $\widetilde{Ham}(M,\omega)$ *into* $\widetilde{Q}$.

Proof. Let F be any Hamiltonian on M. The lift of F is the contact Hamiltonian

$$\widetilde{F} : P \times S^1 \to \mathbb{R}$$

defined by $\widetilde{F} := p^*F$. This contact Hamiltonian generates an element of $\widetilde{Q}$. It is well known that every representative of an element of $\widetilde{Q}$ is generated by a unique contact Hamiltonian which is a lift of a Hamiltonian defined on M. We further remind the reader that Q is a central extension of Ham. That is, we have the short exact sequence

$$1\!\!1 \to S^1 \to Q \to \ Ham \ \to 1\!\!1.$$

This is, from the Lie algebra point of view, the Poisson bracket extension of the algebra of Hamiltonian vector fields. That is,

$$0 \to \mathbb{R} \to C^\infty(M \times S^1) \to \ \mathcal{F} \ \to 0.$$

We first explain how to map the identity element $[1\!\!1]$ of $\widetilde{Ham}$ into $\widetilde{Q}$. Let f_t be any contractible loop representing $[1\!\!1]$, and let F_t be the unique normalized Hamiltonian which generates it. The path $\hat{f}_t$, in Q, which is generated by the lift of F_t, p^*F_t, is a lift of f_t. Nevertheless, it is not necessarily closed (actually it is closed, as we show below). It easy to check that there is a constant, c, such that $p^*F_t + c$ generates a lift, $\hat{h}(t)$, of the loop f_t and this lift is a loop based at the identity of Q. We claim that

a. The loop $\hat{h}(t)$ is contractible.

b. The constant c equals zero.

We first prove **a**. Since the loop f_t is contractible, there is a homotopy

$$K_s(t) : [0,1] \times [0,1] \to Ham$$

such that for every s, $K_s(t)$ is a contractible loop, $K_0 = f_t$ and $K_1 = \mathbb{1}$. We denote by $F_s(t)$ the normalized Hamiltonian which generates the loop K_s. As we explained, for every s there is a constant $c(s)$ such that $p^* F_s + c(s)$ generates a loop, $\hat{h}_s(t)$, in Q which is a lift of K_s.

Now since $\hat{h}_s(t)$ is a lift of the homotopy $K_s(t)$ (which is a homotopy between f_t and $\mathbb{1}$) it is a homotopy between $\hat{h}(t)$ and the identity element of Q. Thus $\hat{h}(t)$ is contractible.

We now prove **b**. First we remind the reader of the Calabi Weinstein invariant, which is well-defined on the group $\pi_1(Q)$. It was proposed by Weinstein in [26]. For a loop $\gamma \in Q$ define the Calabi Weinstein invariant by

$$cw(\gamma) = \int_0^1 dt \int\limits_M F_t \omega^n$$

where $p^* F_t$ is the unique Hamiltonian generating γ.

Now it is clear that

$$cw([\mathbb{1}]) = 0.$$

We conclude that for every s

$$\int_0^1 dt \int\limits_M (F_s(t,x) + c(s)) \omega^n = 0$$

from which we get

$$c(s) vol(M) = -\int_0^1 dt \int\limits_M F_s(t,x) \omega^n.$$

Now since F_s is normalized for every s we get that $c(s) = 0$ as desired.

The map of an arbitrary element of $\widetilde{Ham}$ is defined in the same manner. Let f_t be any representative of an element, $[f]$, of $\widetilde{Ham}$. Let F_t be the unique normalized Hamiltonian generating f_t. Then the image of $[f]$ is the homotopy class of the path, $\hat{f}_t$, generated by the lift of the Hamiltonian F_t, namely $p^* F_t$. By using the same type of argument as above it can be shown that this map is well-defined. Finally, this map is clearly injective.

The universal cover of Q carries a natural normal cone consisting of all elements generated by non-negative contact Hamiltonians. The set of dominants consists of those elements of $\widetilde{Q}$ which are generated by strictly positive Hamiltonian. The normal cone always gives rise to a genuine partial order on $\widetilde{Q}$. Thus one can define the corresponding metric space (Z, d), as was explained in subsection 1.1. We state this as a theorem.

Theorem 3. *Let (M,ω) be a symplectic form such that $\omega \in H^2(M,\mathbb{Z})$. Let (P,ξ) be a prequantization of (M,ω). Then $\widetilde{Q}$, the universal cover of the group of quantomorphisms of Q, is orderable.*

Proof. According to [12] all we need to check is that there are no contractible loops in Q which are generated by a strictly positive contact Hamiltonian. Such contractible loops cannot exist since the Calabi Weinstein invariant of any representative of a contractible loop must be zero, and thus can not be generated by a strictly positive Hamiltonian.

In light of Theorems 1 and 3 we should remark the following.

Remark. Let (M,ω) be the complex projective space $\mathbb{C}P^{n-1}$ endowed with the Fubini Study symplectic form normalized to be integral. Let (P,ξ) be the sphere S^{2n-1} endowed with the standard contact structure, and let $(\mathbb{R}P^{2n-1},\beta)$ be the standard contact real projective space, where β denotes the contact structure. It is well known that these two contact manifolds are prequantizations of (M,ω) (see [16], [25] and [27] for preliminaries on prequantization). Let $Cont(P,\xi)$ and $Cont(\mathbb{R}P^{2n-1},\beta)$ be the groups of all Hamiltonian contact transformations of these manifolds.

The relation of the symplectic linear group to these groups is as follows. The group $Sp(2n,\mathbb{R})$ is the subgroup of all elements of $Cont(P,\xi)$ which commutes with $-1\!\!1$, and $Sp(2n,\mathbb{R})/\pm 1\!\!1$ is a subgroup of $Cont(\mathbb{R}P^{2n-1},\beta)$. Now, it is known that $Cont(\mathbb{R}P^{2n-1},\beta)$ admits a nontrivial partial order (while $Cont(P,\xi)$ does not). The non-orderability of $Cont(P,\xi)$ follows from [11], this is done by using the criterion mentioned above. The orderability of $Cont(\mathbb{R}P^{2n-1},\beta)$ follows from the theory of the nonlinear Maslov index introduced by Givental (see [13]).

In what follows we establish a connection between the partial order on $\widetilde{Q}$ (and its metric space Z) and the Hofer metric. So first let us recall some basic definitions related to the Hofer metric. The *Hofer distance* between $f \in \widetilde{Ham}(M,\omega)$ and $1\!\!1$, is defined by

$$\rho(f,1\!\!1) = \inf_G \int_0^1 \left(\max_{x\in M} G(x,t) - \min_{x\in M} G(x,t) \right) dt, \tag{6}$$

where the infimum is taken over all time 1-periodic Hamiltonian functions G generating paths in Ham which belong to the homotopy class represented by f with fixed end points (see [15], [17] and [20], for further information). Furthermore, we define the *positive and the negative parts of the Hofer distance* as

$$\rho_+(f,1\!\!1) = \inf \int_0^1 \max_{x\in M} G(x,t)dt, \qquad \rho_-(f,1\!\!1) = \inf \int_0^1 - \min_{x\in M} G(x,t)dt, \tag{7}$$

where the infimum is taken as above and $G \in \mathcal{F}$.

Now, for each $f \in \widetilde{Ham}(M,\omega)$, generated by some Hamiltonian, set

$$\|f\|_+ = \inf\{\max F(x,t)\}, \quad \|f\|_- = \inf\{-\min F(x,t)\} \tag{8}$$

where the infimum is taken over all Hamiltonian functions $F \in \mathcal{F}$ generating paths in Ham which belong to the homotopy class represented by f. The following result is due to Polterovich (see [21]).

Lemma 2. *For every $f \in \widetilde{Ham}(M,\omega)$ we have*

$$\rho_+(f) = \|f\|_+ \tag{9}$$

$$\rho_-(f) = \|f\|_- . \tag{10}$$

We further define the positive and negative *asymptotic* parts of the Hofer metric as follows:

$$\|f\|_{+,\infty} := \lim_{n\to\infty} \frac{\|f^n\|_+}{n} \text{ and } \|f\|_{-,\infty} := \lim_{n\to\infty} \frac{\|f^n\|_-}{n}. \tag{11}$$

Before stating our first main result we fix the following notation. Given (P,ξ) (a prequantization of a symplectic manifold (M,ω)) we denote by e^{is} the diffeomorphism of P obtained by rotating the fibers by total angle s. Note that the set $\{e^{is}\}_{s\in\mathbb{R}}$ is a one-parameter subgroup of contact transformations in $\widetilde{Q}$.

Theorem 4. *Let f be the time-1 map of a flow generated by Hamiltonian $F \in \mathcal{F}$. Let $\widetilde{F}$ be the lift of F to the prequantization space and let $\tilde{f}$ be its time-1- map. Take any $s \geq 0$ such that $e^{is}\tilde{f}$ is a dominant. Then*

$$dist_K(\{[e^{it}]\},[e^{is}\tilde{f}]) := \inf_t K([e^{it}],[e^{is}\tilde{f}]) = \frac{1}{2}\log\frac{s+\|f\|_{+,\infty}}{s-\|f\|_{-,\infty}} \tag{12}$$

where $[e^{it}],[e^{is}\tilde{f}] \in \widetilde{Q}$, and f represents an element of $\widetilde{Ham}$.

Before stating our second result we need the following preliminaries.

Definition 4. Let (M,ω) be a closed symplectic manifold. Let L be a closed Lagrangian submanifold of M. We say that L has the *Lagrangian intersection property* if L intersects its image under any exact Lagrangian isotopy.

Definition 5. We say that L *has the stable Lagrangian intersection property* if $L\times\{r=0\}$ has the Lagrangian intersection property in $(M\times T^*S^1, \omega\oplus dr\wedge dt)$ where (r,t) are the standard coordinates on the symplectic manifold T^*S^1 and $M\times T^*S^1$ is considered as a symplectic manifold with the symplectic form $\omega\oplus dr\wedge dt$.

For more information on Lagrangian intersections see [20] Chapter 6.

Consider now the following situation. For (M,ω) a closed symplectic manifold, $[\omega] \in H^2(M,\mathbb{Z})$, assume that we have an open dense subset M_0 of M, such that M_0 is foliated by a family of closed Lagrangian submanifolds $\{L_\alpha\}_{\alpha\in\Lambda}$, and each Lagrangian in this family has the stable Lagrangian intersection property.

We denote by $\mathtt{F}$ the family of autonomous functions in $\mathcal{F}$ defined on M which are constant when restricted to L_α for every α. Note that this means that the elements of $\mathtt{F}$ commutes relative to the Poisson bracket. That is, for every $F,G \in \mathtt{F}$ we have $\{F,G\} = 0$.

We denote by $\{\mathtt{F}, \| \ \|_{max}\}$ the metric space of the family of functions $\mathtt{F}$ endowed with the *max* norm where the *max* norm is $\|F\|_{max} := \max_{x\in M} |F(x)|$.

In the course of the proof we will use the following subspace: Let V be the subspace of $\widetilde{Q}$ constiting of all elements generated by contact Hamiltonians of the form $s + \widetilde{F}$ where $\widetilde{F}$ is a lift of a Hamiltonian $F \in \mathtt{F}$ and $s + \widetilde{F} > 0$. Note that the existence of s is justified due to the fact that the function F, and thus $\widetilde{F}$, attains a minimum.

We now state our second result.

Theorem 5. *Let (M,ω) be a symplectic manifold which admits a subset M_0 foliated by a family of Lagrangians which has the stable intersection property. Let $(\mathtt{F}, \| \ \|_{max})$ be as above. Then there is an isometric injection of the metric space $(\mathtt{F}, \| \ \|_{max})$ into the metric space Z.*

Examples

Example 1. Consider the $2n$-dimensional standard symplectic torus. That is $T^{2n} = \mathbb{R}^{2n}/\mathbb{Z}^{2n}$ with the symplectic structure $dP \wedge dQ = \sum_{i=1}^{n} dp_i \wedge dq_i$. We now foliate the torus by a family of Lagrangians which depends only on the P coordinate. This way we exhibit the torus as a Lagrangian fibration parameterized by an n torus (the P coordinate which in the notation of Theorem 5 is the parameter Λ) and with Lagrangian fibers (spanned by the Q coordinate). It is known that this family of Lagrangian has the stable Lagrangian intersection property (see [20]). Now, define the family of autonomous functions on T^{2n} which depend only on the P coordinate. Thus all the conditions of Theorem 5 are satisfied (we remind the reader that all functions of this family are Poisson commuting). We conclude that the metric space Z of a prequantization space of the standard $2n$-dimensional symplectic torus contains an infinite dimensional metric space. Denote by $C^\infty_{nor}(T^n)$ the space of smooth normalized functions on the n-dimensional torus. Then we have the isometric injection:

$$(C^\infty_{nor}(T^n), \| \ \|_{max}) \hookrightarrow Z.$$

Example 2. Let Σ be any surface of genus greater than or equal than 2. It is known that non-contractible loops on Σ have the stable Lagrangian

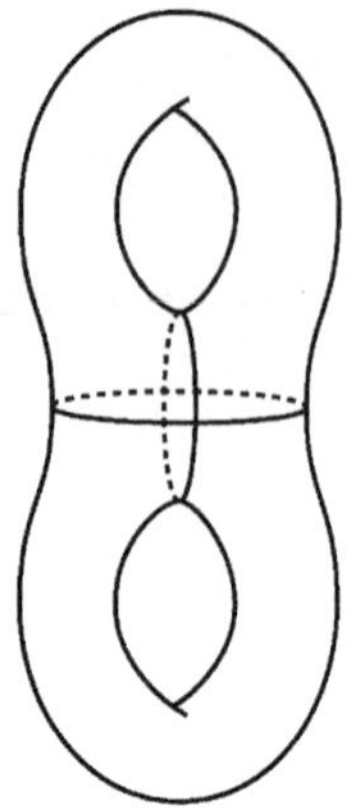

Fig. 1. The curves we remove from Σ

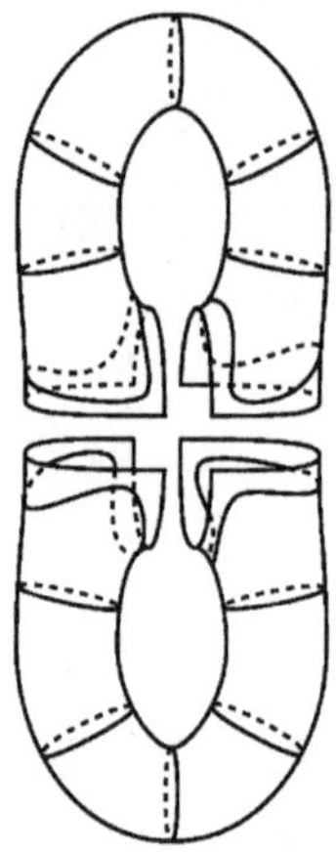

Fig. 2. The foliation of Σ_0

property (see for example [22]). For such surfaces one can foliate a subset, Σ_0, of Σ by a family of disjoint non-contractible closed loops such that Σ_0 is an open dense subset of Σ (actually $\Sigma \setminus \Sigma_0$ is a finite collection of closed arcs, see figures 1 and 2). Note that all the conditions of Theorem 5 are satisfied. In figures (1) and (2) we show an example of such family of Lagrangians for a surface of genus 2.

2 The metric space $\mathbf{Z}(\widetilde{Sp})$

2.1 The metric space $\mathbf{Z}(\widetilde{U}(n))$

In this subsection we prove Theorem 2 for the unitary group $U(n)$. This is a result by itself. Moreover elements of the proof will be used to prove

Theorem 2 for the symplectic case. Here we use occasionally basic properties of the exponential map on the Lie algebra of a matrix Lie group (see [14] Chapter 2).

We denote by $\mathcal{A}$ the universal cover of $U(n)$. We view $\mathcal{A}$ as the set of all paths starting at $\mathbb{1}$, the identity element, up to homotopy relation between paths with the same end point. We will denote an element $[u] \in \mathcal{A}$, with a slight abuse of notation, by u where u is a path representing $[u]$.

As in the symplectic case (see definition 7) we can define for an element $u \in \mathcal{A}$ its *Maslov index* as $\alpha(1) - \alpha(0)$ where $e^{i2\pi\alpha(t)} = \det(u(t))$. Note that it takes its values in $\mathbb{R}$. We will denote it by $\mu(u)$. Using the same definitions as in subsection 1.2 we have the following. An element $u \in \mathcal{A}$ is called *semi-positive* if for some representative, the path of Hermitian matrices $h_u(t)$ defined by the equation

$$\dot{u}u^{-1} = ih_u \tag{13}$$

is semi-positive definite. Note that if we consider $U(n)$ as a subgroup of $Sp(2n, \mathbb{R})$ via the realization

$$A + iB \mapsto \begin{pmatrix} A & -B \\ B & A \end{pmatrix}$$

equation (13) is the same as equation (2). We proceed as in 1.2, define a partial ordering on $\mathcal{A}$, $\geq$, by $f \geq g$ if and only if fg^{-1} is semi-positive. Let $\mathcal{C}$ denote the subset of all semi-positive elements and denote by $\mathcal{C}^+$ the set of positive definite elements of $\mathcal{A}$. These are, respectively, the normal cone and the set of dominants of $\mathcal{A}$ (see subsection 1.1).

We define the function γ_n, the relative growth, and the metric space Z in the same way as in 1.1.

Preliminary basic lemmas

We need the following two elementary lemmas.

Lemma 3. *Let u and v be two paths in $U(n)$ with the same end-points. Then $[u] = [v]$ if and only if $[|u|] = [|v|]$ in $U(1)$.*

Proof. It is well known that the determinant map $U(n) \to U(1)$ induces an isomorphism between the fundamental groups of these spaces. Let $\hat{\gamma}$ be the inverse path to γ and let $*$ be the juxtaposition of paths. Thus $[u] = [v]$ if and only if $u * \hat{v}$ is homotopic to the identity, if and only if

$$|u * \hat{v}| = |u| * \widehat{|v|}$$

is homotopic to the identity, if and only if $[|u|] = [|v|]$.

Lemma 4. *Assume that u, h_u satisfies equation* (13). *Then*

$$\mu(u) = \frac{1}{2\pi}\int_0^1 \mathrm{tr}\ h_u(t)dt. \tag{14}$$

Proof. For a complex path $z(t) \neq 0$ we have of course $\frac{d}{dt}\log z(t) = \frac{\dot{z}(t)}{z(t)}$. Taking the imaginary part we have

$$\frac{d}{dt}\arg z(t) = Im\frac{\dot{z}(t)}{z(t)}.$$

Fix $t = t_0$ in $[0,1]$ and write $u(t) = u(t_0)v(t)$ where $v(t_0) = \mathbb{1}$. Let D be the differentiation with respect to t at $t = t_0$. Note that

$$\dot{v}v^{-1} = ih_u.$$

Then $D|v| = \mathrm{tr}(\dot{v})$ at t_0, which implies that

$$D \arg|u| = \mathrm{Im}\frac{D|u|}{|u|} = \mathrm{Im}\frac{D|v|}{|v|}$$

$$= \mathrm{Im}\ \mathrm{tr}(\dot{v}) = \mathrm{Im}\ \mathrm{tr}(ih_u v) = \mathrm{tr}(h_u).$$

We emphasize that D represents differentiation at $t = t_0$. Also, in the computation above recall that $v(t_0) = \mathbb{1}$ at $t = t_0$. Now the above formula follows since $\arg|u(0)| = 0$.

As a corollary of the lemma we have that if u is semi-positive then $\mu(u) \geq 0$ (since if h_u is semi-positive then $\mathrm{tr} h_u \geq 0$). The converse of the corollary is not true (we will not give an example). Nevertheless we have:

Lemma 5. *Let v be an element of $\mathcal{A}$ (the universal cover of $U(n)$). Then if $\mu(v) \geq 2\pi n$ we have $v \geq \mathbb{1}$. In this case, we can have a representative $v(t) = e^{itA}$ where tr$(A) = \mu(v)$, A is hermitian and positive semidefinite, and if $\lambda_1, \ldots, \lambda_n$ are the eigenvalues of A, then* $\max_{i,j} |\lambda_i - \lambda_j| \leq 2\pi n$.

Proof. Suppose that $\mu(v) \geq 2\pi n$. Then according to lemma 3 and the choice of an appropriate representative we can write $v(t) = e^{itA}$ with A hermitian. The choice of A is not unique, however, $\mathrm{tr}(A) = \mu(v)$ up to a multiple of 2π. Still, we can modify A so that $\mathrm{tr}(A) = \mu(v)$. To see this, diagonalize A by a unitary matrix and then modify the eigenvalues by multiples of 2π (here we are using the fact that if C is an invertible matrix then $e^{CXC^{-1}} = Ce^X C^{-1}$ for any arbitrary matrix X).

In fact we can do better: We may still modify without changing $\mathrm{tr}(A)$ such that now A will be positive semi-definite and $\max_{i,j} |\lambda_i - \lambda_j| \leq 2\pi n$. To see this, after A has been diagonalized, let $\lambda_1, \ldots, \lambda_n$ be the eigenvalues of A (in an increasing order). Then we have

$$\sum_i \lambda_i = \mathrm{tr}(A) = \mu(v) \geq 2\pi n.$$

Now let $0 < \bar{\lambda}_i \leq 2\pi$ be the unique number that is congruent to λ_i modulo $2\pi\mathbb{Z}$. Then we have

$$\sum_i \bar{\lambda}_i = \mathrm{tr}(A) - 2\pi k, \ k > 0.$$

Note that $\max_{i,j} |\bar{\lambda}_i - \bar{\lambda}_j| \leq 2\pi$. We first assume that $k < n$ (the case $n = 1$ is trivial so we assume $n \geq 2$). In this case we change the $\bar{\lambda}_i$s by distributing the k 2πs to the first k $\bar{\lambda}_i$s. It is clear that the new maximal $\bar{\lambda}_i$ will come from the first, new, k $\bar{\lambda}_i$s and that the condition $\max_{i,j} |\bar{\lambda}_i - \bar{\lambda}_j| \leq 2\pi n$ is kept.

Now if $k \geq n$ then write $k = ln + m$ where $l, m \in \mathbb{N}$ and $m < n$. In this case we add 2π to all the $\bar{\lambda}_i$s l times and then we add to the first m $\bar{\lambda}_i$'s the remaining m 2π's. Now we are in the same position as in the first case where $k < n$. The resulting matrix A (before diagonalization) has the desired properties.

Calculating the metric in terms of the Maslov index

Let $f, g \in \mathcal{C}^+$. We wish to compute $\gamma(f, g)$. We have the following theorem.

Theorem 6. *For every non-constant $f, g \in \mathcal{C}^+$ we have:*

$$\gamma(f, g) = \frac{\mu(g)}{\mu(f)}. \tag{15}$$

Proof. We start with the following definition.

$$\gamma^*(f, g) := \inf\{\frac{r}{s} | f^r \geq g^s \ , r \in \mathbb{Z} \ , s \in \mathbb{N}\}. \tag{16}$$

We claim that $\gamma^* = \gamma$. Indeed, denote by T the set of numbers which satisfy the condition of the right hand side of (16). Assume that $\frac{r}{s} \in T$. Then the equivalence of the definitions follows from the inequality

$$r \geq \gamma_s(f, g) \geq s\gamma^*(f, g).$$

We claim that a sufficient condition for $f^r \geq g^s$ is that $r\mu(f) - s\mu(g) \geq 2\pi n$. Indeed, if so then $\mu(f^r g^{-s}) = r\mu(f) - s\mu(g) \geq 2\pi n$, so by Lemma 5,

$$f^r g^{-s} \geq \mathbb{1} \Leftrightarrow f^r \geq g^s.$$

Now, given $\varepsilon > 0$ choose $s > \frac{1}{\varepsilon}$. Let $r \geq 0$ be the smallest integer so that

$$r\mu(f) - s\mu(g) \geq 2\pi n.$$

Thus

$$r\mu(f) - s\mu(g) < 2\pi n + \mu(f)$$

thus

$$\frac{r}{s}\mu(f) - \mu(g) < (2\pi n + \mu(f))\varepsilon$$

from which we get

$$\frac{r}{s} < \frac{\mu(g)}{\mu(f)} + \frac{(2\pi n + \mu(f))\varepsilon}{\mu(f)}.$$

Since ε is arbitrary then

$$\gamma^*(f,g) = \inf\{\frac{r}{s}|f^r \geq g^s\} \leq \frac{\mu(g)}{\mu(f)}.$$

Now

$$\gamma(f,g)\gamma(g,f) \geq 1$$

gives the desired equality:

$$\gamma(f,g) = \frac{\mu(g)}{\mu(f)}.$$

Now we can finish the proof of Theorem 2 in the unitary case. First, we recall the definition of the metric space Z. $Z = \mathcal{C}^+/\sim$ where $f \sim g$ provided $K(f,g) = 0$ and $K(f,g) = \max\{\log\gamma(f,g), \log\gamma(g,f)\}$. Define a map $p : \mathcal{A} \to \mathbb{R}$ by $p(u) = \log(\mu(u))$. We claim that the map p induces an isomorphism of metric spaces, that is,

$$\frac{\mathcal{C}^+}{\sim} = Z \cong \mathbb{R}.$$

Indeed, by Theorem 6 we get

$$|p(f) - p(g)| = \max\{\log\mu(g)/\mu(f), \log\mu(f)/\mu(g)\} = K(f,g).$$

Thus p is an isometry.

Remark. Note also that p preserves order as well. Indeed, $f \geq g$ implies that $0 \leq \mu(fg^{-1}) = \mu(f) - \mu(g)$ which implies that $p(f) \geq p(g)$. See [12] subsection 1.7 for more details on this phenomenon, in a broader context.

2.2 The Symplectic case

We now turn to the symplectic case. Let J be the standard symplectic structure on $\mathbb{R}^{2n}$ (see subsection 1.2). We recall that the symplectic linear group $Sp(2n,\mathbb{R})$ is the group of all matrices A which satisfy $A^TJA = J$. As we already remarked, $U(n)$ is a subgroup of $Sp(2n,\mathbb{R})$. The elements of $U(n)$ are precisely those which commute with J.

We denote by $\mathcal{S}$ the universal cover of this group having as base point the identity matrix $\mathbb{1}$. As before we use the same letter to denote an element in $\mathcal{S}$ and a representing path. However, we shall use capital letters instead.

Now, assume that $X, Y \in \mathcal{S}$. Let H_X, H_Y, H_{XY} be the Hamiltonians generating, respectively, X, Y, XY (see 1.2). For later use we need the following formula. The formula is well known and it is easy to prove.

$$H_{XY} = H_X + {X^{-1}}^T H_Y X^{-1}. \tag{17}$$

The Maslov quasimorphism

Before proving theorem 2 in the symplectic case we define an important property which we use in a crucial way in the course of the proof.

Definition 6. Let G be a group. A *quasimorphism* r on G is a function $r : G \to \mathbb{R}$ which satisfies the homomorphism equation up to a bounded error: there exists $R > 0$ such that

$$|r(fg) - r(f) - r(g)| \leq R$$

for all $f, g \in G$.

Roughly speaking a quasimorphism is a homomorphism up to a bounded error. See [7] for preliminaries on quasimorphisms. A quasimorphism r_h is called *homogeneous* if $r_h(g^m) = m r_h(g)$ for all $g \in G$ and $m \in \mathbb{Z}$. Every quasimorphism r gives rise to a homogeneous one,

$$r_h(g) = \lim_{m \to \infty} \frac{r(g^m)}{m}.$$

As we said, a basic tool in the proof of Theorem 2 is the *Maslov quasimorphism* whose definition is a generalization of the Maslov index from loops to paths in $Sp(2n, \mathbb{R})$.

Definition 7. Let $\Psi(t) = U(t)P(t)$ be a representative of a point in $\widetilde{Sp}$,where $U(t)P(t)$ is the polar decomposition in $\widetilde{Sp}$ of $\Psi(t)$, $U(t)$ is unitary and $P(t)$ is symmetric and positive definite. Choose $\alpha(t)$ such that $e^{i2\pi\alpha(t)} = \det(U(t))$. Define the Maslov quasimorphism μ by $\mu([\Psi]) = \alpha(1) - \alpha(0)$.

Theorem 7. *The Maslov quasimorphism is a quasimorphism.*

For the proof of Theorem 7 see [5] and [10]. Theorem 7 is the reason why we use this extended version of the standard Maslov index for loops (which assumes its values in $\mathbb{Z}$). Note that the Maslov quasimorphism takes its values in $\mathbb{R}$.

We denote by $\widetilde{\mu}$ the homogeneous quasimorphism corresponding to μ:

$$\widetilde{\mu}(X) = \lim_{k \to \infty} \frac{\mu(X^k)}{k}.$$

We further remark that the restriction of the (homogeneous) Maslov quasimorphism to $\mathcal{A}$ (since there is an isomorphism $\pi_1(Sp(2n, \mathbb{R})) \cong \pi_1(U(n))$, we may consider $\mathcal{A}$ as a subset of $\mathcal{S}$) is the Maslov index we have defined in subsection 2.1 on $\mathcal{A}$.

Calculating the metric using the Maslov quasimorphism

We denote by $\mathcal{S}^+$ the set of dominants of $\mathcal{S}$. Note that we have the inclusion $\mathcal{C}^+ \hookrightarrow \mathcal{S}^+$. Recall that $\mathcal{S}^+$ induces a nontrivial partial order on $\mathcal{S}$ (see Theorem 1), thus we can define for every $f, g \in \mathcal{S}^+$ the function $\gamma(f, g)$. Here one must be cautious as this quantity may have two different meanings for unitary f, g. According to the following theorem this causes no difficulties (see the remark following theorem 8).

The key theorem of this subsection is the following.

Theorem 8. *For all $X, Y \in \mathcal{S}^+$ we have*

$$\gamma(X, Y) = \frac{\tilde{\mu}(Y)}{\tilde{\mu}(X)}. \tag{18}$$

Theorem 2 now follows from theorem 8. Indeed, define $p(X) = \log(\tilde{\mu}(X))$ for every $X \in \mathcal{S}^+$. Now repeat verbatim the last part of subsection 2.1.

Remark. Notice that if X and Y are unitary, the theorem shows that their symplectic relative growth is the same as the unitary relative growth, since on unitary paths, $\tilde{\mu} = \mu$.

We now prove Theorem 8 in two steps.

Proof of Theorem 8 Step 1

In step 1 we prove the following lemma.

Lemma 6. *For every positive definite symmetric symplectic **matrix** P, there exists a positive path X connecting $\mathbb{1}$ with P, such that*

$$\mu(X) \leq 4\pi n. \tag{19}$$

Proof. Without loss of generality we may assume that P is diagonal. Otherwise, diagonalize P by a unitary matrix U_0 and notice that for every path Y, we have by a direct calculation that

$$H_{U_0 Y U_0^{-1}} = U_0 H_Y U_0^{-1}$$

where H_Y is a Hamiltonian generating Y. Thus X is positive if and only if $U_0 X U_0^{-1}$ is. Now, assume that the lemma has been proved for the case $n = 1$. Using this assumption we prove the general case.

We first define n different embeddings

$$j_i : Sp(2, \mathbb{R}) \hookrightarrow Sp(2n, \mathbb{R}),\ 1 \leq i \leq n$$

as follows. We copy a 2×2 matrix A to the $(i, n+i)$ block of a $2n \times 2n$ matrix B. That is,

$$B_{ii} := A_{11},\ \ B_{i,i+n} := A_{12}$$

$$B_{i+n,i} := A_{21},\ B_{i+n,i+n} := A_{22}.$$

The rest of the elements of B are defined as follows (we assume of course that $(k,l) \neq (i,i), (i,i+n), (i+n,i), (i+n,i+n)$).

$$B_{kl} := \begin{cases} 0 & k \neq l \\ 1 & k = l\ . \end{cases}$$

These are at least homomorphisms to $GL(2n,\mathbb{R})$, but in fact one can easily check that the image is symplectic. These embeddings preserve the transpose operator and thus symmetry and orthogonality. As a result the embeddings respect the symmetric-unitary decomposition. In particular

$$\mu(X) = \mu(j_i(X)) \tag{20}$$

for every $X \in Sp(2,\mathbb{R})$ and $1 \leq i \leq n$.

Note the following property of the embeddings j_i. For every $i \neq k$ and any $X, Y \in Sp(2,\mathbb{R})$ we have

$$j_i(X)j_k(Y) = j_k(Y)j_i(X). \tag{21}$$

Now, by assumption P is diagonal; let $(\lambda_1, ..., \lambda_n, 1/\lambda_1, ..., 1/\lambda_n)$ be the diagonal of P. Define $P_i = diag(\lambda_i, 1/\lambda_i)$, and notice that

$$P = \prod_{i=1}^{n} j_i(P_i). \tag{22}$$

Having proved the theorem for $n = 1$, we can find paths X_i connecting $\mathbb{1}$ (of $Sp(2,\mathbb{R})$) with P_i, such that $\mu(X_i) \leq 4\pi$. Define

$$X = \prod_{i=1}^{n} j_i(X_i). \tag{23}$$

We write $j_i(X_i) = P_i(X_i)U_i(X_i)$ for the polar decomposition of the matrix $j_i(X_i)$ in $Sp(2n,\mathbb{R})$. Note that according to what we have remarked above, this representation of this polar decomposition is the image of the polar decomposition of X_i with respect to the homomorphism j_i for all i. Anyhow we get

$$X = \prod_{i=1}^{n} P_i(X_i)U_i(X_i). \tag{24}$$

Note that according to (21) all the elements in the product of (24) commute. Thus we can write

$$X = \prod_{i=1}^{n} P_i(X_i) \prod_{i=1}^{n} U_i(X_i). \tag{25}$$

We claim that the r.h.s. of (25) is the polar decomposition of X. Indeed, all the elements in the product $\prod_{i=1}^{n} P_i(X_i)$, according to (21), commute. Thus, this product is a symmetric (positive definite) matrix. The fact that the product $\prod_{i=1}^{n} U_i(X_i)$ is unitary proves our claim. We get that

$$\mu(X) = \mu(\prod_{i=1}^{n} U_i(X_i)) = \sum_{i=1}^{n} \mu(U_i(X_i))$$

$$= \sum_{i=1}^{n} \mu(X_i) \leq 4\pi n.$$

Note that the last equality follows from (20). Finally, H_X is positive definite since it is a direct sum of positive definite Hamiltonians (see formula (17)).

It remains to prove the case $n = 1$. We produce here a concrete example. As argued above, we may assume that $P = diag(\lambda, 1/\lambda)$. Consider the function $f(t) = \tan(\pi/4 + at)$, where $|a| < \pi/4$ is chosen so that $f(1) = \lambda$. We then have

$$|f'| = |a(f^2 + 1)| < f^2 + 1.$$

Now, let

$$U(t) := \begin{pmatrix} \cos(2\pi t) & -\sin(2\pi t) \\ \sin(2\pi t) & \cos(2\pi t) \end{pmatrix}.$$

Define the path

$$X(t) = U(t)F(t)U(t)$$

where $F(t) := diag(f(t), 1/f(t))$. We claim that $\mu(X) = 4\pi$. Indeed, $\mu(X) = \mu(U) + \mu(f) + \mu(U) = 2\pi + 0 + 2\pi = 4\pi$.

Finally we want to show that X is positive.
Using (17) we get:

$$H_X = H_U + UH_FU^{-1} + UF^{-1}H_UF^{-1}U^{-1},$$

and we know that

$$H_U = 2\pi I; \text{ and } H_F = \begin{pmatrix} 0 & f'(t)/f(t) \\ f'(t)/f(t) & 0 \end{pmatrix}.$$

Thus H_X is positive if and only if

$$U^{-1}H_XU = H_U + H_F + F^{-1}H_UF^{-1}$$

$$= 2\pi(I + F^{-2}) + H_F = \begin{pmatrix} 2\pi(1 + f^{-2}) & f'(t)/f(t) \\ f'(t)/f(t) & 2\pi(1 + f^2) \end{pmatrix}$$

-is positive. Since the diagonal entries of the latter matrix are positive, then the matrix is positive if and only if its determinant is positive. Indeed we have:

$$4\pi^2(1+f^2)(1+f^{-2}) - (f'/f)^2 = 4\pi^2(f+1/f)^2 - (f'/f)^2 > 0$$

where in the last inequality we have used the fact that $|f'| < 1 + f^2$. This completes the proof of the lemma.

Proof of Theorem 8 Step 2

In what follows we denote by C the constant which appears in the definition of the Maslov quasimorphism. That is:

$$|\mu(fg) - \mu(f) - \mu(g)| \leq C$$

for all $f, g \in \mathcal{S}$. It is not hard to prove the following fact (the proof appears in the literature, see [7]).

Lemma 7. *Let $\tilde{\mu}$ be the homogeneous quasimorphism of μ. Then*

$$|\tilde{\mu}(X) - \mu(X)| \leq C \textit{ for all } X \in \mathcal{S}.$$

We then have $|\tilde{\mu}(XY) - \tilde{\mu}(X) - \tilde{\mu}(Y)| \leq 4C$ for all $X, Y \in \mathcal{S}$.

We will denote the constant $4C$ by C_1. The main lemma of Step 2 is the following.

Lemma 8. *Every element $Y \in \mathcal{S}$ for which $\mu(Y) \geq 6\pi n + C$ is positive.*

Proof. Let $Y(t) = P(t)V(t)$ be the polar decomposition of Y. Let X be the positive path, given by Lemma 6, ending at $P(1)$. Define

$$Z(t) = X^{-1}(t)P(t).$$

Note that Z is a closed path, moreover we have

$$\mu(Z) \geq \mu(P) - \mu(X) - C \geq -4\pi n - C \tag{26}$$

where in the first inequality we have used the quasimorphism property, and in the second inequality we have used the fact that $\mu(P) = 0$, and for every $X \in \mathcal{S}$ we have $\mu(X) = -\mu(X^{-1})$. From (26) we get

$$\mu(X^{-1}Y) = \mu(X^{-1}PV) = \mu(ZV)$$

$$= \mu(Z) + \mu(V) = \mu(Z) + \mu(Y) > 2\pi n.$$

Now, since $X^{-1}Y(1)$ is unitary, then $X^{-1}Y$ is homotopic to its unitary projection, which we denote by T. So we have $\mu(T) = \mu(X^{-1}Y) > 2\pi n$ and from lemma 5 it follows that $T \geq \mathbb{1}$ in the unitary sense and thus in the symplectic sense. Consequently $X^{-1}Y$is positive and thus $Y = XX^{-1}Y$ is positive.

We are now ready to prove Theorem 8:

We wish to compute $\gamma(X,Y)$. We claim that a sufficient condition for $X^r \geq Y^s$ to hold is

$$r\tilde{\mu}(X) - s\tilde{\mu}(Y) \geq 6\pi n + 2C + C_1. \tag{27}$$

Indeed, if this condition holds, then

$$\mu(X^r Y^{-s}) \geq \tilde{\mu}(X^r Y^{-s}) - C \geq \tilde{\mu}(X^r) + \tilde{\mu}(Y^{-s}) - C - C_1$$
$$= r\tilde{\mu}(X) - s\tilde{\mu}(Y) - C - C_1 \geq 6\pi n + C.$$

Where we have used in the first and the second inequalities lemma 7, and in the third inequality we have used our assumption. Now, by lemma 8, we see that $X^r Y^{-s} \geq 1\!\!1$ or that $X^r \geq Y^s$.

Finally, since the r.h.s. of (27) is independent of r and s, then by similar considerations we have used in the proof of theorem 6, for every $\varepsilon \geq 0$ we can find large r and s, which satisfy (27), such that

$$0 < \frac{r}{s}(\tilde{\mu}(X)) - \tilde{\mu}(Y) < \varepsilon.$$

Since $\gamma^*(X,Y) \leq r/s$, we deduce that

$$\gamma^*(X,Y) = \gamma(X,Y) \leq \frac{\tilde{\mu}(Y)}{\tilde{\mu}(X)}.$$

The fact that $\gamma(X,Y)\gamma(Y,X) \geq 1$ finishes the proof of theorem 8 and thus of theorem 2.

3 The metric of $\mathbf{Z}(\widetilde{Q})$ and the Hofer metric

Before we start proving the main theorems we add here one more basic construction, that is, the construction of vector fields generating the elements of the subgroup of the contactomorphism group, *Quant* (recall that *Quant* is the subgroup of all contactomorphisms which preserve the contact form, see subsection (1.3)). The purpose of this construction is to give an intuitive geometrical description of the group *Quant*.

Given (M,ω) a symplectic manifold, such that $[\omega] \in H^2(M,\mathbb{Z})$, and (P,ξ) its prequantization (as we remarked in 1.3 P carries the structure of a principal S^1 bundle over M) recall that in such a case the manifold P carries a normalized 1-form, α, globally defined on P, such that $\xi = \{v \in T_aP | \alpha(v) = 0, \forall a \in P\}$. Moreover α is a connection form on P such that ω, after normalization, is its curvature. That is $d\alpha = p^*\omega$ where $p : P \to M$ is the fiber bundle projection from P to M (See [18] and [24] Chapter 8 as a source on connections and curvature of principal bundle) This means that for every $a \in P$ we have the smooth decomposition

$$T_aP = \xi_{|a} \oplus V_a \tag{28}$$

where V_a is the vertical subspace of T_aP defined canonically as the subspace tangent to the fiber over the point $p(a)$ at the point a. Equivalently V_a can be defined as the restriction to a of all vector fields which are the image under the Lie algebra homomorphism from the Lie algebra of S^1 into the Lie algebra of vector fields of P under the S^1 action (recall that this Lie algebra homomorphism is induced by the action of the Lie group S^1 on P).

Now, it can be easily verified that for every $x \in M$ and $a \in p^{-1}(x)$ p_{*a} is an isomorphism between T_xM and $\xi_{|a}$. Let p^*F be a contact Hamiltonian on P which is a lift of a Hamiltonian F on M (see subsection 1.3).

Using the decomposition (28) one can describe the contact vector field obtained from p^*F as follows: Denote by $X_F(x)$ the Hamiltonian vector field obtained from F at the point x. Let $a \in p^{-1}(x)$ be a point above x. Then the contact vector vector field X_{p^*F} at the point a is the sum

$$X_{p^*F} = p_*^{-1}(X_F(x)) \oplus v \tag{29}$$

where $p_*^{-1}(X_F(x)) \in \xi_{|a}$ and v is the unique vector in V_a determined by the condition $\alpha(v) = F(x)$. One can say that the horizontal component of the contact vector field is determined by the (symplectic) Hamiltonian vector field and its vertical component is the measure of transversality (determined by F) to the horizontal field ξ.

Take for example any constant time-dependent Hamiltonian $c(t) : M \to \mathbb{R}$. Then in this case we have $p_*^{-1}(X_F(x)) = 0$ and the vertical component of the contact vector field is $\alpha(v(t)) = c(t)$. It is easily verified that the dynamics on P after time t is

$$a \mapsto e^{i\int c(t)dt}(a)$$

where we have used the notation from subsection 1.3.

3.1 The Hofer metric and the partial order

We start with the following lemma in which we establish a connection between the Hofer metric and the partial order.

Lemma 9. *Let f be the time-1-map of a flow generated by a Hamiltonian $F \in \mathcal{F}$. Let $\widetilde{F}$ be its lift to the prequantization space and let $\tilde{f}$ be its time-1-map. Then we have the formulas:*

$$\|f\|_+ = \inf\{s|\ e^{is} \geq \tilde{f}\}, \tag{30}$$

$$\|f\|_- = \inf\{s|\ \tilde{f} \geq e^{-is}\}. \tag{31}$$

Proof. We prove here formula (30), formula (31) is proved along the same lines. Assume that $e^{is} \geq \tilde{f}$, that is, $e^{is}\tilde{f}^{-1} \geq 1\!\!1$. This means that $\exists H_1 \geq 0$ which generates the element $e^{is}\tilde{f}^{-1}$.

Moreover we have that the contact Hamiltonian $H = H_1 - s$ connects $1\!\!1$ to $\tilde{f}^{-1}$ which implies that $-H$ connects $1\!\!1$ to $\tilde{f}$. Now the fact that $H_1 \geq 0$ gives us that $-H \leq s$ if and only if $\max(-H) \leq s$.

Recalling that $-H$ generates $\tilde{f}$ (note that we can assume that $H \in \mathcal{F}$) we have

$$\|f\|_+ \leq \inf\{s|\ e^{is} \geq \tilde{f}\}.$$

On the other hand assume that $s \geq \|f\|_+$. This means that $\exists F$ such that $F \in \mathcal{F}$ and $\max F \leq s$. Define $-H = F$. So we have $H + s \geq 0$ which implies that $e^{is}\tilde{f}^{-1} \geq 1\!\!1$ which implies that $e^{is} \geq \tilde{f}$.

Thus we conclude

$$\|f\|_+ \geq \inf\{s|\ e^{is} \geq \tilde{f}\}$$

which is the desired inequality.

As a result we have the following corollary.

Corollary 1. *Let $F : M \to \mathbb{R}$ be a Hamiltonian. Let $\{f_t\}$ be its flow and let f be the time-1-map of the flow. Let $\widetilde{F}$ be the lift of F and denote by $\tilde{f}$ its time-1-map. Then*

$$e^{is}\tilde{f} \geq 1\!\!1 \Longleftrightarrow \|f\|_- \leq s, \tag{32}$$

and

$$e^{is}\tilde{f} \leq 1\!\!1 \Longleftrightarrow \|f\|_+ \leq -s\ . \tag{33}$$

Proof. We use formula (30) to derive formula (33). The derivation of formula (32) from formula (31) can be shown in the same way.

For both directions note that

$$\|f\|_+ \leq -k$$

if and only if

$$\inf\{s|\ e^{is} \geq \tilde{f}\} \leq -k$$

if and only if

$$1\!\!1 \geq e^{-i(-k)}\tilde{f}$$

if and only if

$$1\!\!1 \geq e^{ik}\tilde{f}.$$

3.2 Proof of Theorem 4

We start with the following formulas ($t, s, \tilde{f}$ are as in the theorem).

$$\gamma(e^{it}, e^{is}\tilde{f}) = \frac{s + \|f\|_{+,\infty}}{t}, \tag{34}$$

$$\gamma(e^{is}\tilde{f}, e^{it}) = \frac{t}{s - \|f\|_{-,\infty}}. \tag{35}$$

We will prove formula (34). Formula (35) is proved along the same lines. First, recall that the function γ can be defined alternatively as in (16), which is the definition we use here. So assume that

$$c := \frac{r}{p} \geq \gamma(e^{it}, e^{is}\tilde{f}),$$

thus,

$$e^{itr} \geq e^{isp}\tilde{f}^p$$

(here we use the fact that all maps commutes, and the alternative definition of γ)

$$\Rightarrow e^{i(tr-sp)} \geq \tilde{f}^p \Rightarrow tr - sp \geq \|f^p\|_+$$

(here we use formula (30))

$$\Rightarrow t\frac{r}{p} \geq \frac{\|f^p\|_+}{p} + s \Rightarrow c \geq \frac{s + \frac{\|f^p\|_+}{p}}{t}.$$

Since p can be as large as we want, we conclude that

$$c \geq \frac{s + \|f\|_{+,\infty}}{t}$$

from which we get

$$\gamma(e^{it}, e^{is}\tilde{f}) \geq \frac{s + \|f\|_{+,\infty}}{t}.$$

On the other hand, assume that

$$c \geq \frac{s + \|f\|_{+,\infty}}{t}.$$

Then for any sequence $\{k_n\}$ of real numbers such that $k_n \to \infty$,

$$c \geq \frac{s + \frac{\|f^{k_n}\|_+}{k_n}}{t}$$

for n sufficiently large.
We conclude that

$$tk_nc - sk_n \geq \|f^{k_n}\|_+ \Rightarrow e^{i(\frac{trk_n}{p} - sk_n)} \geq \tilde{f}^{k_n}$$

from which we get

$$e^{it\frac{rk_n}{p}} \geq (e^{is}\tilde{f})^{k_n}. \tag{36}$$

Now choose a sequence α_n such that $0 \leq \alpha_n \leq 1$ and $\frac{rk_n}{p} + \alpha_n \in \mathbb{N}$. From inequality (36) and the choice of α_n we get

$$e^{it(\frac{rk_n}{p} + \alpha_n)} \geq (e^{is}\tilde{f})^{k_n}.$$

From the last inequality and the definition of γ we use here we get

$$\frac{\frac{rk_n}{p} + \alpha_n}{k_n} \geq \gamma(e^{it}, e^{is}\tilde{f}),$$

thus

$$\frac{r}{p} + \frac{\alpha_n}{k_n} \geq \gamma(e^{it}, e^{is}\tilde{f}).$$

Now, since $\lim_{n\to\infty} \frac{\alpha_n}{k_n} = 0$ we get that

$$c = \frac{r}{p} \geq \gamma(e^{it}, e^{is}\tilde{f})$$

which is what we need.

At this point we remark that due to (34) and (35) and the fact that $\gamma(f,g)\gamma(g,f) \geq 1$ for every f and g, we infer that the r.h.s. of (12) is defined.

Now we can actually calculate $K(e^{it}, e^{is}\tilde{f})$. From this calculation we will derive formula (12).

By the very definition of K and formulas (34) and (35) we have

$$\begin{aligned}
K(e^{it}, e^{is}\tilde{f}) &= \max\{\log(s + \|f\|_{+,\infty}) - \log t,\ \log t - \log(s - \|f\|_{-,\infty})\} \\
&= \frac{\log(s + \|f\|_{+,\infty}) - \log(s - \|f\|_{-,\infty})}{2} \\
&+ \frac{|2\log t - (\log(s + \|f\|_{+,\infty}) + \log(s - \|f\|_{-,\infty}))|}{2}.
\end{aligned}$$

Now clearly this expression attains its infimum when

$$t = e^{\frac{\log(s+\|f\|_{+,\infty}) + \log(s - \|f\|_{-,\infty})}{2}}.$$

Thus we conclude that the l.h.s. of (12) equals

$$= \frac{\log(s + \|f\|_{+,\infty}) - \log(s - \|f\|_{-,\infty})}{2}$$

or simply

$$\frac{1}{2}\log\frac{s + \|f\|_{+,\infty}}{s - \|f\|_{-,\infty}}$$

which is the desired expression.

3.3 A simple expression for the Hofer metric

We begin with the following lemma. For the definition of F see the preliminaries, in subsection 1.3, of Theorem 5.

Lemma 10. *Let $F \in \mathtt{F}$ be a Hamiltonian. Let f be the time-1-map of the flow generated by F. Then we have the formulas:*

$$\max F = \|f\|_+, \tag{37}$$

and

$$-\min F = \|f\|_-. \tag{38}$$

Proof. Our starting point is the following fact, due to Polterovich, which can be easily deduced from [21].

Fact. Let (M, ω) be a symplectic manifold. Let L be a closed Lagrangian in M with the stable Lagrangian intersection property. Moreover, let F be an autonomous Hamiltonian defined on M such that $F \in \mathcal{F}$ (see subsection 1.3) and for some positive constant C we have $F_{|L} \geq C$. Denote by f the time-1-map defined by F. Then we have

$$\|f\|_+ \geq C. \tag{39}$$

Now let L_α be a Lagrangian of the family of Lagrangians which foliate M_0 as in theorem 5. Assume that $F_{|L_\alpha} \geq C_\alpha$. Then using (39) we have the following double inequality.

$$\max F \geq \|f\|_+ \geq C_\alpha, \,\forall \alpha$$

where in the first inequality we have used the very definition of the norm $\| \ \|_+$. Now due to the fact that the set of Lagrangians foliates an open dense set in M we get that

$$\max F = \max_\alpha \max_{L_\alpha} F.$$

From this we get we get formula (37).

In the same manner we obtain formula (38).

3.4 Proof of Theorem 5

We now calculate the relative growth of elements of V (see subsection 1.3 for the definition of V).

Let $F, G \in \mathtt{F}$. Denote by $\varphi_F$the time-1- map of F and by φ_G the time-1-map of G. Let $\widetilde{\varphi}_F$, $\widetilde{\varphi}_G$ be their lifts to the prequantization space. Now let s, t be any real numbers such that

$$e^{is}\widetilde{\varphi}_F, e^{it}\widetilde{\varphi}_G \geq \mathbb{1}.$$

Note that this means that $e^{is}\widetilde{\varphi}_F, e^{it}\widetilde{\varphi}_G$ are in V and of course all elements of V can be characterized in this way. Now, by definition of the relative growth, we have

$$\gamma(e^{is}\widetilde{\varphi}_F, e^{it}\widetilde{\varphi}_G) = \lim_{n\to\infty} \frac{\gamma_n(e^{is}\widetilde{\varphi}_F, e^{it}\widetilde{\varphi}_G)}{n}.$$

We calculate γ by a direct calculation of the functions γ_n on elements of V.

By the definition of γ_n we have

$$\begin{aligned}\gamma_n(e^{is}\widetilde{\varphi}_F, e^{it}\widetilde{\varphi}_G) &= \inf\{m|\ e^{ims}\widetilde{\varphi}_F^m \geq e^{int}\widetilde{\varphi}_G^n\}\\ &= \inf\{m|\ e^{i(ms-nt)}\widetilde{\varphi}_{mF-nG} \geq 1\!\!1\}\\ &= \inf\{m|\ \|\varphi_{mF-nG}\|_- \leq ms-nt\}\\ &= \inf\{m|\ -\min(mF-nG) \leq ms-nt\}\end{aligned}$$

where in the second equality we have used the fact that the functions of the union $\mathsf{F}\cup\{e^{is}\}_{s\in\mathbb{R}}$ are all Poisson commute, in the third equality we have used Corollary 1, and the fourth equality follows from formula (38).

So we need m that will satisfy

$$\max(nG-mF) \leq ms-nt$$

if and only if

$$nG-mF \leq ms-nt$$

if and only if

$$n(G+t) \leq m(F+s).$$

Dividing both sides of the last inequality by the positive function $F+s$ (recall that $F+s$ generates the dominant $e^{is}\widetilde{\varphi}_F$) we get the equivalent condition

$$\max(\frac{G+t}{F+s})n \leq m.$$

Thus we have

$$\gamma_n(e^{is}\widetilde{\varphi}_F, e^{it}\widetilde{\varphi}_G) - 1 \leq \max(\frac{G+t}{F+s})n \leq \gamma_n(e^{is}\widetilde{\varphi}_F, e^{it}\widetilde{\varphi}_G),$$

thus

$$\frac{\gamma_n(e^{is}\widetilde{\varphi}_F, e^{it}\widetilde{\varphi}_G) - 1}{n} \leq \max(\frac{G+t}{F+s}) \leq \frac{\gamma_n(e^{is}\widetilde{\varphi}_F, e^{it}\widetilde{\varphi}_G)}{n}.$$

So we have the formula

$$\gamma(e^{is}\widetilde{\varphi}_F, e^{it}\widetilde{\varphi}_G) = \max(\frac{G+t}{F+s}). \tag{40}$$

Now, let $\widetilde{f}, \widetilde{g} \in V$ generated by the Hamiltonians $\widetilde{F}, \widetilde{G}$ respectively. Then according to formula (40) we have

$$K(\widetilde{f}, \widetilde{g}) = \max |\log \widetilde{F} - \log \widetilde{G}| \tag{41}$$

Using formula (41), for K, we define the isometry of $\{\mathtt{F}, \| \ \|_{max}\}$ to Z.

Let $\widetilde{f} \in V$ generated by the Hamiltonian $\widetilde{F}$. Then the correspondence $\widetilde{f} \leftrightarrow \log \widetilde{F}$ clearly induces the required isometric imbedding of $\{\mathtt{F}, \| \ \|_{max}\}$ into Z. This conclude the proof of the theorem.

Acknowledgments. I want to thank deeply Leonid Polterovich for reaching a helping hand each time it was needed, in all aspects of academic research activity: support, ideas, and encouragement. I also want to thank Yasha Eliashberg for inviting me to Stanford University and for sharing with me his opinion on this work, and his kind hospitality. The first part of this work started at the beginning of my Ph.D. studies and benefited a lot from many discussions with Assaf Goldberger. I thank him very much for his help.

References

1. Abraham,R.,Marsden,J. *Foundations of Mechanics*, second edition, 1994. Addison-Wesley Publishing Company.
2. Arnold,V.I. *Mathematical Methods of Classical Mechanics*, second edition, Graduate Texts in Mathematics; 60, 1989.
3. Arnold,V.I.,Givental,A. Symplectic Geometry, in, *Encyclopaedia of Mathematical Sciences, Vol. 4, Dynamical systems IV,* V. Arnold, S. Novikov eds., Springer, 1990, pp. 1-136.
4. Banyaga, A.,*Sur la structure du groupe des difféomorphismes qui préservent une forme symplectique,* Comm. Math. Helv. **53**:2 (1978), 174-227.
5. Barge,J.,Ghys,E. *Cocycle D'Euler et de Maslov* Math.Ann. **294**:2 (1992) 235-265.
6. Bates,S.,Weinstein,A. *Lectures on the Geometry of Quantization.* AMS, Berkeley Center for Pure and Applied Mathematics, 1997.
7. Bavard, C. *Longueur stable des commutateurs,* L'Enseign. Math. **37**:1-2 (1991), 109-150.
8. Bott,R.,Tu,L.W. *Differential Forms in algebraic Topology,* Springer, 1982.
9. Carter,R.,Segal,G.,MacDonald,I. *Lectures on Lie Groups and Lie Algebras.* London Mathematical Society Student Texts 32, Cambridge University Press.
10. Dupont,J. *Bounds for the chercteristic numbers of flat bundles.* Notes Math.,vol 763 p.109-119 Berlin, Heidelberg, New York, Springer, 1979.
11. Eliashberg,Y., Kim,S.K., Polterovich,P. *Geometry of contact transformations and domains: orderability vs. squeezing,* Preprint math.SG/0511658, 2005.
12. Eliashberg,Y.;Polterovich,L. *Partially ordered groups and geometry of contact transformations.* Geom.Funct.Anal., **10**(2000), 6, 1448-1476.
13. Givental A.B. *Nonlinear generalization of the Maslov index.* Theory of singularities and its applications, Adv. Soviet Math., vol,1. American Mathematical Society, Rhode Island,1990, p.71-103.

14. Hall,B.C. *Lie groups, Lie Algebras, and Representations: An Elementary Introduction.*, Graduate Texts in Mathematics; 222, Springer, 2003.
15. Hofer,H.,Zehnder,E. *Symplectic Invariants and Hamiltonian Dynamics*, Birkhäuser Advanced Texts, Birkhäuser Verlag, 1994.
16. Kirillov,A. Geometric quantization, in *Encyclopaedia of Mathematical Sciences, Vol. 4, Dynamical systems IV,* V. Arnold, S. Novikov eds., Springer, 1990, pp. 137-172.
17. McDuff,D., Salamon, D., *Introduction to symplectic topology*, second edition, Oxford University Press, Oxford, 1998.
18. Nakahara,M. *Geometry, Topology and Physics*, Institute of Physics Publishing Bristol and Philadelphia, 1990.
19. Olshanskii,G.I. *Invariant orderings in simple Lie groups. The solution to E.B Vinberg's problem.* Functional Analysis Appl. 16 (1982), 4, 311-313.
20. Polterovich L. *The geometry of the group of symplectic diffeomorphisms.* Lectures in Mathematics ETH Zürich, Birkhäuser Verlag, Basel, 2001.
21. Polterovich L. *Hofer's diameter and Lagrangian intersections,.* Int.Math.Res.Notices **4** (1998), 217-223.
22. Polterovich,L.;Siburg,K.F.;*On the asymptotic geometry of area preserving maps* .Math.Res.Lett. **7.** 233-243 (2000).
23. Robbin, J.W. and Salamon, D.A. *The Maslov index for paths*, Topology, **32**, (1993), 827-844.
24. Spivak,M. *A Comprehensive Introduction to Differential Geometry, Volume 2,* third edition, 1999, Publish or Perish,Inc.
25. Weinstein,A. *Lectures on Symplectic manifolds*, Regional Conference Series in Mathematics **29**, Amer. Math. Soc., 1977.
26. Weinstein,A. *Cohomology of Symplectomorphism Groups and Critical Values of Hamiltonians.* Math.Zeit., 201, (1989), 75-82. .
27. Woodhouse,N.M.J. *Geometric Quantization*, second edition, Oxford University Press, 1991.

Towards Quantum Cohomology of Real Varieties

Özgür Ceyhan

Max-Planck-Institute for Mathematics, Bonn, Germany
ceyhan@mpim-bonn.mpg.de

Summary. This chapter is devoted to a discussion of Gromov–Witten–Welschinger (GWW) classes and their applications. In particular, Horava's definition of quantum cohomology of real algebraic varieties is revisited by using GWW classes and is introduced as a differential graded operad. In light of this definition, we speculate about mirror symmetry for real varieties.

AMS Classification Codes (2000). 14N35, 14P99, 81T30

The strangeness and absurdity of these replies arise from the fact that modern history, like a deaf man, answers questions no one has asked.
Leo Tolstoy, *War and Peace.*

1 Introduction

1.1 Quantum cohomology of complex varieties

Let $X(\mathbb{C})$ be a projective algebraic variety, and let $\overline{M}_{\mathbf{S}}(\mathbb{C})$ be the moduli space of $\mathbf{S}$-pointed complex stable curves of genus zero.

In their seminal work [24], Kontsevich and Manin define the Gromov–Witten (GW) classes of the variety $X(\mathbb{C})$ as a collection of linear maps

$$\{I^X_{\mathbf{S},\beta} : \bigotimes_{\mathbf{S}} H^*(X(\mathbb{C})) \to H^*\overline{M}_{\mathbf{S}}(\mathbb{C})\}$$

that are expected to satisfy a series of formal and geometric properties. These invariants actually give appropriate enumerations of rational curves in $X(\mathbb{C})$ satisfying certain incidence conditions.

The quantum cohomology of $X(\mathbb{C})$ is a formal deformation of its cohomology ring. The parameters of this deformation are coordinates on the space

Ö. Ceyhan et al. (eds.), *Arithmetic and Geometry Around Quantization*,
Progress in Mathematics 279, DOI: 10.1007/978-0-8176-4831-2_4,

$H^*(X(\mathbb{C}))$, and the structure constants are the third derivatives of a formal power series Φ whose coefficients are the top-dimensional GW classes of $X(\mathbb{C})$.

The formal solution Φ of associativity, or WDVV, equation has provided a source for solutions to problems in complex enumerative geometry.

A refined algebraic picture: Homology operad of $\overline{M}_{\mathbf{S}}(\mathbb{C})$. By dualizing $\{I^X_{\mathbf{S},\beta}\}$, a more refined algebraic structure on $H^*(X(\mathbb{C}))$ is obtained:

$$\mathbf{Y_S} : H_*\overline{M}_{\mathbf{S}\cup\{s\}}(\mathbb{C}) \to Hom(\bigotimes_{\mathbf{S}} H^*X(\mathbb{C}), H^*X(\mathbb{C})).$$

In this picture, any homology class in $\overline{M}_{\mathbf{S}\cup\{s\}}(\mathbb{C})$ is interpreted as an n-ary operation on $H^*X(\mathbb{C})$. The additive relations in $H_*\overline{M}_{\mathbf{S}\cup\{s\}}(\mathbb{C})$ become identities between these operations. Therefore, $H^*X(\mathbb{C})$ carries a structure of an algebra over the cyclic operad $H_*\overline{M}_{\mathbf{S}\cup\{s\}}(\mathbb{C})$.

One of the remarkable basic results in the theory of the associativity equations (or Frobenius manifolds) is the fact that their formal solutions are the same as cyclic algebras over the homology operad $H_*\overline{M}_{\mathbf{S}\cup\{s\}}(\mathbb{C})$.

1.2 Quantum cohomology of real varieties

Physicists have long suspected that there should exist analogous algebraic structures in open-closed string theory arising from the enumeration of real rational curves/discs with Lagrangian boundary condition. By contrast with the spectacular achievements of closed string theory in complex geometry, the effects of open-closed string theory in real algebraic geometry remained deficient due to the lack of suitable enumerative invariants. The main obstacle with defining real enumerative invariants is that the number of real objects usually varies along the parameter space.

The perspective for the real situation radically changed after the discovery of Welschinger invariants. In a series of papers [39]–[42], J.Y. Welschinger introduced a set of invariants for real varieties that give lower bounds on the number of real solutions.

Welschinger invariants. Welschinger has defined a set of invariants counting, with appropriate weight ± 1, real rational J-holomorphic curves intersecting a generic real configuration (i.e., real or conjugate pairs) of marked points. Unlike the usual homological definition of Gromov–Witten invariants, Welschinger invariants are originally defined by assigning signs to individual curves based on certain geometric-topological criteria (such as number of solitary double points in the case of surfaces, self-linking numbers in the case of 3-folds etc.). A homological interpretation of Welschinger invariants has been recently given by J. Solomon (see [36]):

Let $(X(\mathbb{C}), c_X)$ be a real variety and $X(\mathbb{R}) := \mathrm{Fix}(c_X)$ be its real part. Let $\overline{R}^{\sigma}_{\mathbf{S}}(X, c_X, \mathbf{d})$ be the moduli space of real stable maps which are invariant under relabeling by the involution σ.

Let μ_* and α_* be Poincare duals of point classes (respectively in $X(\mathbb{C})$ and $X(\mathbb{R})$). Solomon showed that Welschinger invariants $N^{\sigma}_{\mathbf{S},\mathbf{d}}$ can be defined in terms of the (co)homology of the moduli space of real maps $\overline{R}^{\sigma}_{\mathbf{S}}(X, c_X, \mathbf{d})$:

$$N^{\sigma}_{\mathbf{S},\mathbf{d}} := \int_{[\overline{R}^{\sigma}_{\mathbf{S}}(X,c_X,\mathbf{d})]} \left\{ \bigwedge_{\{s,\bar{s}\}\subset \mathbf{Perm}(\sigma)} ev_s^*(\mu_s) \wedge \bigwedge_{s\in \mathbf{Fix}(\sigma)} ev_s^*(\alpha_s) \right\}. \quad (1)$$

Here, it is important to note that the moduli space $\overline{R}^{\sigma}_{\mathbf{S}}(X, c_X, \mathbf{d})$ has codimension one boundaries which is in fact the source of one of the main difficulties of the definition of open Gromov–Witten invariants. That is why earlier studies on open Gromov–Witten invariants focus on homotopy invariants instead of actual enumerative invariants (see, for instance [11]).

Quantum cohomology of real varieties. Quantum cohomology for real algebraic varieties was introduced surprisingly early, in 1993, by P. Horava in [16]. In his paper, Horava describes a $\mathbb{Z}_2$-equivariant topological sigma model of a real variety $(X(\mathbb{C}), c_X)$ whose set of physical observables (closed and open string states) is a direct sum of the cohomologies of $X(\mathbb{C})$ and $X(\mathbb{R})$;

$$\mathcal{H}_c \oplus \mathcal{H}_o := H^*(X(\mathbb{C})) \oplus H^*(X(\mathbb{R})).$$

The analogue of the quantum cohomology ring in Horava's setting is a structure of $H^*(X(\mathbb{C}))$-module on $H^*(X(\mathbb{C})) \oplus H^*(X(\mathbb{R}))$ obtained by deforming cup and mixed products.

Solomon's homological interpretation of Welschinger invariants opens a new gate to reconsider the quantum cohomology of real varieties. In this chapter, we define the *Welschinger classes* using Gromov-Witten theory as a guideline: By considering the following diagram

$$\begin{array}{ccc} \overline{R}^{\sigma}_{\mathbf{S}}(X, c_X, \mathbf{d}) & \xrightarrow{ev} & \prod_{s\in \mathbf{Perm}(\sigma)/\sigma} X(\mathbb{C}) \times \prod_{s\in \mathbf{Fix}(\sigma)} X(\mathbb{R}) \\ {\scriptstyle\nu}\downarrow & & \\ \overline{M}^{\sigma}_{\mathbf{S}}(\mathbb{R}) & & \end{array}$$

we give Welschinger classes as a family of linear maps

$$W^X_{\mathbf{S},\mathbf{d}} : \bigotimes_{\mathbf{Perm}(\sigma)/\sigma} H^*(X(\mathbb{C})) \bigotimes_{\mathbf{Fix}(\mathbf{S})} H^*(X(\mathbb{R}); \det(TX(\mathbb{R}))) \to H^*(\overline{M}^{\sigma}_{\mathbf{S}}(\mathbb{R}), \mathfrak{D})$$

where $\overline{M}^{\sigma}_{\mathbf{S}}(\mathbb{R})$ is the moduli space of pointed real stable curves and $\mathfrak{D}$ is the union of its substrata of codimension one or higher.

In a recent paper [4], we have shown that homology of the moduli space $\overline{M}^{\sigma}_{\mathbf{S}}(\mathbb{R})$ is isomorphic to the homology of a combinatorial graph complex which is generated by the strata of $\overline{M}^{\sigma}_{\mathbf{S}}(\mathbb{R})$. By using a reduced version of this graph complex $\mathcal{C}_\bullet$, in this chapter, we introduce *quantum cohomology of a real variety* $(X(\mathbb{C}), c_X)$. We construct a differential graded (partial) operad

$$\{\mathbf{Z}^\sigma_{\mathbf{S}} : \mathcal{C}_\bullet \to Hom(\bigotimes_{\mathbf{Perm}(\sigma)/\sigma} \mathcal{H}_c \bigotimes_{\mathbf{Fix}(\sigma)} \mathcal{H}_o, \mathcal{H}_o)\}$$

along with an algebra over $H_*\overline{M}_{\mathbf{S}\cup\{s\}}$-operad

$$\{\mathbf{Y}_{\mathbf{S}} : H_*\overline{M}_{\mathbf{S}\cup\{s\}} \to Hom(\bigotimes_{\mathbf{S}} \mathcal{H}_c, \mathcal{H}_c)\}$$

which serve as the *quantum cohomology of the real variety* $(X(\mathbb{C}), c_X)$.

A reconstruction theorem for Welschinger invariants. Until very recently, the only calculation technique for Welschinger invariants was Mikhalkin's method which is based on tropical algebraic geometry (see [31,32]). By using Mikhalkin's technique, Itenberg, Kharlamov, and Shustin proved a recursive formula for Welschinger invariants [17]. In their work, they also give a definition of higher genus Welschinger invariants in the tropical setting.

However, a real version of Kontsevich's recursive formula (for Welschinger invariants) was still missing. In [37], Jake Solomon announced a differential equation satisfied by the generating function of Welschinger invariants and gave a recursion relation between Welschinger invariants of $\mathbb{P}^2(\mathbb{C})$ with standard real structure.

Solomon's method is very similar to the calculations of Gromov–Witten invariants in the complex situation i.e., considering certain degenerations and homological relation between degenerate loci. However, the cycle in the moduli of real maps which he considers is quite intriguing. We recently noticed that the cycle, which leads to Solomon's formula, is a combination of certain cycles satisfying A_∞ and Cardy type relations.

The enumerative aspects of quantum cohomology of real varieties and its relation to Solomon's formula will be presented in a subsequent paper [5].

'Realizing' mirror symmetry. Kontsevich's conjecture of homological mirror symmetry and our definition of quantum cohomology of real varieties as DG-operads, suggest a correspondence between two open-closed homotopy algebras (in an appropriate sense):

Symplectic side X (A-side)	$\Longleftrightarrow$	Complex side Y (B-side)
$(\mathcal{H}_c, \mathcal{H}_o) =$ $(H^*(X), HF^*(X(\mathbb{R}), X(\mathbb{R})))$	$\Leftrightarrow$	$(\widehat{\mathcal{H}}_c, \widehat{\mathcal{H}}_o) =$ $(H^{*,*}(Y), H^{0,*}(Y))$

A brief account of the real version of mirror symmetry is given in the final section of this chapter.

Acknowledgements. I take this opportunity to express my deep gratitude to Yuri Ivanovic Manin and Matilde Marcolli. They have been a mathematical inspiration for me ever since I met them, but as I've gotten to know them better, what have impressed me most are their characters. Their interest, encouragement, and suggestions have been invaluable to me.

I also would like to thank D. Auroux and M. Abouzaid. The last part of this chapter containing the discussion on mirror symmetry was first inspired during a discussion with them at Gökova Conference.

I am thankful to O. Cornea, C. Faber, F. Lalonde, A. Mellit, S. Merkulov, D. Radnell, and A. Wand for their comments and suggestions.

This paper was conceived during my stay at Max-Planck-Institut für Mathematik (MPI), Bonn, its early draft was written at Centre de Recherhes Mathématiques (CRM), Montreal, and this final version was prepared during my stay at Mittag-Leffler Institute, Stockholm. Thanks are also due to these three institutes for their hospitality and support.

Table of contents:

Part I: Moduli spaces of pointed complex and real curves

In order to define correlators of open-closed string theories, we need explicit descriptions of the homologies of both the moduli space of pointed complex curves and moduli space of pointed real curves. This section reviews the basic facts on pointed complex/real curves of genus zero, their moduli spaces, and the homologies of these moduli spaces.

Notation/Convention

We denote a finite set $\{s_1, \ldots, s_n\}$ by $\mathbf{S}$, and the symmetric group consisting of all permutations of $\mathbf{S}$ by $\mathbb{S}_n$. We denote the set $\{1, \ldots, n\}$ simply by $\underline{\mathbf{n}}$.

In this chapter, the genus of all curves is zero except when the contrary is stated. Therefore, we usually omit mentioning the genus of curves.

2 Moduli space of pointed complex curves

An $\mathbf{S}$-*pointed curve* $(\Sigma;\mathbf{p})$ is a connected complex algebraic curve Σ with distinct, smooth, *labeled points* $\mathbf{p}=(p_{s_1},\dots,p_{s_n})\subset\Sigma$ satisfying the following conditions:

- Σ has only nodal singularities.
- The arithmetic genus of Σ is equal to zero.

The nodal points and labeled points are called *special* points.

A *family of* $\mathbf{S}$-*pointed curves* over a complex manifold $B(\mathbb{C})$ is a proper, flat holomorphic map $\pi_B:U_B(\mathbb{C})\to B(\mathbb{C})$ with n sections $p_{s_1},\dots,p_{s_n}$ such that each geometric fiber $(\Sigma(b);\mathbf{p}(b))$ is an $\mathbf{S}$-pointed curve.

Two $\mathbf{S}$-pointed curves $(\Sigma;\mathbf{p})$ and $(\Sigma';\mathbf{p}')$ are *isomorphic* if there exists a bi-holomorphic equivalence $\Phi:\Sigma\to\Sigma'$ mapping p_s to p'_s for all $s\in\mathbf{S}$.

An $\mathbf{S}$-pointed curve is *stable* if its automorphism group is trivial (i.e., on each irreducible component, the number of singular points plus the number of labeled points is at least three).

Graphs

A *graph* γ is a pair of finite sets of *vertices* $\mathbf{V}_\gamma$ and *flags* (or *half edges*) $\mathbf{F}_\gamma$ with a boundary map $\boldsymbol{\partial}_\gamma:\mathbf{F}_\gamma\to\mathbf{V}_\gamma$ and an involution $\mathbf{j}_\gamma:\mathbf{F}_\gamma\to\mathbf{F}_\gamma$ ($\mathbf{j}_\gamma^2=\mathbf{id}$). We call $\mathbf{E}_\gamma=\{(f_1,f_2)\in\mathbf{F}_\gamma^2\mid f_1=\mathbf{j}_\gamma(f_2)\ \&\ f_1\neq f_2\}$ the set of *edges*, and $\mathbf{T}_\gamma=\{f\in\mathbf{F}_\gamma\mid f=\mathbf{j}_\gamma(f)\}$ the set of *tails*. For a vertex $v\in\mathbf{V}_\gamma$, let $\mathbf{F}_\gamma(v)=\boldsymbol{\partial}_\gamma^{-1}(v)$ and $|v|=|\mathbf{F}_\gamma(v)|$ be the *valency* of v.

We think of a graph γ in terms of its *geometric realization* $||\gamma||$ as follows: Consider the disjoint union of closed intervals $\bigsqcup_{f_i\in\mathbf{F}_\gamma}[0,1]\times f_i$, and identify $(0,f_i)$ with $(0,f_j)$ if $\boldsymbol{\partial}_\gamma(f_i)=\boldsymbol{\partial}_\gamma(f_j)$, and identify (t,f_i) with $(1-t,\mathbf{j}_\gamma(f_i))$ for $t\in]0,1[$ and $f_i\neq\mathbf{j}_\gamma(f_i)$. The geometric realization of γ has a piecewise linear structure.

A *tree* is a graph whose geometric realization is connected and simply-connected. If $|v|>2$ for all vertices, then such a tree is called *stable*.

There are only finitely many isomorphism classes of stable trees whose set of tails $\mathbf{T}_\gamma$ is equal to $\mathbf{S}$ (see, [28] or [2]). We call the isomorphism classes of such trees $\mathbf{S}$-*trees*.

Dual trees of S-pointed curves

Let $(\Sigma;\mathbf{p})$ be an $\mathbf{S}$-pointed stable curve and $\eta:\hat{\Sigma}\to\Sigma$ be its normalization. Let $(\hat{\Sigma}_v;\hat{\mathbf{p}}_v)$ be the following pointed stable curve: $\hat{\Sigma}_v$ is a component of $\hat{\Sigma}$, and $\hat{\mathbf{p}}_v$ is the set of points consisting of the pre-images of special points on $\Sigma_v:=\eta(\hat{\Sigma}_v)$. The points $\hat{\mathbf{p}}_v=(p_{f_1},\dots,p_{f_{|v|}})$ on $\hat{\Sigma}_v$ are ordered by the elements f_* in the set $\{f_1,\dots,f_{|v|}\}$.

The *dual tree* of an **S**-pointed stable curve $(\Sigma; \mathbf{p})$ is an **S**-tree γ where

- $\mathbf{V}_\gamma$ is the set of components of $\hat{\Sigma}$.
- $\mathbf{F}_\gamma(v)$ is the set consisting of the pre-images of special points in Σ_v.
- $\boldsymbol{\partial}_\gamma : f \mapsto v$ if and only if $p_f \in \hat{\Sigma}_v$.
- $\mathbf{j}_\gamma : f \mapsto f$ if and only if $\eta(p_f)$ is a labeled point, and $\mathbf{j}_\gamma : f_1 \mapsto f_2$ if and only if $p_{f_1} \in \hat{\Sigma}_{v_1}$ and $p_{f_2} \in \hat{\Sigma}_{v_2}$ are the pre-images of a nodal point $\Sigma_{v_1} \cap \Sigma_{v_2}$.

Combinatorics of degenerations: Contraction morphisms of S-trees

Let $(\Sigma; \mathbf{p})$ be an **S**-pointed stable curve whose dual tree is γ. Consider the deformations of a nodal point of $(\Sigma; \mathbf{p})$. Such a deformation of $(\Sigma; \mathbf{p})$ gives a *contraction* of an edge of γ: Let $e = (f_e, f^e) \in \mathbf{E}_\gamma$ be the edge corresponding to the nodal point which is deformed, and let $\boldsymbol{\partial}_\gamma(f_e) = v_e, \boldsymbol{\partial}_\gamma(f^e) = v^e$. Consider the equivalence relation $\sim$ on the set of vertices, defined by: $v \sim v$ for all $v \in \mathbf{V}_\gamma \setminus \{v_e, v^e\}$, and $v_e \sim v^e$. Then, there is an **S**-tree γ/e whose set of vertices is $\mathbf{V}_\gamma/\sim$ and whose set of flags is $\mathbf{F}_\gamma \setminus \{f_e, f^e\}$. The boundary map and involution of γ/e are the restrictions of $\boldsymbol{\partial}_\gamma$ and $\mathbf{j}_\gamma$.

We use the notation $\gamma < \tau$ to indicate that the **S**-tree τ is obtained by contracting a set of edges of γ.

2.1 Moduli space of S-pointed curves

The moduli space $\overline{M}_\mathbf{S}(\mathbb{C})$ is the space of isomorphism classes of **S**-pointed stable curves. This space is stratified according to the degeneration types of **S**-pointed stable curves. The degeneration types of **S**-pointed stable curves are combinatorially encoded by **S**-trees. In particular, the principal stratum $M_\mathbf{S}(\mathbb{C})$ parameterizes **S**-pointed irreducible curves, i.e., it is associated to the one-vertex **S**-tree. The principal stratum $M_\mathbf{S}(\mathbb{C})$ is the quotient of the product $(\mathbb{P}^1(\mathbb{C}))^n$ minus the diagonals $\Delta = \bigcup_{k<l}\{(p_{s_1}, \ldots, p_{s_n}) | p_{s_k} = p_{s_l}\}$ by $Aut(\mathbb{P}^1(\mathbb{C})) = PSL_2(\mathbb{C})$.

Theorem 1 (Knudsen & Keel [22, 26]) *(a) For any* $|\mathbf{S}| \geq 3$, $\overline{M}_\mathbf{S}(\mathbb{C})$ *is a smooth projective algebraic variety of (real) dimension* $2|\mathbf{S}| - 6$.

(b) Any family of **S***-pointed stable curves over* $B(\mathbb{C})$ *is induced by a unique morphism* $\kappa : B(\mathbb{C}) \to \overline{M}_\mathbf{S}(\mathbb{C})$. *The universal family of curves* $\overline{U}_\mathbf{S}(\mathbb{C})$ *of* $\overline{M}_\mathbf{S}(\mathbb{C})$ *is isomorphic to* $\overline{M}_{\mathbf{S}\cup\{s_{n+1}\}}(\mathbb{C})$.

(c) For any **S***-tree* γ, *there exists a quasi-projective subvariety* $D_\gamma(\mathbb{C}) \subset \overline{M}_\mathbf{S}(\mathbb{C})$ *parameterizing the* **S***-pointed curves whose dual tree is given by* γ. *The subvariety* $D_\gamma(\mathbb{C})$ *is isomorphic to* $\prod_{v\in\mathbf{V}_\gamma} M_{\mathbf{F}_\gamma(v)}(\mathbb{C})$. *Its (real) codimension (in* $\overline{M}_\mathbf{S}(\mathbb{C})$*) is* $2|\mathbf{E}_\gamma|$.

(d) $\overline{M}_\mathbf{S}(\mathbb{C})$ *is stratified by pairwise disjoint subvarieties* $D_\gamma(\mathbb{C})$. *The closure* $\overline{D}_\gamma(\mathbb{C})$ *of any stratum* $D_\gamma(\mathbb{C})$ *is stratified by* $\{D_{\gamma'}(\mathbb{C}) \mid \gamma' \leq \gamma\}$.

Forgetful morphisms

We say that $(\Sigma; p_{s_1}, \ldots, p_{s_{n-1}})$ is obtained by forgetting the labeled point p_{s_n} of an $\mathbf{S}$-pointed curve $(\Sigma; p_{s_1}, \ldots, p_{s_n})$. However, the resulting pointed curve may well be unstable. This happens when the component Σ_v of Σ supporting p_{s_n} has only two additional special points. In this case, we contract this component to its intersection point(s) with the components adjacent to Σ_v. With this *stabilization*, we extend this map to the whole space and obtain $\pi_{\{s_n\}} : \overline{M}_{\mathbf{S}}(\mathbb{C}) \to \overline{M}_{\mathbf{S}'}(\mathbb{C})$ where $\mathbf{S}' = \mathbf{S} \setminus \{s_n\}$. There exists a canonical isomorphism $\overline{M}_{\mathbf{S}}(\mathbb{C}) \to \overline{U}_{\mathbf{S}'}(\mathbb{C})$ commuting with the projections to $\overline{M}_{\mathbf{S}'}(\mathbb{C})$. In other words, $\pi_{\{s_n\}} : \overline{M}_{\mathbf{S}}(\mathbb{C}) \to \overline{M}_{\mathbf{S}'}(\mathbb{C})$ can be identified with the universal family of curves.

A very detailed study on the moduli space $\overline{M}_{\mathbf{S}}(\mathbb{C})$ can be found in chapter 3.2 and 3.3 in [28], and also in [22, 26].

2.2 Intersection ring of $\overline{M}_{\mathbf{S}}(\mathbb{C})$

In [22], Keel gave a construction of the moduli space $\overline{M}_{\mathbf{S}}(\mathbb{C})$ via a sequence of blowups of $\overline{M}_{\mathbf{S}\setminus\{s_n\}}(\mathbb{C}) \times \mathbb{P}^1(\mathbb{C})$ along certain (complex) codimension two subvarieties. This inductive construction of $\overline{M}_{\mathbf{S}}(\mathbb{C})$ allowed him to calculate the intersection ring in terms of the divisor classes $[\overline{D}_\gamma(\mathbb{C})]$. Note that the divisors $D_\gamma(\mathbb{C})$ parameterize $\mathbf{S}$-pointed curves whose dual trees have only one edge.

For $|\mathbf{S}| \geq 4$, choose $i, j, k, l \in \mathbf{S}$, and let $\gamma, \tau \in \mathcal{T}ree$ such that $\tau \not\approx \gamma$ and $|\mathbf{E}_\tau| = |\mathbf{E}_\gamma| = 1$. We write $ij\gamma kl$ if tails labeled by i, j and k, l belongs to different vertices of γ. We call γ and τ *compatible* if there is no $\{i, j, k, l\} \subset \mathbf{S}$ such that simultaneously $ij\gamma kl$ and $ik\tau jl$.

From now on, we denote the divisor classes $[\overline{D}_\gamma(\mathbb{C})]$ simply by $[\overline{D}_\gamma]$.

Theorem 2 (Keel [22]) *For* $|\mathbf{S}| \geq 3$,

$$H_*(\overline{M}_{\mathbf{S}}(\mathbb{C}); \mathbb{Z}) = \mathbb{Z}[\ [\overline{D}_\gamma] \mid \gamma \in \mathcal{T}ree,\ |\mathbf{E}_\gamma| = 1]/I_{\mathbf{S}}$$

is a graded polynomial ring, deg $[\overline{D}_\gamma] = 1$. *The ideal* $I_{\mathbf{S}}$ *is generated by the following relations:*

1. *For any distinct four elements* $i, j, k, l \in \mathbf{S}$:

$$\sum_{ij\gamma kl} [\overline{D}_\gamma] - \sum_{ik\tau jl} [\overline{D}_\tau] = 0.$$

2. $[\overline{D}_\gamma] \cdot [\overline{D}_\tau] = 0$ *unless* γ *and* τ *are compatible.*

Additive and multiplicative structures of $H_*(\overline{M}_{\mathbf{S}}(\mathbb{C}))$

The precise description of homogeneous elements in $H_*(\overline{M}_{\mathbf{S}}(\mathbb{C}),\mathbb{Z})$ is given by Kontsevich and Manin in [25]. The monomial $[\overline{D}_{\gamma_1}]\cdots[\overline{D}_{\gamma_d}]$ is called *good*, if $|\mathbf{E}_{\gamma_i}|=1$ for all i, and γ_i's are pairwise compatible. Consider any $\mathbf{S}$-tree γ. Any edge $e\in\mathbf{E}_\gamma$ defines an $\mathbf{S}$-tree $\gamma(e)$ which is obtained by contracting all edges of γ but e. Then, we can associate a good monomial

$$[\overline{D}_\gamma]:=\prod_{e\in\mathbf{E}_\gamma}[\overline{D}_{\gamma(e)}]$$

of degree $|\mathbf{E}_\gamma|$ to γ. The map $\gamma\mapsto[\overline{D}_\gamma]$ establishes a bijection between the good monomials of degree d in $H_*(\overline{M}_{\mathbf{S}}(\mathbb{C}),\mathbb{Z})$, and $\mathbf{S}$-trees γ with $|\mathbf{E}_\gamma|=d$ (see [25]). Since boundary divisors intersect transversally when their trees are pairwise compatible, the good monomials are represented by the corresponding closed strata.

Theorem 3 (Kontsevich and Manin [25]) *The classes of good monomials linearly generate $H_*(\overline{M}_{\mathbf{S}}(\mathbb{C});\mathbb{Z})$.*

Multiplication in $H_*(\overline{M}_{\mathbf{S}}(\mathbb{C}))$

Let $\tau,\gamma\in\mathcal{T}ree$ and $|\mathbf{E}_\tau|=1$. In [25], a product formula of $[\overline{D}_\tau]\cdot[\overline{D}_\gamma]$ is given in three distinguished cases:

A. Suppose that there exists an $e\in\mathbf{E}_\gamma$ such that $\gamma(e)$ and τ are not compatible (i.e., $\overline{D}_\tau(\mathbb{C})\cap\overline{D}_{\gamma(e)}(\mathbb{C})=\emptyset$). Then $[\overline{D}_\tau]\cdot[\overline{D}_\gamma]=0$.
B. Suppose that $[\overline{D}_\tau]\cdot[\overline{D}_\gamma]$ is a good monomial, i.e., $\tau,\gamma(e)$'s are pairwise compatible for all $e\in\mathbf{E}_\gamma$. Then there exists a unique $\mathbf{S}$-tree τ' with $e'\in\mathbf{E}_{\tau'}$ such that $\tau'/e'=\gamma$, $\tau'(e)=\tau$, and $[\overline{D}_\tau]\cdot[\overline{D}_\gamma]=[\overline{D}_{\tau'}]$.
C. Suppose now that there exists an $e\in\mathbf{E}_\gamma$ such that $\gamma(e)=\tau$ i.e., $[\overline{D}_\tau]$ divides $[\overline{D}_\gamma]$. For a given quadruple $\{i,j,k,l\}$ such that $ij\tau kl$, we have

$$\sum_{ij\tau_1 kl}[\overline{D}_{\tau_1}]\cdot[\overline{D}_\gamma]-\sum_{ik\tau_2 jl}[\overline{D}_{\tau_2}]\cdot[\overline{D}_\gamma]=0.$$

Since the elements of the second sum are not compatible with D^γ, we have

$$[\overline{D}_\tau]\cdot[\overline{D}_\gamma]=-\sum_{\substack{\tau_1\neq\tau\\ ij\tau_1 kl}}[\overline{D}_{\tau_1}]\cdot[\overline{D}_\gamma].$$

Here, $[\overline{D}_{\tau_1}]\cdot[\overline{D}_\gamma]$ are good monomials, so they can be computed as in the previous case (B).

Additive relations in $H_*(\overline{M}_{\mathbf{S}}(\mathbb{C}))$

It remains to give the linear relations between degree d monomials. In [25], these relations are given in the following way. Consider an $\mathbf{S}$-tree γ with $|\mathbf{E}_\gamma| = d-1$, and a vertex $v \in \mathbf{V}_\gamma$ with $|v| \geq 4$. Let $f_1, f_2, f_3, f_4 \in \mathbf{F}_\gamma(v)$ be pairwise distinct flags. Put $\mathbf{F} = \mathbf{F}_\gamma(v) \setminus \{f_1, f_2, f_3, f_4\}$ and let $\mathbf{F}_1, \mathbf{F}_2$ be two disjoint subsets of $\mathbf{F}$ such that $\mathbf{F} = \mathbf{F}_1 \bigcup \mathbf{F}_2$. We define two $\mathbf{S}$-trees γ_1, γ_2. The $\mathbf{S}$-tree γ_1 is obtained by inserting a new edge $e = (f_e, f^e)$ to γ at v with boundary $\boldsymbol{\partial}_{\gamma_1}(e) = \{v_e, v^e\}$ and flags $\mathbf{F}_{\gamma_1}(v_e) = \mathbf{F}_1 \bigcup \{f_1, f_2, f_e\}$ and $\mathbf{F}_{\gamma_1}(v^e) = \mathbf{F}_2 \bigcup \{f_3, f_4, f^e\}$. The $\mathbf{S}$-tree γ_2 is also obtained by inserting an edge e to γ at the same vertex v, but the flags are distributed differently on vertices $\boldsymbol{\partial}_{\gamma_2}(e) = \{v_e, v^e\}$: $\mathbf{F}_{\gamma_2}(v_e) = \mathbf{F}_1 \bigcup \{f_1, f_3, f_e\}$ and $\mathbf{F}_{\gamma_2}(v^e) = \mathbf{F}_2 \bigcup \{f_2, f_4, f^e\}$. Put

$$R(\gamma, v; f_1, f_2, f_3, f_4) := \sum_{\gamma_1} [\overline{D}_{\gamma_1}] - \sum_{\gamma_2} [\overline{D}_{\gamma_2}] \tag{2}$$

where summation is taken over all possible γ_1 and γ_2 for fixed set of flags $\{f_1, \ldots, f_4\}$ as given above.

Theorem 4 (Kontsevich and Manin [25]) *All linear relations between good monomials of degree d are spanned by $R(\gamma, v; f_1, f_2, f_3, f_4)$ with $|\mathbf{E}_\gamma| = d-1$.*

A very detailed study of the moduli space $\overline{M}_{\mathbf{S}}(\mathbb{C})$ can be found in chapter 3.2 and 3.3 in [28], and also in [22, 26].

3 Moduli space of pointed real curves

In this section, we review some basic facts on the moduli spaces of $\mathbf{S}$-pointed real curves of genus zero and their homology groups.

3.1 Real structures of complex varieties

A *real structure* on a complex variety $X(\mathbb{C})$ is an anti-holomorphic involution $c_X : X(\mathbb{C}) \to X(\mathbb{C})$. The fixed point set $X(\mathbb{R}) := \mathrm{Fix}(c_X)$ of the involution c_X is called the *real part* of the variety $X(\mathbb{C})$ (or of the real structure c_X).

Notation/Convention

For an involution $\sigma \in \mathbb{S}_n$, we denote the subset $\{s \in \mathbf{S} \mid s = \sigma(s)\}$ by $\mathbf{Fix}(\sigma)$ and its complement $\mathbf{S} \setminus \mathbf{Fix}(\sigma)$ by $\mathbf{Perm}(\sigma)$.

Welschinger invariants for $\mathbb{Z}_2$-equivariant point configurations with no real points (i.e., when $\mathbf{Fix}(\sigma) = \emptyset$) are known to be zero (see [39, 40]). Moreover, the homology of the moduli space for these cases requires slightly different presentation. Therefore, we exclude these special cases and assume that the involution σ has fixed points i.e., $\mathbf{Fix}(\sigma) \neq \emptyset$.

3.2 σ-Invariant curves and their families

An **S**-pointed stable curve $(\Sigma; \mathbf{p})$ is called *σ-invariant* if it admits a real structure $c_\Sigma : \Sigma \to \Sigma$ such that $c_\Sigma(p_s) = p_{\sigma(s)}$ for all $s \in \mathbf{S}$.

Let $\pi_B : U_B(\mathbb{C}) \to B(\mathbb{C})$ be a family of **S**-pointed stable curves with a pair of real structures

$$\begin{array}{ccc} U_B(\mathbb{C}) & \xrightarrow{c_U} & U_B(\mathbb{C}) \\ {\scriptstyle \pi_B}\downarrow & & \downarrow{\scriptstyle \pi_B} \\ B(\mathbb{C}) & \xrightarrow{c_B} & B(\mathbb{C}). \end{array}$$

Such a family is called *σ-equivariant* if the following conditions are met;

- if $\pi^{-1}(b) = \Sigma$, then $\pi^{-1}(c_B(b)) = \overline{\Sigma}$ for every $b \in B$;
- $c_U : z \in \Sigma = \pi^{-1}(b) \mapsto z \in \overline{\Sigma} = \pi^{-1}(c_B(b))$.

Here a complex curve Σ is regarded as a pair $\Sigma = (C, J)$, where C is the underlying two-dimensional manifold, J is a complex structure on C, and $\overline{\Sigma} = (C, -J)$ is its complex conjugated pair.

Remark 1. If $\pi_B : U_B(\mathbb{C}) \to B(\mathbb{C})$ is a σ-equivariant family, then each $(\Sigma(b), \mathbf{p}(b))$ for $b \in B(\mathbb{R})$ is σ-invariant. This follows from the fact that the group of automorphisms of **S**-pointed stable curves is trivial.

Remark 2. Since we have set $\mathbf{Fix}(\sigma) \neq \emptyset$, all σ-invariant curves are of type 1, i.e., their real parts $\Sigma(\mathbb{R})$ are trees of real projective spaces having nodal singularities. This follows from the fact that real parts of σ-invariant curves cannot be the empty set (i.e., they can't be type 2) and all special points must be distinct (i.e., they can not have real isolated nodal points).

By contrast, σ-invariant curves can be of both type 1, type 2 or can have real isolated node when $\mathbf{Fix}(\sigma) = \emptyset$ (see [4]).

Combinatorial types of σ-invariant curves

σ-Invariant curves inherit additional structures on their sets of special points. In this subsection, we first introduce the 'oriented' versions of these structures.

Let $(\Sigma; \mathbf{p})$ be a σ-invariant curve, and let γ be its dual tree. We denote the set of real components $\{v \mid c_\Sigma(\Sigma_v) = \Sigma_v\}$ of $(\Sigma; \mathbf{p})$ by $\mathbf{V}_\gamma^{\mathbb{R}}$.

Oriented combinatorial types

Let $(\hat{\Sigma}; \hat{\mathbf{p}})$ be the normalization of a σ-invariant curve $(\Sigma; \mathbf{p})$. By identifying $\hat{\Sigma}_v$ with $\Sigma_v \subset \Sigma$, we obtain a real structure on $\hat{\Sigma}_v$ for a real component Σ_v. The real part $\hat{\Sigma}_v(\mathbb{R})$ of this real structure divides $\hat{\Sigma}_v$ into two halves: two 2-dimensional open discs, Σ_v^+ and Σ_v^-, having $\hat{\Sigma}_v(\mathbb{R})$ as their common boundary in $\hat{\Sigma}_v$. The real structure $c_\Sigma : \hat{\Sigma}_v \to \hat{\Sigma}_v$ interchanges $\Sigma_v^\pm$, and the complex orientations of $\Sigma_v^{\pm}$ induce two opposite orientations on $\hat{\Sigma}_v(\mathbb{R})$, called its *complex orientations.*

If we fix a labeling of halves of $\hat{\Sigma}_v$ by $\Sigma_v^{\pm}$ (or equivalently, if we orient $\hat{\Sigma}_v(\mathbb{R})$ with one of the complex orientations), then the set of pre-images of special points $\hat{\mathbf{p}}_v \in \hat{\Sigma}_v$ admits the following structures:

- *An oriented cyclic ordering on the set of special points lying in* $\Sigma_v(\mathbb{R})$: For any point $p_{f_i} \in (\hat{\mathbf{p}}_v \cap \hat{\Sigma}_v(\mathbb{R}))$, there is a unique $p_{f_{i+1}} \in (\hat{\mathbf{p}}_v \cap \hat{\Sigma}_v(\mathbb{R}))$ which follows the point p_{f_i} in the positive direction of $\hat{\Sigma}_v(\mathbb{R})$ (the direction which is determined by the complex orientation induced by the orientation of Σ_v^{+}).
- *An ordered two-partition of the set of special points lying in* $\Sigma_v \setminus \Sigma_v(\mathbb{R})$: The subsets $\hat{\mathbf{p}}_v \cap \Sigma_v^{\pm}$ of $\hat{\mathbf{p}}_v$ give an ordered partition of $\hat{\mathbf{p}}_v \cap (\Sigma_v \setminus \Sigma_v(\mathbb{R}))$ into two disjoint subsets.

The relative positions of the special points lying in $\hat{\Sigma}_v(\mathbb{R})$ and the complex orientation of $\hat{\Sigma}_v(\mathbb{R})$ give an *oriented cyclic ordering* on the corresponding labeling set $\mathbf{F}_\gamma^{\mathbb{R}}(v) := (\hat{\mathbf{p}}_v \cap \hat{\Sigma}_v(\mathbb{R}))$. Moreover, the partition $\{p_f \in \Sigma_v^{\pm}\}$ gives an *ordered two-partition* $\mathbf{F}_\gamma^{\pm}(v) := \{f \mid p_f \in \Sigma_v^{\pm}\}$ of $\mathbf{F}_\gamma(v) \setminus \mathbf{F}_\gamma^{\mathbb{R}}(v)$.

The *oriented combinatorial type* of a real component Σ_v with a fixed complex orientation is the following set of data:

$$o_v := \{\text{two partition } \mathbf{F}_\gamma^{\pm}(v); \text{oriented cyclic ordering on } \mathbf{F}_\gamma^{\mathbb{R}}(v)\}.$$

If we consider a σ-invariant curve $(\Sigma; \mathbf{p})$ with a fixed complex orientation at each real component, then the set of oriented combinatorial types of real components

$$o := \{o_v \mid v \in \mathbf{V}_\gamma^{\mathbb{R}}\}$$

is called an *oriented combinatorial type* of $(\Sigma; \mathbf{p})$.

Un-oriented combinatorial types

The definition of oriented combinatorial types requires additional choices (such as complex orientations) which are not determined by real structures of σ-invariant curves. By identifying the oriented combinatorial types for such different choices, we obtain *unoriented combinatorial types* of σ-invariant curves. Namely, for each real component Σ_v of a σ-invariant curve $(\Sigma; \mathbf{p})$, there are two possible ways of choosing Σ_v^{+} in $\hat{\Sigma}_v$. These two different choices give the *opposite* oriented combinatorial types o_v and $\bar{o}_v$: The oriented combinatorial type $\bar{o}_v$ is obtained from o_v by reversing the cyclic ordering of $\mathbf{F}_\gamma^{\mathbb{R}}(v)$ and swapping $\mathbf{F}_\gamma^{+}(v)$ and $\mathbf{F}_\gamma^{-}(v)$.

The *unoriented combinatorial type* of a real component Σ_v of $(\Sigma; \mathbf{p})$ is the pair of opposite oriented combinatorial types $u_v := \{o_v, \bar{o}_v\}$. The set of unoriented combinatorial types of real components

$$u := \{u_v \mid v \in \mathbf{V}_\gamma^{\mathbb{R}}\}$$

is called the *unoriented combinatorial type* of $(\Sigma; \mathbf{p})$.

Dual trees of σ-invariant curves

Let $(\Sigma;\mathbf{p})$ be an σ-invariant curve and let γ be its dual tree.

O-planar trees

An *oriented locally planar (o-planar) structure* on γ is a set of data which encodes an oriented combinatorial type of $(\Sigma;\mathbf{p})$. O-planar structures are explicitly given as follows:

- $\mathbf{V}^{\mathbb{R}}_\gamma \subset \mathbf{V}_\gamma$ is the set of real components of Σ (i.e., the set of *real vertices*).
- $\mathbf{F}^{\mathbb{R}}_\gamma(v) \subset \mathbf{F}_\gamma(v)$ is the set of the pre-images of special points in $\Sigma_v(\mathbb{R})$ (i.e., the set of *real flags* adjacent to the real vertex $v \in \mathbf{V}^{\mathbb{R}}_\gamma$).
- $\mathbf{F}^{\mathbb{R}}_\gamma(v)$ carries an oriented cyclic ordering for every $v \in \mathbf{V}^{\mathbb{R}}_\gamma$.
- $\mathbf{F}_\gamma(v) \setminus \mathbf{F}^{\mathbb{R}}_\gamma(v)$ carries an ordered two-partition for every $v \in \mathbf{V}^{\mathbb{R}}_\gamma$.

We denote $\mathbf{S}$-trees γ, τ, μ with o-planar structures by $(\gamma, o), (\tau, o), (\mu, o)$, or by bold Greek characters with tilde $\widetilde{\boldsymbol{\gamma}}, \widetilde{\boldsymbol{\tau}}, \widetilde{\boldsymbol{\mu}}$. When it is necessary to indicate different o-planar structures on the same $\mathbf{S}$-tree, we use indices in parentheses (e.g., $\widetilde{\boldsymbol{\tau}}_{(i)}$).

Notations

For each vertex $v \in \mathbf{V}^{\mathbb{R}}_\gamma$ (resp. $v \in \mathbf{V}_\gamma \setminus \mathbf{V}^{\mathbb{R}}_\gamma$) of an o-planar tree $\widetilde{\boldsymbol{\gamma}}$, we associate a subtree $\widetilde{\boldsymbol{\gamma}}_v$ (resp. γ_v) which is given by $\mathbf{V}_{\gamma_v} = \{v\}, \mathbf{F}_{\gamma_v} = \mathbf{F}_\gamma(v)$, $\mathbf{j}_{\gamma_v} = \mathbf{id}$, $\boldsymbol{\partial}_{\gamma_v} = \boldsymbol{\partial}_\gamma$, and by the o-planar structure o_v of $\widetilde{\boldsymbol{\gamma}}$ at the vertex $v \in \mathbf{V}^{\mathbb{R}}_\gamma$.

A pair of vertices $v, \bar{v} \in \mathbf{V}_\gamma \setminus \mathbf{V}^{\mathbb{R}}_\gamma$ is said to be *conjugate* if $c_\Sigma(\Sigma_v) = \Sigma_{\bar{v}}$. Similarly, we call a pair of flags $f, \bar{f} \in \mathbf{F}_\gamma \setminus \mathbf{F}^{\mathbb{R}}_\gamma$ *conjugate* if c_Σ swaps the corresponding special points.

To each o-planar tree $\widetilde{\boldsymbol{\gamma}}$, we associate the subsets of vertices $\mathbf{V}^{\pm}_\gamma$ and flags $\mathbf{F}^{+}_\gamma$ as follows: Let $v_1 \in \mathbf{V}_\gamma \setminus \mathbf{V}^{\mathbb{R}}_\gamma$, and let $v_2 \in \mathbf{V}^{\mathbb{R}}_\gamma$ be the closest vertex to v_1 in $||\gamma||$. Let $f(v_1) \in \mathbf{F}_\gamma(v_2)$ be in the shortest path connecting the vertices v_1 and v_2. The sets $\mathbf{V}^{\pm}_\gamma$ are the subsets of vertices $v_1 \in \mathbf{V}_\gamma \setminus \mathbf{V}^{\mathbb{R}}_\gamma$ such that the corresponding flags $f(v_1)$ are respectively in $\mathbf{F}^{\pm}_\gamma(v_2)$. The subsets of flags $\mathbf{F}^{\pm}_\gamma$ are defined as $\boldsymbol{\partial}^{-1}_\gamma(\mathbf{V}^{\pm}_\gamma)$.

U-planar trees

A u-planar structure on the dual tree γ of $(\Sigma;\mathbf{p})$ is the set of data encoding the unoriented combinatorial type of $(\Sigma;\mathbf{p})$. It is given by

$$u := \{(\gamma_v, o_v), (\gamma_v, \bar{o}_v) \mid v \in \mathbf{V}^{\mathbb{R}}_\gamma\}$$

We denote $\mathbf{S}$-trees γ, τ, μ with u-planar structures by $(\gamma, u), (\tau, u), (\mu, u)$, or simply by bold Greek characters $\boldsymbol{\gamma}, \boldsymbol{\tau}, \boldsymbol{\mu}$. O-planar planar trees $\widetilde{\boldsymbol{\gamma}}, \widetilde{\boldsymbol{\tau}}, \widetilde{\boldsymbol{\mu}}$ give representatives of u-planar trees $\boldsymbol{\gamma}, \boldsymbol{\tau}, \boldsymbol{\mu}$ respectively.

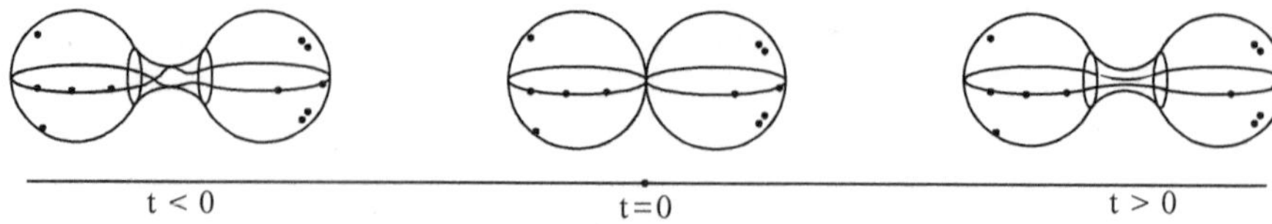

Fig. 1. *Two possible deformations of a real nodal point*

Contraction morphism of o/u-planar trees

Contraction morphism of o-planar trees

Consider a family of σ-invariant curves which is a deformation of a real node of the central fiber $(\Sigma(b_0), \mathbf{p}(b_0))$ with a given oriented combinatorial type. Let $\widetilde{\boldsymbol{\tau}}, \widetilde{\gamma}$ be the o-planar trees associated respectively to generic fibers $(\Sigma(b), \mathbf{p}(b))$ and the central fiber $(\Sigma(b_0), \mathbf{p}(b_0))$ of this family. Let e be the edge corresponding to the nodal point that is deformed. We say that $\widetilde{\boldsymbol{\tau}}$ is obtained by *contracting* the edge e of $\widetilde{\gamma}$, and to indicate this we use the notation $\widetilde{\gamma} < \widetilde{\boldsymbol{\tau}}$.

Contraction morphism of u-planar trees

The definition of contraction morphisms of u-planar trees is the same as that of contraction morphisms of o-planar trees. By contrast, the contraction of an edge of a u-planar tree is not a well-defined operation: We can think of a deformation of a real node as the family $\{x \cdot y = t \mid t \in \mathbb{R}\}$. According to the sign of the deformation parameter t, we obtain two different unoriented combinatorial types of σ-invariant curves, see Figure 1. Different u-planar trees $\boldsymbol{\gamma}_{(i)}$ that are obtained by contraction of the same edge of $\boldsymbol{\tau}$ correspond to different signs of deformation parameters.

Further details of contraction morphisms for o/u-planar trees can be found in [2].

3.3 Moduli space of σ-invariant curves

The moduli space $\overline{M}_{\mathbf{S}}(\mathbb{C})$ comes equipped with a natural real structure $c : (\Sigma; \mathbf{p}) \mapsto (\overline{\Sigma}; \mathbf{p})$.

On the other hand, the permutation group $\mathbb{S}_n$ acts on $\overline{M}_{\mathbf{S}}(\mathbb{C})$ via relabeling: For each $\varrho \in \mathbb{S}_n$, there is an holomorphic map ψ_ϱ defined by $(\Sigma; \mathbf{p}) \mapsto (\Sigma; \varrho(\mathbf{p})) := (\Sigma; p_{\varrho(s_1)}, \ldots, p_{\varrho(s_n)})$. For each involution $\sigma \in \mathbb{S}_n$, we have an additional real structure $c_\sigma := c \circ \psi_\sigma : (\Sigma; \mathbf{p}) \mapsto (\overline{\Sigma}; \sigma(\mathbf{p}))$ of $\overline{M}_{\mathbf{S}}(\mathbb{C})$. The real part $\overline{M}^\sigma_{\mathbf{S}}(\mathbb{R})$ of the real structure $c_\sigma : \overline{M}_{\mathbf{S}}(\mathbb{C}) \to \overline{M}_{\mathbf{S}}(\mathbb{C})$ gives the moduli space of σ-invariant curves:

Theorem 5 (Ceyhan [2]) *(a) For any $|\mathbf{S}| \geq 3$, $\overline{M}^\sigma_{\mathbf{S}}(\mathbb{R})$ is a smooth projective real manifold of dimension $|\mathbf{S}| - 3$.*

(b) Any σ-equivariant family $\pi_B : U_B(\mathbb{C}) \to B(\mathbb{C})$ of **S***-pointed stable curves is induced by a unique pair of real morphisms*

$$\begin{array}{ccc} U_B(\mathbb{C}) & \xrightarrow{\hat{\kappa}} & \overline{U}_{\mathbf{S}}(\mathbb{C}) \\ {\scriptstyle \pi_S}\downarrow & & \downarrow{\scriptstyle \pi} \\ B(\mathbb{C}) & \xrightarrow{\kappa} & \overline{M}_{\mathbf{S}}(\mathbb{C}). \end{array}$$

(c) Let $\mathfrak{M}_\sigma(\mathbb{C})$ be the contravariant functor that sends each real variety $(B(\mathbb{C}), c_B)$ to the set of σ-equivariant families of curves over $B(\mathbb{C})$. The moduli functor $\mathfrak{M}_\sigma(\mathbb{C})$ is represented by the real variety $(\overline{M}_{\mathbf{S}}(\mathbb{C}), c_\sigma)$.

(d) Let $\mathfrak{M}_\sigma(\mathbb{R})$ be the contravariant functor that sends each real analytic manifold R to the set of families of σ-invariant curves over R. The moduli functor $\mathfrak{M}_\sigma(\mathbb{R})$ is represented by the real part $\overline{M}^\sigma_{\mathbf{S}}(\mathbb{R})$ of $(\overline{M}_{\mathbf{S}}(\mathbb{C}), c_\sigma)$.

Remark 3. The group of holomorphic automorphisms of $\overline{M}_{\mathbf{S}}(\mathbb{C})$ that respect its stratification is isomorphic to $\mathbb{S}_n$. Therefore, c_σ's are the real structures preserving the stratification of $\overline{M}_{\mathbf{S}}(\mathbb{C})$ (see [2]).

However, we don't know whether there exist real structures other than c_σ, since the whole group of holomorphic automorphisms $Aut(\overline{M}_{\mathbf{S}}(\mathbb{C}))$ is not necessarily isomorphic to $\mathbb{S}_n$. For example, the automorphism group of $\overline{M}_{\mathbf{S}}(\mathbb{C})$ is $PSL_2(\mathbb{C})$ when $|\mathbf{S}| = 4$.

It is believed that $Aut(\overline{M}_{\mathbf{S}}(\mathbb{C})) \cong \mathbb{S}_n$ for $|\mathbf{S}| \geq 5$. In fact, it is true for $|\mathbf{S}| = 5$ and a proof can be found in [9]. To the best of our knowledge, there is no systematic exposition of $Aut(\overline{M}_{\mathbf{S}}(\mathbb{C}))$ for $|\mathbf{S}| > 5$.

3.4 A stratification of moduli space $\overline{M}^\sigma_{\mathbf{S}}(\mathbb{R})$

A stratification for $\overline{M}^\sigma_{\mathbf{S}}(\mathbb{R})$ can be obtained by using the stratification of $\overline{M}_{\mathbf{S}}(\mathbb{C})$ given in Theorem 1.

Lemma 1. *Let γ and $\overline{\gamma}$ be the dual trees of $(\Sigma; \mathbf{p})$ and $(\overline{\Sigma}; \sigma(\mathbf{p}))$ respectively.*

(a) If γ and $\overline{\gamma}$ are not isomorphic, then the restriction of c_σ on the union of complex strata $D_\gamma(\mathbb{C}) \bigcup D_{\overline{\gamma}}(\mathbb{C})$ gives a real structure with empty real part.

(b) If γ and $\overline{\gamma}$ are isomorphic, then the restriction of c_σ on $D_\gamma(\mathbb{C})$ gives a real structure whose corresponding real part $D_\gamma(\mathbb{R})$ is the intersection of $\overline{M}^\sigma_{\mathbf{S}}(\mathbb{R})$ with $D_\gamma(\mathbb{C})$.

A tree γ is called *σ-invariant* if it is isomorphic to $\bar{\gamma}$. We denote the set of σ-invariant **S**-trees by $\mathcal{T}ree(\sigma)$.

Theorem 6 (Ceyhan [2]) *$\overline{M}^\sigma_{\mathbf{S}}(\mathbb{R})$ is stratified by real analytic subsets*

$$D_\gamma(\mathbb{R}) = \prod_{v_r \in \mathbf{V}^{\mathbb{R}}_\gamma} M^{\hat{\sigma}}_{\mathbf{F}_\gamma(v_r)}(\mathbb{R}) \times \prod_{\{v,\overline{v}\} \subset \mathbf{V}_\gamma \setminus \mathbf{V}^{\mathbb{R}}_\gamma} M_{\mathbf{F}_\gamma(v)}(\mathbb{C})$$

where $\gamma \in \mathcal{T}ree(\sigma)$ and $\hat{\sigma}$ is the involution determined by the action of real structure c_Σ on the special points labeled by $\mathbf{F}_\gamma(v_r)$ for $v_r \in \mathbf{V}^{\mathbb{R}}_\gamma$.

Although the notion of σ-invariant trees leads us to a combinatorial stratification of $\overline{M}^{\sigma}_{\mathbf{S}}(\mathbb{R})$ as given in Theorem 6, it does not give a stratification in terms of connected strata. For a σ-invariant γ, $D_{\gamma}(\mathbb{R})$ has many connected components. In the next paragraph, we refine this stratification by using the spaces of $\mathbb{Z}_2$-equivariant point configurations in the projective line $\mathbb{P}^1(\mathbb{C})$.

$\mathbb{Z}_2$-equivariant configurations in $\mathbb{P}^1(\mathbb{C})$

Let $z := [z : 1]$ be an affine coordinate on $\mathbb{P}^1(\mathbb{C})$. Consider the upper half-plane $\mathbb{H}^+ = \{z \in \mathbb{P}^1(\mathbb{C}) \mid \Im(z) > 0\}$ (resp. lower half plane $\mathbb{H}^- = \{z \in \mathbb{P}^1(\mathbb{C}) \mid \Im(z) < 0\}$) as a half of the $\mathbb{P}^1(\mathbb{C})$ with respect to $z \mapsto \bar{z}$, and the real part $\mathbb{P}^1(\mathbb{R})$ as its boundary. Denote by $\mathbb{H}$ the compactified disc $\mathbb{H}^+ \cup \mathbb{P}^1(\mathbb{R})$.

The configuration space of $k = |\mathbf{Perm}(\sigma)|/2$ distinct pairs of conjugate points in $\mathbb{H}^+ \bigsqcup \mathbb{H}^-$ and $l = |\mathbf{Fix}(\sigma)|$ distinct points in $\mathbb{P}^1(\mathbb{R})$ is

$$\begin{aligned}\widetilde{Conf}_{(\mathbf{S},\sigma)} := \{(z_{s_1},\dots,z_{s_{2k}};x_{r_1},\dots,x_{r_l}) \mid & z_s \in \mathbb{H}^+ \bigsqcup \mathbb{H}^- \text{ for } s \in \mathbf{Perm}(\sigma),\\ & z_s = z_{s'} \Leftrightarrow s = s', z_s = \bar{z}_{s'} \Leftrightarrow s = \sigma(s') \ \&\\ & x_r \in \mathbb{P}^1(\mathbb{R}) \text{ for } r \in \mathbf{Fix}(\sigma),\ x_r = x_{r'} \Leftrightarrow r = r'\}.\end{aligned}$$

The number of connected components of $\widetilde{Conf}_{(\mathbf{S},\sigma)}$ is $2^k(l-1)!$ when $l \geq 2$, and 2^k when $l = 1$. They are all pairwise diffeomorphic; natural diffeomorphisms are given by σ-invariant relabelings.

The action of $SL_2(\mathbb{R})$ on $\mathbb{H}$ is given by

$$SL_2(\mathbb{R}) \times \mathbb{H} \to \mathbb{H}, \quad (\Lambda, z) \mapsto \Lambda(z) = \frac{az+b}{cz+d}, \quad \Lambda = \begin{pmatrix} a & b \\ c & d \end{pmatrix} \in SL_2(\mathbb{R}).$$

It induces an isomorphism $SL_2(\mathbb{R})/\pm I \to Aut(\mathbb{H})$. The automorphism group $Aut(\mathbb{H})$ acts on $\widetilde{Conf}_{(\mathbf{S},\sigma)}$ by

$$\Lambda : (z_{s_1},\dots,z_{s_{2k}};x_{r_1},\dots,x_{r_l}) \mapsto (\Lambda(z_{s_1}),\dots,\Lambda(z_{s_{2k}});\Lambda(x_{r_1}),\dots,\Lambda(x_{r_l})).$$

This action preserves each of the connected components of $\widetilde{Conf}_{(\mathbf{S},\sigma)}$. It is free when $2k + l \geq 3$, and it commutes with diffeomorphisms given by σ-invariant relabelings. Therefore, the quotient space $\widetilde{C}_{(\mathbf{S},\sigma)} := \widetilde{Conf}_{(\mathbf{S},\sigma)}/Aut(\mathbb{H})$ is a manifold of dimension $2k + l - 3$ whose connected components are pairwise diffeomorphic.

In addition to the automorphisms considered above, there is a diffeomorphism of $\widetilde{Conf}_{(\mathbf{S},\sigma)}$ which is given in affine coordinates as follows.

$$e : (z_{s_1},\dots,z_{s_{2k}};x_{r_1},\dots,x_{r_l}) \mapsto (-z_{s_1},\dots,-z_{s_{2k}};-x_{r_1},\dots,-x_{r_l}).$$

Consider the quotient space $Conf_{(\mathbf{S},\sigma)} = \widetilde{Conf}_{(\mathbf{S},\sigma)}/(e)$. The diffeomorphism e commutes with each ϱ-invariant relabeling and normalizing action of $Aut(\mathbb{H})$.

Therefore, the quotient space $C_{(\mathbf{S},\sigma)} := Conf_{(\mathbf{S},\sigma)}/Aut(\mathbb{H})$ is a manifold of dimension $2k+l-3$, its connected components are diffeomorphic to the components of $\widetilde{C}_{(\mathbf{S},\sigma)}$, and, moreover, the quotient map $\widetilde{C}_{(\mathbf{S},\sigma)} \to C_{(\mathbf{S},\sigma)}$ is a trivial double covering.

Connected Components of $M_{\mathbf{S}}^{\sigma}(\mathbb{R})$

Each connected component of $C_{(\mathbf{S},\sigma)}$ is associated to an unoriented combinatorial type of σ-invariant curves, and each unoriented combinatorial type is given by a one-vertex u-planar tree γ. We denote the connected components of $C_{(\mathbf{S},\sigma)}$ by C_γ.

Every $\mathbb{Z}_2$-equivariant point configuration defines a σ-invariant curve. Hence, we define $\Xi : \bigsqcup_{\gamma:|\mathbf{V}_\gamma|=1} C_\gamma \to M_{\mathbf{S}}^{\sigma}(\mathbb{R})$ which maps $\mathbb{Z}_2$-equivariant point configurations to the corresponding isomorphisms classes of irreducible σ-invariant curves.

Lemma 2. *(a) The map Ξ is a diffeomorphism.*

(b) Let $|\mathbf{Perm}(\sigma)| = 2k$ and $\mathbf{Fix}(\sigma) = l$. The configuration space C_γ is diffeomorphic to

- $((\mathbb{H}^+)^k \setminus \Delta) \times \mathbb{R}^{l-3}$ *when* $l > 2$,
- $((\mathbb{H}^+ \setminus \{\sqrt{-1}\})^{k-1} \setminus \Delta) \times \mathbb{R}^{l-1}$ *when* $l = 1, 2$,

where Δ is the union of all diagonals where $z_s = z_{s'}$.

Refinement of the stratification

We associate a product of configuration spaces $C_{\widetilde{\gamma}_v}$ and moduli spaces of pointed complex curves $\overline{M}_{\mathbf{F}_\gamma(v)}(\mathbb{C})$ to each o-planar tree $\widetilde{\gamma}$:

$$C_{\widetilde{\gamma}} = \prod_{v\in\mathbf{V}_\gamma^{\mathbb{R}}} C_{\gamma_v} \times \prod_{v\in\mathbf{V}_\gamma^{+}} M_{\mathbf{F}_\gamma(v)}(\mathbb{C}).$$

For each u-planar γ, we first choose an o-planar representative $\widetilde{\gamma}$, and then we set $C_\gamma := C_{\widetilde{\gamma}}$. Note that C_γ does not depend on the o-planar representative.

Theorem 7 (Ceyhan [2]) *(a) $\overline{M}_{\mathbf{S}}^{\sigma}(\mathbb{R})$ is stratified by C_γ.*

(b) A stratum C_γ is contained in the boundary of $\overline{C}_{\boldsymbol{\tau}}$ if and only if $\boldsymbol{\tau}$ is obtained by contracting an invariant set of edges of γ. The codimension of C_γ in $\overline{C}_{\boldsymbol{\tau}}$ is $|\mathbf{E}_\gamma| - |\mathbf{E}_\tau|$.

3.5 Graph homology of $\overline{M}_{\mathbf{S}}^{\sigma}(\mathbb{R})$

In this section, we summarize the results from [4]. We a give a combinatorial complex generated by the strata of $\overline{M}_{\mathbf{S}}^{\sigma}(\mathbb{R})$ whose homology is isomorphic to the homology of $\overline{M}_{\mathbf{S}}^{\sigma}(\mathbb{R})$.

A graph complex of $\overline{M}^{\sigma}_{\mathbf{S}}(\mathbb{R})$

Let $\sigma \in \mathbb{S}_n$ be an involution such that $\mathbf{Fix}(\sigma) \neq \emptyset$. We define a graded group

$$\mathcal{G}_d = \left(\bigoplus_{\gamma:|\mathbf{E}_\gamma|=|\mathbf{S}|-d-3} H_{\dim(C_\gamma)}(\overline{C}_\gamma, Q_\gamma; \mathbb{Z}) \right) / I_d, \tag{3}$$

$$= \left(\bigoplus_{\gamma:|\mathbf{E}_\gamma|=|\mathbf{S}|-d-3} \mathbb{Z}\,[\overline{C}_\gamma] \right) / I_d \tag{4}$$

where $[\overline{C}_\gamma]$ are the (relative) fundamental classes of the strata $\overline{C}_\gamma$ of $\overline{M}^{\sigma}_{\mathbf{S}}(\mathbb{R})$. Here, Q_γ denotes the union of the substrata of $\overline{C}_\gamma$ of codimension one and higher.

For $|\mathbf{Perm}(\sigma)| < 4$, the subgroup I_d (for degree d) is the trivial subgroup. In all other cases (i.e., for $|\mathbf{Perm}(\sigma)| \geq 4$), the subgroup I_d is generated by the following elements.

The generators of the subgroup of relations of the graph complex.

The following paragraphs $\mathfrak{R}$-1 and $\mathfrak{R}$-2 describe the generators of the subgroup of relations of the graph complex.

$\mathfrak{R}$-1. Degeneration of a real vertex

Consider an o-planar representative $\widetilde{\gamma}$ of a u-planar tree γ such that $|\mathbf{E}_\gamma| = |\mathbf{S}|-d-5$, and consider one of its vertices $v \in \mathbf{V}^{\mathbb{R}}_\gamma$ with $|v| \geq 5$ and $|\mathbf{F}^+_\gamma(v)| \geq 2$. Let $f_i, \bar{f}_i \in \mathbf{F}_\gamma \setminus \mathbf{F}^{\mathbb{R}}_\gamma$ be conjugate pairs of flags for $i = 1, 2$ such that $f_1, f_2 \in \mathbf{F}^+_\gamma(v)$ of $\widetilde{\gamma}$, and let $f_3 \in \mathbf{F}^{\mathbb{R}}_{\gamma}$. Put $\mathbf{F} = \mathbf{F}_\gamma(v) \setminus \{f_1, \bar{f}_1, f_2, \bar{f}_2, f_3\}$

We define two u-planar trees γ_1 and γ_2 as follows.

The o-planar representative $\widetilde{\gamma}_1$ of γ_1 is obtained by inserting a pair of conjugate edges $e = (f_e, f^e)$, $\bar{e} = (f_{\bar{e}}, f^{\bar{e}})$ into $\widetilde{\gamma}$ at v in such a way that $\widetilde{\gamma}_1$ gives $\widetilde{\gamma}$ when we contract the edges $e, \bar{e}$. Let $\boldsymbol{\partial}_{\gamma_1}(e) = \{\tilde{v}, v^e\}$, $\boldsymbol{\partial}_{\gamma_1}(\bar{e}) = \{\tilde{v}, v^{\bar{e}}\}$. Then, the distribution of flags of $\widetilde{\gamma}_1$ is given by $\mathbf{F}_{\gamma_1}(\tilde{v}) = \mathbf{F}_1 \cup \{f_3, f_e, f_{\bar{e}}\}$, $\mathbf{F}_{\gamma_1}(v^e) = \mathbf{F}_2 \cup \{f_1, f_2, f^e\}$ and $\mathbf{F}_{\gamma_1}(v^{\bar{e}}) = \overline{\mathbf{F}}_2 \cup \{\bar{f}_1, \bar{f}_2, f^{\bar{e}}\}$ where $(\mathbf{F}_1, \mathbf{F}_2, \overline{\mathbf{F}}_2)$ is an equivariant partition of $\mathbf{F}$.

The o-planar representative $\widetilde{\gamma}_2$ of γ_2 is obtained by inserting a pair of real edges $e_1 = (f_{e_1}, f^{e_1})$, $e_2 = (f_{e_2}, f^{e_2})$ into $\widetilde{\gamma}$ at v in such a way that $\widetilde{\gamma}_2$ produces $\widetilde{\gamma}$ when we contract the edges e_1, e_2. Let $\boldsymbol{\partial}_{\gamma_2}(e_1) = \{\tilde{v}, v^{e_1}\}$, $\boldsymbol{\partial}_{\gamma_2}(e_2) = \{\tilde{v}, v^{e_2}\}$. The sets of flags of $\widetilde{\gamma}_2$ are $\mathbf{F}_{\gamma_2}(\tilde{v}) = \widetilde{\mathbf{F}}_1 \cup \{f_3, f_{e_1}, f_{e_2}\}$, $\mathbf{F}_{\gamma_2}(v^{e_1}) = \widetilde{\mathbf{F}}_2 \cup \{f_1, \bar{f}_1, f^{e_1}\}$, $\mathbf{F}_{\gamma_2}(v^{e_2}) = \widetilde{\mathbf{F}}_3 \cup \{f_2, \bar{f}_2, f^{e_2}\}$ where $(\widetilde{\mathbf{F}}_1, \widetilde{\mathbf{F}}_2, \widetilde{\mathbf{F}}_3)$ is an equivariant partition of $\mathbf{F}$.

The u-planar trees γ_1, γ_2 are the equivalence classes represented by $\widetilde{\gamma}_1, \widetilde{\gamma}_2$ given above.

Then, we define

$$\mathcal{R}(\gamma; v, f_1, f_2, f_3) := \sum_{\gamma_1} [\overline{C}_{\gamma_1}] - \sum_{\gamma_2} [\overline{C}_{\gamma_2}], \tag{5}$$

where the summation is taken over all possible $\gamma_i, i = 1, 2$ for a fixed set of flags $\{f_1, \bar{f}_1, f_2, \bar{f}_2, f_3\}$.

$\mathfrak{R}$-*2. Degeneration of a conjugate pair of vertices*

Consider an o-planar representative $\widetilde{\gamma}$ of a u-planar tree γ such that $|\mathbf{E}_\gamma| = |\mathbf{S}| - d - 5$, and a pair of its conjugate vertices $v, \bar{v} \in \mathbf{V}_\gamma \setminus \mathbf{V}_\gamma^{\mathbb{R}}$ with $|v| = |\bar{v}| \geq 4$. Let $f_i \in \mathbf{F}_\gamma(v), i = 1, \dots, 4$ and $\bar{f}_i \in \mathbf{F}_\gamma(\bar{v})$ be the flags conjugate to $f_i, i = 1, \dots, 4$. Put $\mathbf{F} = \mathbf{F}_\gamma(v) \setminus \{f_1, \dots, f_4\}$. Let $(\mathbf{F}_1, \mathbf{F}_2)$ be a two-partition of $\mathbf{F}$, and $\overline{\mathbf{F}}_1, \overline{\mathbf{F}}_2$ be the sets of flags that are conjugate to the flags in $\mathbf{F}_1, \mathbf{F}_2$ respectively.

We define two u-planar trees γ_1 and γ_2 as follows.

The o-planar representative $\widetilde{\gamma}_1$ of γ_1 is obtained by inserting a pair of conjugate edges $e = (f_e, f^e)$, $\bar{e} = (f_{\bar{e}}, f^{\bar{e}})$ in $\widetilde{\gamma}$ at $v, \bar{v}$ in such a way that $\widetilde{\gamma}_1$ produces $\widetilde{\gamma}$ when we contract the edges $e, \bar{e}$. Let $\boldsymbol{\partial}_{\gamma_1}(e) = \{v_e, v^e\}$, $\boldsymbol{\partial}_{\gamma_1}(\bar{e}) = \{v_{\bar{e}}, v^{\bar{e}}\}$. The sets of flags of $\widetilde{\gamma}_1$ are $\mathbf{F}_{\gamma_1}(v_e) = \mathbf{F}_1 \cup \{f_1, f_2, f_e\}$, $\mathbf{F}_{\gamma_1}(v^e) = \mathbf{F}_2 \cup \{f_3, f_4, f^e\}$ and $\mathbf{F}_{\gamma_1}(v_{\bar{e}}) = \overline{\mathbf{F}}_1 \cup \{\bar{f}_1, \bar{f}_2, f_{\bar{e}}\}$, $\mathbf{F}_{\gamma_1}(v^{\bar{e}}) = \overline{\mathbf{F}}_2 \cup \{\bar{f}_3, \bar{f}_4, f^{\bar{e}}\}$.

The o-planar representative $\widetilde{\gamma}_2$ of γ_2 is also obtained by inserting a pair of conjugate edges into $\widetilde{\gamma}$ at the same vertices $v, \bar{v}$, but the flags are distributed differently on vertices. Let $\boldsymbol{\partial}_{\gamma_2}(e) = \{v_e, v^e\}$, $\boldsymbol{\partial}_{\gamma_2}(\bar{e}) = \{v_{\bar{e}}, v^{\bar{e}}\}$. Then, the distribution of the flags of $\widetilde{\gamma}_2$ is given by $\mathbf{F}_{\gamma_2}(v_e) = \mathbf{F}_1 \cup \{f_1, f_3, f_e\}$, $\mathbf{F}_{\gamma_2}(v^e) = \mathbf{F}_2 \cup \{f_2, f_4, f^e\}$, $\mathbf{F}_{\gamma_2}(v_{\bar{e}}) = \overline{\mathbf{F}}_1 \cup \{\bar{f}_1, \bar{f}_3, f_{\bar{e}}\}$, and $\mathbf{F}_{\gamma_2}(v^{\bar{e}}) = \overline{\mathbf{F}}_2 \cup \{\bar{f}_2, \bar{f}_4, f^{\bar{e}}\}$.

The u-planar trees γ_1, γ_2 are the equivalence classes represented by $\widetilde{\gamma}_1, \widetilde{\gamma}_2$ given above.

We define

$$\mathcal{R}(\gamma; v, f_1, f_2, f_3, f_4) := \sum_{\gamma_1} [\overline{C}_{\gamma_1}] - \sum_{\gamma_2} [\overline{C}_{\gamma_2}], \tag{6}$$

where the summation is taken over all γ_1, γ_2 for a fixed set of flags $\{f_1, \dots, f_4\}$.

The subgroup I_d is generated by $\mathcal{R}(\gamma; v, f_1, f_2, f_3)$ and $\mathcal{R}(\gamma; v, f_1, f_2, f_3, f_4)$ for all γ and v satisfying the required conditions above.

The boundary homomorphism of the graph complex

We define the *graph complex* $\mathcal{G}_\bullet$ of the moduli space $\overline{M}_{\mathbf{S}}^\sigma(\mathbb{R})$ by introducing a boundary map $\partial : \mathcal{G}_d \to \mathcal{G}_{d-1}$

$$\partial\ [\overline{C}_{\boldsymbol{\tau}}] \ = \ \sum_{\gamma} \ \pm\ [\overline{C}_\gamma], \tag{7}$$

where the summation is taken over all u-planar trees γ which give $\boldsymbol{\tau}$ after contracting one of their real edges.

Theorem 8 (Ceyhan [4]) *The homology of the graph complex $\mathcal{G}_\bullet$ is isomorphic to the singular homology $H_*(\overline{M}^\sigma_{\mathbf{S}}(\mathbb{R});\mathbb{Z})$ for* $\mathbf{Fix}(\sigma) \neq \emptyset$.

Remark 4. In [4], the graph homology is defined and a similar theorem is proved for $\mathbf{Fix}(\sigma) = \emptyset$ case. The generators of the ideal are slightly different in this case.

Remark 5. If $|\mathbf{S}| > 4$ and $|\mathbf{Fix}(\sigma)| \neq 0$, then the moduli space $\overline{M}^\sigma_{\mathbf{S}}(\mathbb{R})$ is not orientable. A combinatorial construction of the orientation double covering of $\overline{M}^\sigma_{\mathbf{S}}(\mathbb{R})$ is given in [2]. A stratification of the orientation cover is given in terms of certain equivalence classes of o-planar trees. By following the same ideas above, it is possible to construct a graph complex generated by fundamental classes of the strata that calculates the homology of the orientation double cover of $\overline{M}^\sigma_{\mathbf{S}}(\mathbb{R})$.

Remark 6. In their recent preprint [10], Etingof *et al* calculated the cohomology algebra $H^*(\overline{M}^\sigma_{\mathbf{S}}(\mathbb{R});\mathbb{Q})$ in terms of generators and relations for the $\sigma = id$ case. Until this work, little was known about the topology of $\overline{M}^\sigma_{\mathbf{S}}(\mathbb{R})$ (except [6, 8, 13, 21]).

The graph homology, in a sense, treats the homology of the moduli space $\overline{M}^\sigma_{\mathbf{S}}(\mathbb{R})$ in the complementary directions to [10]: The graph complex provides a recipe to calculate the homology of $\overline{M}^\sigma_{\mathbf{S}}(\mathbb{R})$ in $\mathbb{Z}$ coefficient for all possible involutions σ, and it reduces to cellular complex of the moduli space $\overline{M}^\sigma_{\mathbf{S}}(\mathbb{R})$ for $\sigma = id$ (see [8, 21]). Moreover, our presentation is based on stratification of $\overline{M}^\sigma_{\mathbf{S}}(\mathbb{R})$ suited for investigating Gromov–Witten–Welschinger classes.

The graph homology provides a set of homological relations between the classes of $\overline{D}_\gamma(\mathbb{R})$ that are essential for quantum cohomology of real varieties:

Corollary 1. *If τ is a σ-invariant tree, then*

$$\partial\, [\overline{D}_\tau(\mathbb{R})] = \sum_{\substack{\text{all possible} \\ \text{u-planar str. of } \gamma}} \sum_{\gamma} \pm[\overline{C}_\gamma] = \sum_{\gamma} [\overline{D}_\gamma(\mathbb{R})] \tag{8}$$

is homologous to zero.

Corollary 2. *If γ is an o-planar tree satisfying the condition required in $\mathfrak{R}$-1, then the sum*

$$\sum_{\substack{\text{all possible} \\ \text{u-planar str. of } \gamma}} \mathcal{R}(\gamma; v, f_1, f_2, f_3) = \sum_{\gamma_1} [\overline{D}_{\gamma_1}(\mathbb{R})] - \sum_{\gamma_2} [\overline{D}_{\gamma_2}(\mathbb{R})] \tag{9}$$

is homologous to zero.

Part II: Quantum cohomology of real varieties

In this part, we recall the definition of Gromov–Witten classes and quantum cohomology. Then, we introduce Welschinger classes and quantum cohomology of real varieties. Our main goal is to revise Horava's attempt of quantum cohomology of real varieties by investigating the properties of Gromov–Witten and Welschinger invariants.

4 Gromov–Witten classes

Let $X(\mathbb{C})$ be a projective algebraic manifold, and let $\beta \in H_2(X(\mathbb{C});\mathbb{Z})$ such that $(L \cdot \beta) \geq 0$ for all Kähler L. Let $\overline{M}_{\mathbf{S}}(X,\beta)$ be the set of isomorphism classes of $\mathbf{S}$-pointed stable maps $(\Sigma;\mathbf{p};f)$ where Σ is a projective nodal complex curve of genus zero, $p_{s_1},\ldots,p_{s_n}$ are distinct smooth labeled points of Σ, and $f:\Sigma\to X$ is a morphism satisfying $f_*([\Sigma])=\beta$.

For each labelled point $s\in\mathbf{S}$, the moduli space of stable maps $\overline{M}_{\mathbf{S}}(X,\beta)$ inherits a canonical *evaluation* map

$$ev_s:\overline{M}_{\mathbf{S}}(X,\beta)\to X(\mathbb{C})$$

defined for $(\Sigma;\mathbf{p};f)\in\overline{M}_{\mathbf{S}}(X,\beta)$ by:

$$ev_s:(\Sigma;\mathbf{p},f)\mapsto f(p_s).$$

Given classes $\mu_{s_1},\cdots,\mu_{s_n}\in H^*(X(\mathbb{C}))$, a product is determined in the ring $H^*\overline{M}_{\mathbf{S}}(X,\beta)$ by:

$$ev_{s_1}^*(\mu_{s_1})\cup\cdots\cup ev_{s_n}^*(\mu_{s_n}). \tag{10}$$

If $\sum\operatorname{codim}(\mu_s)=\dim(\overline{M}_{\mathbf{S}}(X,\beta))$, then the product (10) can be evaluated on the (virtual) fundamental class of $\overline{M}_{\mathbf{S}}(X,\beta)$. In this case, the *Gromov–Witten* invariant is defined as the degree of the evaluation maps:

$$\int_{[\overline{M}_{\mathbf{S}}(X,\beta)]} ev_{s_1}^*(\mu_{s_1})\cup\cdots\cup ev_{s_n}^*(\mu_{s_n}).$$

This gives an appropriate counting of parametrized curves of genus zero, lying in a given homology class β and satisfying certain incidence conditions encoded by μ_s's.

The above definition of Gromov–Witten invariants leads to more general Gromov–Witten invariants in $H^*\overline{M}_{\mathbf{S}}(\mathbb{C})$: Given classes $\mu_{s_1},\ldots,\mu_{s_n}$, we have

$$I^X_{\mathbf{S},\beta}(\mu_{s_1}\otimes\cdots\otimes\mu_{s_n})=\nu_*(ev_{s_1}^*(\mu_{s_1})\cup\cdots\cup ev_{s_n}^*(\mu_{s_n}))$$

where $\nu:\overline{M}_{\mathbf{S}}(X,\beta)\to\overline{M}_{\mathbf{S}}$ is the projection that forgets the morphisms $f:\Sigma\to X$ and stabilizes the resultant pointed curve (if necessary). The set of multilinear maps

$$\{I^X_{\mathbf{S},\beta} : \bigotimes_{\mathbf{S}} H^*X(\mathbb{C}) \to H^*\overline{M}_{\mathbf{S}}(\mathbb{C})\}$$

is called the *(tree-level) system of Gromov–Witten classes.*

4.1 Cohomological field theory

A two-dimensional cohomological field theory (CohFT) with coefficient field $\mathbb{K}$ consists of the following data:

- A $\mathbb{K}$-linear superspace (of fields) $\mathcal{H}$, endowed with an even non-degenerate pairing g.
- A family of even linear maps (correlators)

$$\{I_{\mathbf{S}} : \bigotimes_{\mathbf{S}} \mathcal{H} \to H^*\overline{M}_{\mathbf{S}}(\mathbb{C})\} \tag{11}$$

  defined for all $|\mathbf{S}| \geq 3$.

These data must satisfy the following axioms:

A. $\mathbb{S}_n$-covariance: The maps $I_{\mathbf{S}}$ are compatible with the actions of $\mathbb{S}_n$ on $\bigotimes_{\mathbf{S}} \mathcal{H}$ and on $\overline{M}_{\mathbf{S}}(\mathbb{C})$.

B. Splitting: Let $\{\mu_a\}$ denote a basis of $\mathcal{H}$, $\Delta := \sum g^{ab}\mu_a \otimes \mu_b$ is the Casimir element of the pairing.

Let $\mathbf{S}_a = \{s_{a_1}, \ldots, s_{a_m}\}$, $\mathbf{S}_b = \{s_{b_1}, \ldots, s_{b_n}\}$ and $\mathbf{S} = \mathbf{S}_1 \sqcup \mathbf{S}_2$. Let γ be an $\mathbf{S}$-tree such that $\mathbf{V}_\gamma = \{v_a, v_b\}$, and the set of flags $\mathbf{F}_\gamma(v_a) = \mathbf{S}_a \cup \{s_a\}$, $\mathbf{F}_\gamma(v_b) = \mathbf{S}_b \cup \{s_b\}$, and let

$$\varphi_\gamma : \overline{M}_{\mathbf{F}_\gamma(v_a)}(\mathbb{C}) \times \overline{M}_{\mathbf{F}_\gamma(v_b)}(\mathbb{C}) \to \overline{M}_{\mathbf{S}}(\mathbb{C})$$

be the map which assigns to the pointed curves $(\Sigma_a; p_{s_{a_1}}, \ldots, p_{s_{a_m}}, p_{s_a})$ and $(\Sigma_b; p_{s_{b_1}}, \ldots, p_{s_{b_n}}, p_{s_b})$, their union $\Sigma_a \cup \Sigma_b$ identified at p_{s_a} and p_{s_b}.

The Splitting Axiom reads:

$$\begin{aligned} &\varphi^*_\gamma(I_{\mathbf{S}}(\mu_{s_1} \cdots \mu_{s_n})) \\ &= \varepsilon(\gamma) \sum_{a,b} g^{ab}\, I_{\mathbf{F}_\gamma(v_a)}(\bigotimes_{\mathbf{S}_a} \mu_{s_*} \otimes \mu_{s_a})) \otimes I_{\mathbf{F}_\gamma(v_a)}(\bigotimes_{\mathbf{S}_b} \mu_{s_*} \otimes \mu_{s_b})) \end{aligned} \tag{12}$$

where $\varepsilon(\gamma)$ is the sign of the permutation $(\mathbf{S}_1, \mathbf{S}_2)$ on $\{\mu_*\}$ of odd dimension.

CohFTs are abstract generalizations of tree-level systems of Gromov–Witten classes: any system of GW classes for a manifold $X(\mathbb{C})$ satisfying appropriate assumptions gives rise to the cohomological field theory where $\mathcal{H} = H^*(X(\mathbb{C}); \mathbb{K})$.

4.2 Homology operad of $\overline{M}_{\mathbf{S}}(\mathbb{C})$

There is another useful reformulation of CohFT. By dualizing (11), we obtain a family of maps

$$\{\mathbf{Y_S} : H_*\overline{M}_{\mathbf{S}\cup\{s\}}(\mathbb{C}) \to Hom(\bigotimes_{\mathbf{S}} \mathcal{H}, \mathcal{H})\}. \tag{13}$$

Therefore, any homology class in $\overline{M}_{\mathbf{S}\cup\{s\}}(\mathbb{C})$ is interpreted as an n-ary operation on $\mathcal{H}$. The additive relations given in Section 2.2 and the splitting axiom (12) provide identities between these operations, i.e., $\mathcal{H}$ carries the structure of an algebra over the cyclic operad $H_*\overline{M}_{\mathbf{S}\cup\{s\}}(\mathbb{C})$.

$Comm_\infty$-algebras

Each n-ary operation is a composition of k-ary operations corresponding to fundamental classes of $\overline{M}_{\mathbf{F}\cup\{s\}}(\mathbb{C})$ for some $\mathbf{F}$ of order $k \leq n$. Rephrasing an algebra over $H_*\overline{M}_{\mathbf{S}\cup\{s\}}(\mathbb{C})$-operad by using additive relations in $H_*\overline{M}_{\mathbf{S}\cup\{s\}}(\mathbb{C})$ provides us a *$Comm_\infty$-algebra.*

The structure of *$Comm_\infty$-algebra* on $(\mathcal{H}, g)$ is a sequence of even polylinear maps $\mathbf{Y}_k : \mathcal{H}^{\otimes k} \to \mathcal{H}, k \geq 2$, satisfying the following conditions:

A′. Higher commutativity: $\mathbf{Y}_k$ are $\mathbb{S}_k$-symmetric.

B′. Higher associativity: For all $m \geq 0$, and $\mu_\alpha, \mu_\beta, \mu_\delta, \mu_{i_1}, \cdots, \mu_{i_m} \in A$, we have

$$\sum_{\substack{\tau:\mathbf{S}_1\sqcup\mathbf{S}_2=\underline{\mathbf{n}} \\ \alpha,\beta,j_*\in\mathbf{S}_1\&\delta,k_*\in\mathbf{S}_2}} (-1)^{\varepsilon(\tau)}\mathbf{Y}_{|\mathbf{S}_2|}(\mathbf{Y}_{|\mathbf{S}_1|}(\mu_\alpha, \mu_\beta, \mu_{j_1}, \cdots, \mu_{j_r}), \mu_\delta, \mu_{k_1}, \cdots, \mu_{k_s})$$

$$= \sum_{\substack{\gamma:\mathbf{S}'_1\sqcup\mathbf{S}'_2=\underline{\mathbf{n}} \\ \beta,\delta,j_*\in\mathbf{S}'_2\&\alpha,k_*\in\mathbf{S}'_1}} (-1)^{\varepsilon(\gamma)}\mathbf{Y}_{|\mathbf{S}'_2|}(\mathbf{Y}_{|\mathbf{S}'_1|}(\mu_\beta, \mu_\delta, \mu_{j_1}, \cdots, \mu_{j_r}), \mu_\alpha, \mu_{k_1}, \cdots, \mu_{k_s})$$

Here, the summations runs over all partitions of $\underline{\mathbf{n}}$ into two disjoint subsets $(\mathbf{S}_1, \mathbf{S}_2)$ and $(\mathbf{S}'_1, \mathbf{S}'_2)$ satisfying required conditions. Note that these relations are consequences of the homology relations given in Theorem 2 by Keel.

Since we identify the bases with n-corollas of the $H_*\overline{M}_{\mathbf{S}\cup\{s\}}(\mathbb{C})$-operad (see Fig. 2 for an n-corolla), it is represented as a linear span of all possible $\mathbf{S}'$-trees modulo the higher associativity relations for $\mathbf{S}' = \mathbf{S}\cup\{s\}$ (see Fig. 3).

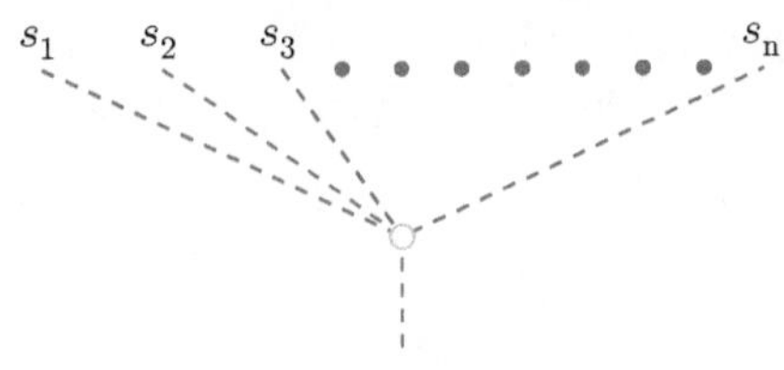

Fig. 2. *n-corolla corresponding to the fundamental class of* $\overline{M}_{\mathbf{S}\cup\{s\}}(\mathbb{C})$

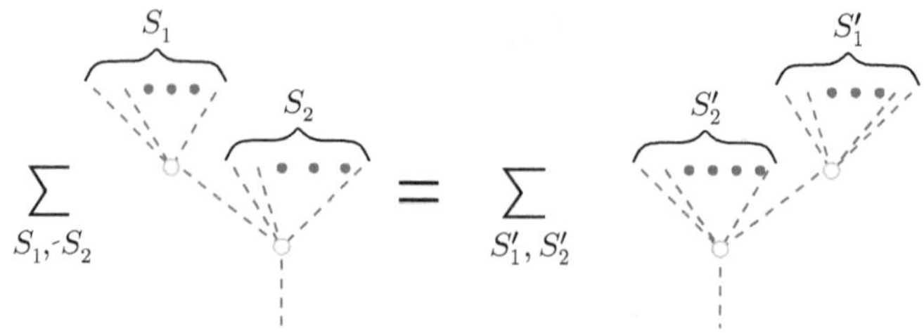

Fig. 3. *Higher associativity relations*

We call $Comm_\infty$-algebra $(\mathcal{H}, g, \mathbf{Y}_*)$ *cyclic* if the tensors

$$\eta_{k+1} : \mathcal{H}^{\otimes k+1} \to \mathbb{K},$$
$$\eta_{k+1}(\mu_1, \cdots, \mu_k, \mu_{k+1}) := g(\mathbf{Y}_k(\mu_1, \ldots, \mu_k), \mu_{k+1})$$

are $\mathbb{S}_{k+1}$-symmetric .

4.3 Gromov–Witten potential and quantum product

Let $X(\mathbb{C})$ be a manifold equipped with a system of tree level GW classes. Put

$$\langle \mu_{s_1} \cdots \mu_{s_n} \rangle_\beta = \langle I^X_{\mathbf{S},\beta} \rangle (\mu_{s_1} \otimes \cdots \otimes \mu_{s_n}) := \int_{[\overline{M}_{\mathbf{S}}(\mathbb{C})]} I^X_{\mathbf{S},\beta}(\mu_{s_1} \otimes \cdots \otimes \mu_{s_n}).$$

We define Φ as a formal sum depending on a variable point $\Gamma \in H^*(X(\mathbb{C}))$:

$$\Phi(\Gamma) := \sum_{|\mathbf{S}| \geq 3} \sum_{\beta} \langle I^X_{\mathbf{S},\beta} \rangle (\Gamma^{\otimes |\mathbf{S}|}) \frac{q^\beta}{|\mathbf{S}|!}.$$

The quantum multiplication is defined as a product on tangent space of homology space $H^*(X(\mathbb{C}))$ by

$$\partial_a * \partial_b = \sum_{c,d} \Phi_{abc}\, \partial_d$$

where Φ_{abc} denotes the third derivative $\partial_a \partial_b \partial_c(\Phi)$ of the potential function.

Theorem 9 (Kontsevich & Manin [24]) *This definition makes the tangent sheaf of the homology space $H^*(X(\mathbb{C}), \mathbb{C})$ a Frobenius manifold.*

The main property, the associativity of the quantum product, is a consequence of the splitting axiom of GW classes and the defining relations of the homology ring of $\overline{M}_{\mathbf{S}}(\mathbb{C})$ (see Theorem 2).

5 Gromov–Witten–Welschinger classes

In this section, we introduce the Welschinger classes by following the same principles and steps which lead to Gromov–Witten theory: Firstly, we introduce the real enumerative invariants in terms of homology of moduli space of suitable maps. Then, we transfer this definition to a new one in terms of the homology of $\overline{M}^{\sigma}_{\mathbf{S}}(\mathbb{R})$. We then further generalize our definition to open-closed CohFTs and study these algebraic structures by using $H_*\overline{M}^{\sigma}_{\mathbf{S}}(\mathbb{R})$, i.e., introduce the quantum cohomology of real varieties.

5.1 Moduli space of stable real maps

Let $(X(\mathbb{C}), c_X)$ be a projective real algebraic manifold. Let $\mathbf{d} := (\beta, d) \in H_2(X(\mathbb{C})) \oplus H_1(X(\mathbb{R}))$.

We call a stable map $(\Sigma; \mathbf{p}; f) \in \overline{M}_{\mathbf{S}}(X, \beta)$ *σ-invariant* if Σ admits a real structure $c_\Sigma : \Sigma \to \Sigma$ such that $c_\sigma(p_s) = p_{\sigma(s)}$, $c_X \circ f = f \circ c_\Sigma$ and $f_*([\Sigma(\mathbb{R})]) = \pm d \in H_1(X(\mathbb{R}))$.

We denote the moduli space of σ-invariant stable maps by $\overline{R}^{\sigma}_{\mathbf{S}}(X, c_X, \mathbf{d})$. The open stratum $R^{g}_{\mathbf{S}}(X, c_X, \mathbf{d})$ of $\overline{R}^{\sigma}_{\mathbf{S}}(X, c_X, \mathbf{d})$ is the subspace where the domain curve Σ is irreducible. It is important to note that the moduli space $\overline{R}^{\sigma}_{\mathbf{S}}(X, c_X, \mathbf{d})$ has boundaries (since we have fixed the homology class d represented by the real parts of the curves $f(\Sigma)$). The σ-invariant maps $(\Sigma; \mathbf{p}; f)$ lying in the boundaries have at least two components of the domain curve Σ which are not contracted to a point by morphism $f : \Sigma \to X$.

Obviously, $\overline{R}^{\sigma}_{\mathbf{S}}(X, c_X, \mathbf{d})$ is a subspace of the real part of the real structure

$$c_\sigma : \overline{M}_{\mathbf{S}}(X, \beta) \to \overline{M}_{\mathbf{S}}(X, \beta); \qquad (\Sigma; \mathbf{p}; f) \mapsto (\overline{\Sigma}; \sigma(\mathbf{p}); c_X \circ f).$$

Therefore, the restrictions of the evaluation maps provide us

$$\begin{aligned} ev_s &: \overline{R}^{\sigma}_{\mathbf{S}}(X, c_X, \mathbf{d}) \to X(\mathbb{C}), \ \text{ for } s \in \mathbf{Perm}(\sigma), \\ ev_s &: \overline{R}^{\sigma}_{\mathbf{S}}(X, c_X, \mathbf{d}) \to X(\mathbb{R}), \ \text{ for } s \in \mathbf{Fix}(\sigma). \end{aligned}$$

These moduli spaces have been extensively studied in the open Gromov–Witten invariants context. The compactification of these moduli spaces was studied in (see [11,27,36]). The orientibility of the open stratum $R^{g}_{\mathbf{S}}(X, c_X, \mathbf{d})$ has been noted first by Fukaya and his collaborators (for $\sigma = id$ case), Liu and (implicitly) by Welschinger in [11,27,39,40]. Solomon treated this problem in most general setting (see [36]).

5.2 Welschinger invariants as degrees of evaluation maps

In a series of papers [39]–[42], Welschinger defined a set of invariants counting, with appropriate weight ± 1, real rational J-holomorphic curves intersecting a generic σ-invariant collection of marked points. Unlike the usual homological definition of Gromov–Witten invariants, Welschinger invariants are originally defined by assigning signs to individual curves based on certain geometric-topological criteria. A homological interpretation of Welschinger invariants has been given by J. Solomon very recently (see [36]):

Let $H^*(X; \det(TX(\mathbb{R})))$ denote the cohomology of $X(\mathbb{R})$ with coefficients in the flat line bundle $\det(TX(\mathbb{R}))$. Let $\alpha_s \in \Omega^*(X(\mathbb{R}), TX(\mathbb{R}))$ be representatives of a set classes in $H^*(X; \det(TX(\mathbb{R})))$ for $s \in \mathbf{Fix}(\sigma)$. Furthermore, for $\{s, \bar{s}\} \subset \mathbf{Perm}(\sigma)$, $\mu_s \in \Omega^* X(\mathbb{C})$ represent a set of classes in $H^*(X(\mathbb{C}))$. Then, a product is determined in $\Omega^* R^g_{\mathbf{S}}(X, c_X, \mathbf{d})$ by

$$\bigwedge_{\{s,\bar{s}\}\subset \mathbf{Perm}(\sigma)} ev_s^*(\mu_s) \wedge \bigwedge_{s\in \mathbf{Fix}(\sigma)} ev_s^*(\alpha_s). \tag{14}$$

If $\sum \operatorname{codim}(\alpha_*) + \sum \operatorname{codim}(\mu_*) = \dim(R^g_{\mathbf{S}}(X, c_X, \mathbf{d}))$, the form (14) extends along the critical locus of evaluation map except σ-invariant maps with reducible domain curve. In other words, it provides a relative cohomology class in $H^*(\overline{R}^\sigma_{\mathbf{S}}(X, c_X, \mathbf{d}), \mathscr{D})$. Here, $\mathscr{D}$ is the subset of $\overline{R}^\sigma_{\mathbf{S}}(X, c_X, \mathbf{d})$ whose elements $(\Sigma; \mathbf{p}; f)$ have more than one component. This extension of the differential form (14) is a direct consequence of Welschinger's theorems.

If μ_* and α_* are Poincare duals of point classes (respectively in $X(\mathbb{C})$ and $X(\mathbb{R})$), then one can define

$$N^\sigma_{\mathbf{S},\mathbf{d}} := \int_{[\overline{R}^\sigma_{\mathbf{S}}(X, c_X, \mathbf{d})]} \left\{ \bigwedge_{\{s,\bar{s}\}\subset \mathbf{Perm}(\sigma)} ev_s^*(\mu_s) \wedge \bigwedge_{s\in \mathbf{Fix}(\sigma)} ev_s^*(\alpha_s) \right\}. \tag{15}$$

Theorem 10 (Solomon [36]) *The sum*

$$\sum_d N^\sigma_{\mathbf{S},\mathbf{d}} \tag{16}$$

which is taken over all possible homology classes $d \in H_1(X(\mathbb{R}))$ realized by σ-invariant maps, is equal to Welschinger invariants.

5.3 Welschinger classes

Let $\nu : \overline{R}^\sigma_{\mathbf{S}}(X, c_X, \mathbf{d}) \to \overline{M}^\sigma_{\mathbf{S}}(\mathbb{R})$ be the restriction of the contraction map $\nu : \overline{M}_{\mathbf{S}}(X, \beta) \to \overline{M}_{\mathbf{S}}(\mathbb{C})$. Let $\mathfrak{D}$ be the subspace of $\overline{M}^\sigma_{\mathbf{S}}(\mathbb{R})$ whose elements $(\Sigma; \mathbf{p})$ have at least two irreducible components. Note that the contraction morphism ν maps $\mathscr{D} \subset \overline{R}^\sigma_{\mathbf{S}}(X, c_X, \mathbf{d})$ onto $\mathfrak{D}$.

By using the contraction morphism as in the definition of Gromov–Witten classes, Solomon's definition of Welschinger invariants can be put into a more general setting:

$$W^X_{\mathbf{S},\mathbf{d}}\Big(\bigotimes_{\{s,\bar{s}\}\subset \mathbf{Perm}(\sigma)} \mu_s \otimes \bigotimes_{s\in \mathbf{Fix}(\sigma)} \alpha_s \Big)$$
$$:= \nu_*\Big(\bigwedge_{\{s,\bar{s}\}\subset \mathbf{Perm}(\sigma)} ev_s^*(\mu_s) \wedge \bigwedge_{s\in \mathbf{Fix}(\sigma)} ev_s^*(\alpha_s)\Big).$$

The set of multilinear maps

$$W^X_{\mathbf{S},\mathbf{d}} : \bigotimes_{\mathbf{Perm}(\sigma)/\sigma} H^*(X(\mathbb{C})) \bigotimes_{\mathbf{Fix}(\mathbf{S})} H^*(X(\mathbb{R}); \det(TX(\mathbb{R}))) \to H^*(\overline{M}^\sigma_{\mathbf{S}}(\mathbb{R}), \mathfrak{D})$$

is called the *system Welschinger classes.*

5.4 Open-closed cohomological field theory

An open-closed CohFT with coefficient field $\mathbb{K}$ consists of the following data:

- A pair of superspaces (of closed and open states) $\mathcal{H}_c$ and $\mathcal{H}_o$ endowed with even non-degenerate pairings g and η respectively.
- Two sets of linear maps (open & closed correlators)

$$\{W^\sigma_{\mathbf{S}} : \bigotimes_{\mathbf{Perm}(\sigma)/\sigma} \mathcal{H}_c \otimes \bigotimes_{\mathbf{Fix}(\sigma)} \mathcal{H}_o \to H^*(\overline{M}^\sigma_{\mathbf{S}}(\mathbb{R}), \mathfrak{D})\} \tag{17}$$

$$\{I_{\mathbf{S}} : \bigotimes_{\mathbf{S}} \mathcal{H}_c \to H^* \overline{M}_{\mathbf{S}}(\mathbb{C})\}$$

defined for all $|\mathbf{S}| \geq 3$.

These data must satisfy the following axioms:

A′. CohFT of closed states: The set of maps $\{I_{\mathbf{S}} : \bigotimes_{\mathbf{S}} \mathcal{H}_c \to \overline{M}_{\mathbf{S}}(\mathbb{C})\}$ forms a CohFT.

B′. Covariance: The maps $W^\sigma_{\mathbf{S}}$ are compatible with the actions of σ-invariant relabelling on $\bigotimes_{\mathbf{Perm}(\sigma)/\sigma} \mathcal{H}_c \otimes \bigotimes_{\mathbf{Fix}(\sigma)} \mathcal{H}_o$ and on $\overline{M}^\sigma_{\mathbf{S}}(\mathbb{R})$.

C′. Splitting 1: Let $\{\alpha_a\}$ denote a basis of $\mathcal{H}_o$, and let Δ_o be $\sum \eta^{ef} \alpha_e \otimes \alpha_f$.

Let γ be a σ-invariant tree with $\mathbf{V}_\gamma = \mathbf{V}^{\mathbb{R}}_\gamma - \{v_e, v_f\}$, and let $\mathbf{F}_\gamma(v_e) = \mathbf{S}_e \cup \{s_e\}$, $\mathbf{F}_\gamma(v_f) = \mathbf{S}_f \cup \{s_f\}$ and $\mathbf{S} = \mathbf{S}_e \cup \mathbf{S}_f$. We denote the restriction of σ onto $\mathbf{S}_e, \mathbf{S}_f$ by σ_e, σ_f respectively. Let

$$\phi_\gamma : \overline{D}_\gamma(\mathbb{R}) := \overline{M}^{\sigma_e}_{\mathbf{F}_\gamma(v_e)}(\mathbb{R}) \times \overline{M}^{\sigma_f}_{\mathbf{F}_\gamma(v_f)}(\mathbb{R}) \hookrightarrow \overline{M}^\sigma_{\mathbf{S}}(\mathbb{R})$$

be the embedding of the real divisor $\overline{D}_\gamma(\mathbb{R})$.

The Splitting Axiom reads:

$$\begin{aligned}
&\phi^*_\gamma(W^\sigma_{\mathbf{S}}(\bigotimes_{\mathbf{Perm}(\sigma)/\sigma} \mu_* \otimes \bigotimes_{\mathbf{Fix}(\sigma)} \alpha_*)) \\
&\quad = \varepsilon(\gamma) \sum_{a,b} \eta^{ef} \; W^{\sigma_e}_{\mathbf{F}_\gamma(v_e)}(\bigotimes_{\mathbf{Perm}(\sigma_e)/\sigma_e} \mu_* \bigotimes_{\mathbf{Fix}(\sigma_e)\setminus\{s_e\}} \alpha_* \otimes \alpha_e) \\
&\qquad \otimes W^{\sigma_f}_{\mathbf{F}_\gamma(v_f)}(\bigotimes_{\mathbf{Perm}(\sigma_f)/\sigma_f} \mu_* \bigotimes_{\mathbf{Fix}(\sigma_f)\setminus\{s_f\}} \alpha_* \otimes \alpha_f).
\end{aligned}$$

D′. Splitting 2: Let γ be a σ-invariant tree with $\mathbf{V}_\gamma = \{v_r, v, \bar{v}\}$ and $\mathbf{V}_\gamma^{\mathbb{R}} = \{v_r\}$, and let $\mathbf{F}_\gamma(v_r) = \mathbf{S}_r \cup \{s_e, \bar{s}_e\}$, $\mathbf{F}_\gamma(v) = \mathbf{S}_f \cup \{s_f\}$, $\mathbf{F}_\gamma(\bar{v}) = \overline{\mathbf{S}}_f \cup \{\bar{s}_f\}$ and $\mathbf{S} = \mathbf{S}_r \cup \mathbf{S}_f \cup \overline{\mathbf{S}}_f$. We denote the restriction of σ onto $\mathbf{S}_r$ by σ_r. Let

$$\phi_\gamma : \overline{D}_\gamma(\mathbb{R}) := \overline{M}^{\sigma_r}_{\mathbf{F}_\gamma(v_r)}(\mathbb{R}) \times \overline{M}_{\mathbf{F}_\gamma(v)}(\mathbb{C}) \hookrightarrow \overline{M}^{\sigma}_{\mathbf{S}}(\mathbb{R})$$

be the embedding of the subspace $\overline{D}_\gamma(\mathbb{R})$.

Then, the Splitting Axiom reads:

$$\phi_\gamma^*(W^\sigma_{\mathbf{S}}(\bigotimes_{\mathbf{Perm}(\sigma)/\sigma} \mu_* \bigotimes_{\mathbf{Fix}(\sigma)} \alpha_* \otimes)) = \varepsilon(\gamma)$$

$$\sum_{a,b} g^{ab}\ W^{\sigma_r}_{\mathbf{F}_\gamma(v_r)}(\mu_a \bigotimes_{\mathbf{Perm}(\sigma_r)/\sigma_r} \mu_* \bigotimes_{\mathbf{Fix}(\sigma_r)} \alpha_* \otimes) \otimes I_{\mathbf{F}_\gamma(v)}(\mu_b \otimes \bigotimes_{\mathbf{S}_f} \mu_*).$$

Open-closed CohFTs are abstract generalizations of systems of Gromov–Witten–Welschinger classes: any system of GW classes and Welschinger classes for a real variety $(X(\mathbb{C}), c_X)$ satisfying appropriate assumptions gives rise to the open-closed CohFT where $\mathcal{H}_c = H^*X(\mathbb{C})$ and $\mathcal{H}_o = H^*(X(\mathbb{R}); \det(TX(\mathbb{R})))$.

5.5 Quantum cohomology of real varieties: A DG-operad

In this section, we first give a brief account of Horava's attempt at defining quantum cohomology of real varieties. Then, we define quantum cohomology for real varieties as a DG-operad of the *reduced graph complex* of the moduli space $\overline{M}^{\sigma}_{\mathbf{S}\cup\{s\}}(\mathbb{R})$. The definition of this DG-operad is based on Gromov–Witten–Welschinger classes.

Quantum cohomology of real varieties: Horava's approach

The quantum cohomology for real algebraic varieties was introduced surprisingly early, in 1993, by P. Horava in [16]. In his paper, Horava describes a $\mathbb{Z}_2$-equivariant topological sigma model of a real variety $(X(\mathbb{C}), c_X)$ whose set of physical observables is a direct sum of the cohomologies of $X(\mathbb{C})$ and $X(\mathbb{R})$:

$$\mathcal{H}_c \oplus \mathcal{H}_o := H^*(X(\mathbb{C})) \oplus H^*(X(\mathbb{R})). \tag{18}$$

In the case of usual topological sigma models, the OPE-algebra is known to give a deformation of the cohomology ring of $X(\mathbb{C})$. One may thus wonder what the classical structure of $\mathcal{H}_c \oplus \mathcal{H}_o$ is, of which the quantum OPE-algebra can be expected to be a deformation. Notice first that there is a natural structure of an $\mathcal{H}_c$-module on $\mathcal{H}_c \oplus \mathcal{H}_o$. Indeed, to define a product of a cohomology class α of $X(\mathbb{R})$ with a cohomology class μ of $X(\mathbb{C})$, we will pull back the cohomology class μ by $i : X(\mathbb{R}) \hookrightarrow X(\mathbb{C})$, and take the wedge product with α. Equipped with this natural structure, the module (18) associated with the pair represented by a manifold $X(\mathbb{C})$ and a real structure of it can be chosen as a (non-classical) equivariant cohomology ring in the sense of [1]. Horava required this equivariant cohomology theory to be recovered in the classical limit of the OPE-algebra of the equivariant topological sigma model. Note the important point that the equivariant cohomology theory considered in Horava's study is not the same as Borel's G-equivariant cohomology theory based on the classifying spaces BG of G. It is quite evident that classical equivariant cohomology theory wouldn't provide that rich structure since the classifying space of $\mathbb{Z}_2$ is $\mathbb{P}^\infty(\mathbb{R})$ i.e., it mainly contains two torsion elements.

By ignoring the degeneracy phenomenon, the correlation functions of such a model count the number of equivariant holomorphic curves $f : \Sigma \to X(\mathbb{C})$ meeting a number of subvarieties of $X(\mathbb{C})$ and $X(\mathbb{R})$ in images of prescribed points of Σ. The analogue of the quantum cohomology ring is now a structure of $H^*(X(\mathbb{C}))$-module on $H^*(X(\mathbb{C})) \oplus H^*(X(\mathbb{R}))$ deforming cup and mixed products.

However, the degeneracy phenomenon is the key ingredient which encodes the relations that deformed product structure should respect. In the following paragraph, we define quantum cohomology of real variety as a DG-operad in order to recover these relations.

Reduced graph complex

Let $\mathcal{G}_\bullet$ be the graph complex of $\overline{M}^\sigma_{\mathbf{S}}(\mathbb{R})$. The reduced graph complex $\mathcal{C}_\bullet$ of the moduli space $\overline{M}^\sigma_{\mathbf{S}}(\mathbb{R})$ is a complex similar to $\mathcal{G}_\bullet$ given as follows:

- The graded group is
$$\mathcal{C}_d = \left(\bigoplus_{\substack{\gamma \text{ are } \sigma-\text{invariant} \\ \& |\mathbf{E}_\gamma| = |\mathbf{S}| - d - 3}} \mathbb{Z}\left[\overline{D}_\gamma(\mathbb{R})\right] \right) / \mathcal{I}_d \tag{19}$$
where $[\overline{D}_\gamma(\mathbb{R})]$ are the (relative) fundamental class of the subspaces $\overline{D}_\gamma(\mathbb{R})$ of $\overline{M}^\sigma_{\mathbf{S}}(\mathbb{R})$. The subgroup of relations $\mathcal{I}_d$ is generated by (9).
- The boundary homomorphism of the reduced graph complex is given by the restriction of the boundary homomorphism $\partial : \mathcal{G}_d \to \mathcal{G}_{d-1}$ of the graph complex $\mathcal{G}_\bullet$.

Remark 7. A priori, the dualization of (17) requires the whole homology group $H_*(\mathcal{G}_\bullet)$ rather than $H_*(\mathcal{C}_\bullet)$. However, the covariance property of $W^\sigma_{\mathbf{S}}$ allows us to restrict ourselves to subspaces $\overline{D}_\gamma(\mathbb{R})$ instead of the strata considered in Theorem 7.

DG-operad of reduced graph homology

Let $\mathcal{C}_\bullet$ be the reduced graph complex of $\overline{M}^\sigma_{\mathbf{S}\cup\{s\}}(\mathbb{R})$. By dualizing (17), we obtain a family of maps

$$\{\mathbf{Z}^{g}_{\mathbf{S}} : \mathcal{C}_{\bullet} \to Hom(\bigotimes_{\mathbf{Perm}(\sigma)/\sigma} \mathcal{H}_c \bigotimes_{\mathbf{Fix}(\sigma)} \mathcal{H}_o, \mathcal{H}_o)\} \tag{20}$$

along with an algebra over $H_*\overline{M}_{\mathbf{S}\cup\{s\}}(\mathbb{C})$-operad

$$\{\mathbf{Y}_{\mathbf{S}} : H_*\overline{M}_{\mathbf{S}\cup\{s\}}(\mathbb{C}) \to Hom(\bigotimes_{\mathbf{S}} \mathcal{H}_c, \mathcal{H}_c)\}. \tag{21}$$

Therefore, the homology class of the subspace $\overline{D}_\gamma(\mathbb{R})$ in $\overline{M}^{\sigma}_{\mathbf{S}\cup\{s\}}(\mathbb{R})$ is interpreted as an (k,l)-ary operation on $(\mathcal{H}_o, \mathcal{H}_c)$ where $k = |\mathbf{Perm}(\sigma)|/2$ and $l = |\mathbf{Fix}(\sigma)|$.

Remark 8. The σ-invariant trees which are used in Lemma 1 are different than the trees depicting (k,l)-ary operation. These new trees are in fact quotients of σ-invariant trees with respect to their symmetries arising from the real structures of the corresponding σ-invariant curves (see Fig. 4).

In addition to the higher associativity $\mathbf{Y}_{\mathbf{S}}$, the additives relations in $\mathcal{C}_{\bullet}$ require additional conditions on (k,l)-ary operations. The families of maps $\mathbf{Z}^{g}_{\mathbf{S}}$ and $\mathbf{Y}_{\mathbf{S}}$ satisfy the following conditions:

1. *Higher associativity relations:* $(\mathcal{H}_c, \eta_c, \mathbf{Y}_*)$ form a $Comm_\infty$-algebra structure.

2. *A_∞-type of structure:* The additive relations arising from the image of the boundary homomorphism $\partial : \mathcal{C}_* \to \mathcal{C}_{*-1}$ (i.e., from the identity (8)) and the splitting axiom ($\mathbf{C}'$) become the identities between (k,l)-ary operations depicted in Fig. 5.

It is easy to see that $(\mathcal{H}_o, \eta, \mathbf{Z}^{g}_{\mathbf{S}})$ reduces to an A_∞-algebra when $\sigma = id$.

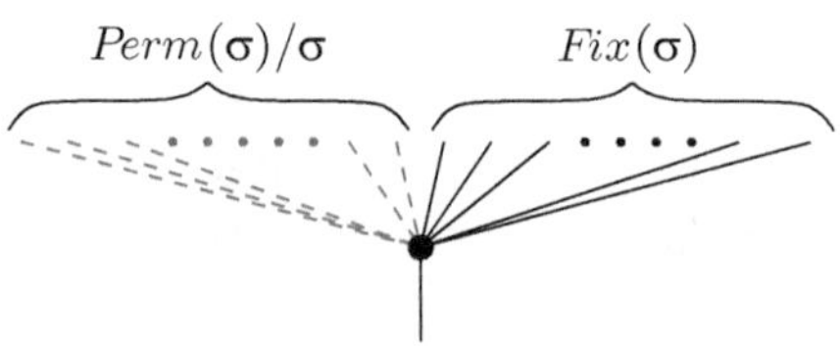

Fig. 4. *(k,l)-ary operation corresponding to the fundamental class of* $\overline{M}^{\sigma}_{\mathbf{S}\cup\{s\}}(\mathbb{R})$

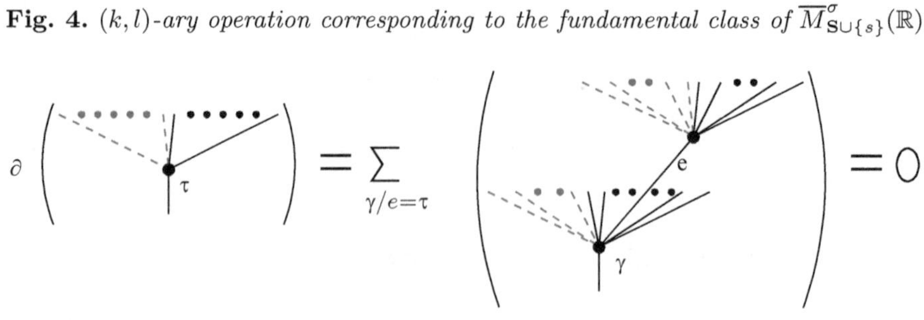

Fig. 5. *The image of boundary map provides A_∞-type of relations*

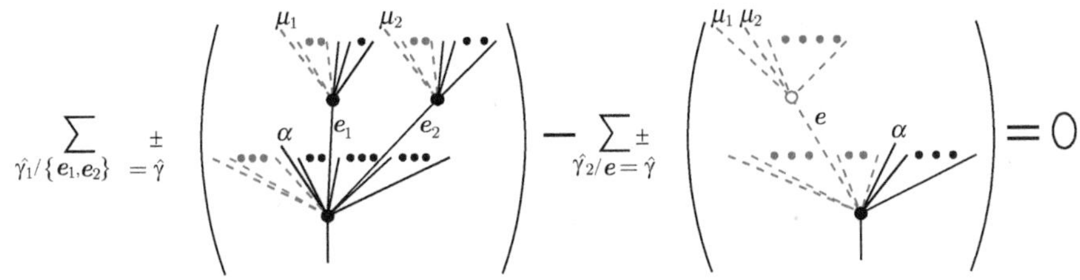

Fig. 6. *Cardy type of relations arising from the subgroup* $\mathcal{I}_*$

3. *Cardy type of relations:* The additive relations arising from the subgroup $\mathcal{I}_d$ of the graph complex $\mathcal{C}_\bullet$ (i.e., from the identity (9)) and the splitting axiom (**D′**) become the identities between (k,l)-ary operations which are depicted in Fig. 6.

Here, $\hat{\gamma}_1.\hat{\gamma}_2$ are the quotients of γ_1, γ_2 in (9) with respect to their $\mathbb{Z}_2$-symmetries induced by real structures of corresponding σ-invariant curves. In this picture, μ_1, μ_2 correspond to the flags f_1, f_2, and α corresponds to the real flag f_3.

Remark 9. The (partial) DG-operad defined above has close relatives which have appeared in the literature.

The first one is the Swiss-cheese operad which is introduced by Voronov in [38]. In fact, Swiss-cheese operads are the chain operad of the spaces of $\mathbb{Z}_2$-equivariant configuration in $\mathbb{P}^1(\mathbb{C})$ (see Section 3.4).

Kajiura and Stasshefff recently introduced *open-closed homotopy algebras* which are amalgamations of L_∞-algebras with A_∞-algebras, see [18]–[20]. In their setting, the compactification of the configuration spaces are different than ours; therefore, the differential and the corresponding relations are different. In particular, there is no Cardy type of relation in their setting.

Very recently, Merkulov studied the 'operad of formal homogeneous space' in [30]. His approach is motivated by studies on relative obstruction theory and its applications (see [34,35]). It seems that a version of mirror symmetry which takes the real structures into account has a connection with this approach.

Part III: Yet again, mirror symmetry!

In his seminal paper [23], Kontsevich proposed that mirror symmetry is a (non-canonical) equivalence between the bounded derived category of coherent sheaves $D^bCoh(Y)$ on a complex variety Y and the derived Fukaya category $DFuk(X)$ of its mirror, a symplectic manifold (X,ω). This first category consists of chain complexes of holomorphic bundles, with quasi-isomorphisms and their formal inverses. Roughly speaking, the derived Fukaya category should be constructed from Lagrangian submanifolds $L \subset X$ (carrying flat

$U(1)$-connection A). Morphisms are given by Floer cohomology of Lagrangian submanifolds. One important feature is that, whereas the mirror of a Calabi–Yau variety is another Calabi–Yau variety, the mirrors of Fano varieties are Landau–Ginzburg models; i.e., affine varieties equipped with a map called the superpotential.

Kontsevich also conjectured that the equivalence of derived categories should imply numerical predictions: For the complex side (B-side), consider the diagonal subvariety $\Delta_Y \subset Y \times Y$, and define the Hochschild cohomology of Y to be the endomorphism algebra of Δ_Y regarded as an object of $D^bCoh(X \times X)$ [23]. Kontsevich interpreted the last definition of $HH^*(\mathcal{O}_Y)$ ($\cong H^*(Y, \bigwedge^* TY)$) as computing the space of infinitesimal deformations of the bounded derived category of coherent sheaves on Y in the class of A_∞-categories. On the other hand, for the symplectic side (A-side), the diagonal $\Delta_X \subset X \times X$ is a Lagrangian submanifold of $(X \times X, (\omega, -\omega))$. The Floer cohomology of the diagonal is canonically isomorphic to $H^*(X)$. Roughly, the above picture suggests that the deformation of $H^*(X)$ constructed using Gromov–Witten invariants should correspond to variations of Hodge structure in the B-side.

'Realizing' mirror symmetry

Let $X \to B$ be a Lagrangian torus fibration and $c_X : X \to X$ be an anti-symplectic involution of (X, ω) whose real part $X(\mathbb{R})$ is a section of $X \to B$.

To recover more about real geometry from Kontsevich's conjecture, we consider the structure sheaf $\mathcal{O}_Y$. Calculating its self-Hom's

$$Ext^i(\mathcal{O}_Y, \mathcal{O}_Y) \cong H^{0,i}(Y)$$

shows that if $\mathcal{O}_Y$ is mirror to the Lagrangian submanifold $X(\mathbb{R})$, then we must have

$$HF^*(X(\mathbb{R}), X(\mathbb{R})) \cong H^{0,*}(Y)$$

as graded vector spaces.

The Kontsevich's conjecture above and the description of quantum cohomology of real varieties that we have discussed in the previous section suggest a correspondence between two 'open-closed homotopy algebras' (in an appropriate sense):

Symplectic side X (A-side)	$\Longleftrightarrow$	Complex side Y (B-side)
$(\mathcal{H}_c, \mathcal{H}_o) =$ $(H^*(X), HF^*(X(\mathbb{R}), X(\mathbb{R})))$	$\Leftrightarrow$	$(\widehat{\mathcal{H}}_c, \widehat{\mathcal{H}}_o) =$ $(H^{*,*}(Y), H^{0,*}(Y))$

If we restrict ourselves to three point operations, we should obtain $\mathcal{H}_c$ and $\widehat{\mathcal{H}}_c$-module structure in respective sides (which recalls Horava's version of quantum cohomology of real varieties).

For the A-side of the mirror correspondence, the DG-operad structure that we have discussed in the previous section provides a promising candidate for extension of this structure. However, the B-side of the story is more intriguing: Open-closed strings in the B-model, their boundary conditions, etc. are quite unclear to us. S. Merkulov pointed out possible connections with Ran's work on relative obstructions, Lie atoms, and their deformations [34, 35].

References

1. G. Bredon, *Equivariant Cohomology Theories.* Lecture Notes in Mathematics 34, Springer Verlag, Berlin, 1967.
2. Ö. Ceyhan, *On moduli of pointed real curves of genus zero. Proceedings 13th Gokova Geometry Topology Conference* 2006, 1–38.
3. Ö. Ceyhan, *On moduli of pointed real curves of genus zero.* Ph.D. Thesis. IRMA Preprints. Ref: 06019.
4. Ö. Ceyhan, *The graph homology of the moduli space of pointed real curves of genus zero. Selecta Math.* 13 (2007), no. 2, 203–237.
5. Ö. Ceyhan, *On Solomon's reconstruction theorem for Gromov-Witten-Welshinger classes.* in preperation.
6. M. Davis, T. Januszkiewicz, R. Scott, *Fundamental groups of blow-ups.* Adv. Math. 177 (2003), no. 1, 115–179.
7. A. Degtyarev, I. Itenberg, V. Kharlamov, *Real enriques surfaces,* Lecture Notes in Mathmetics 1746, Springer-Verlag, Berlin, 2000. xvi+259 pp.
8. S.L. Devadoss, *Tessellations of moduli spaces and the mosaic operad.* in 'Homotopy invariant algebraic structures' (Baltimore, MD, 1998), 91–114, Contemp. Math., 239, Amer. Math. Soc., Providence, RI, 1999.
9. I. Dolgachev, *Topics in Classical Algebraic Geometry. Part I.* lecture notes, available authors webpage *http://www.math.lsa.umich.edu/ idolga/topics1.pdf*
10. P. Etingof, A. Henriques, J. Kamnitzer, E. Rains, *The cohomology ring of the real locus of the moduli space of stable curves of genus 0 with marked points.* preprint math.AT/0507514.
11. K. Fukaya, Y.G. Oh, H. Ohta, K. Ono, *Lagrangian intersection Floer theory: anomaly and obstruction.* preprint 2000.
12. P. Griffiths, J. Harris, *Principles of algebraic geometry.* Pure and Applied Mathematics. Wiley-Interscience [John Wiley and Sons], New York, (1978). xii+813 pp.
13. A.B. Goncharov, Yu.I. Manin, *Multiple ζ-motives and moduli spaces $\overline{M}_{0,n}(\mathbb{C})$.* Compos. Math. 140 (2004), no. 1, 1–14.
14. J. Harris, I. Morrison, *Moduli of curves.* Graduate Texts in Mathematics vol 187, Springer-Verlag, NewYork, (1998), 366 pp.
15. A. Henriques, J. Kamnitzer, *Crystals and coboundary categories.* Duke Math. J. 132 (2006), no. 2, 191–216.
16. P. Horava, *Equivariant topological sigma models.* Nuclear Phys. B 418 (1994), no. 3, 571–602.
17. I. Itenberg, V. Kharlamov, E. Shustin, *A Caporaso-Harris type formula for Welschinger invariants of real toric Del Pezzo surfaces.* preprint math/0608549.
18. H. Kajiura, J. Stasheff, *Jim Homotopy algebras inspired by classical open-closed string field theory.* Comm. Math. Phys. 263 (2006), no. 3, 553–581.

19. H. Kajiura, J. Stasheff, *Open-closed homotopy algebra in mathematical physics.* J. Math. Phys. 47 (2006), no. 2, 023506, 28 pp.
20. H. Kajiura, J. Stasheff, *Homotopy algebra of open-closed strings.* hep-th/0606283.
21. M. Kapranov, *The permutoassociahedron, MacLane's coherence theorem and asymptotic zones for KZ equation.* J. Pure Appl. Algebra 85 (1993), no. 2, 119–142.
22. S. Keel, *Intersection theory of moduli space of stable N-pointed curves of genus zero*, Trans. Amer. Math. Soc. 330 (1992), no. 2, 545–574.
23. M. Kontsevich, it Homological algebra of mirror symmetry. Proceedings of the International Congress of Mathematicians, Vol. 1, 2 (Zürich, 1994), 120–139, Birkhäuser, Basel, 1995.
24. M. Kontsevich, Yu.- I. Manin, *Gromov-Witten classes, quantum cohomology, and enumerative geometry.* Comm. Math. Phys. 164 (1994), no. 3, 525–562.
25. M. Kontsevich, Yu.-I. Manin (with appendix by R. Kaufmann), *Quantum cohomology of a product.* Invent. Math. 124 (1996), no. 1–3, 313–339.
26. F.F. Knudsen, *The projectivity of the moduli space of stable curves. II. The stacks* $M_{g,n}$. Math. Scand. 52 (1983), no. 2, 161–199.
27. C.C.M. Liu, *Moduli J-holomorphic curves with Lagrangian boundary conditions and open Gromov-Witten invariants for an* S^1*-equivariant pair.* preprint math.SG/0210257.
28. Yu.-I. Manin, *Frobenius manifolds, quantum cohomology and moduli spaces.* AMS Colloquium Publications vol 47, Providence, RI, (1999), 303 pp.
29. Yu.-I. Manin, *Gauge Field Theories and Complex Geometry.* Grundlehren der Mathematischen Wissenschaften 289, Springer-Verlag, Berlin, (1997), 346 pp.
30. S. A. Merkulov, *Operad of formal homogeneous spaces and Bernoulli numbers.* math/0708.0891.
31. G. Mikhalkin, *Counting curves via lattice paths in polygons.* C. R. Math. Acad. Sci. Paris 336 (2003), no. 8, 629–634.
32. G. Mikhalkin, *Enumerative tropical algebraic geometry in* $\mathbb{R}^2$. J. Amer. Math. Soc. 18 (2005), no. 2, 313–377.
33. R. Pandharipande, J. Solomon, J. Walcher, *Disk enumeration on the quintic 3-fold.* math.SG/0610901.
34. Z. Ran, *Lie Atoms and their deformation theory.* math/0412204.
35. Z. Ran, *Bernoulli numbers and deformations of schemes and maps.* math/0609223.
36. J. Solomon, *Intersection theory on the moduli space of holomorphic curves with Lagrangian boundary conditions.* math.SG/0606429.
37. J. Solomon, *A differential equation for the open Gromov-Witten potential.* a talk given at *Real, tropical, and complex enumerative geometry workshop* at CRM-Montreal.
38. A. Voronov, *The Swiss-cheese operad.* Homotopy invariant algebraic structures (Baltimore, MD, 1998), 365–373, Contemp. Math., 239, Amer. Math. Soc., Providence, RI, 1999.
39. J.Y. Welschinger, *Invariants of real symplectic 4-manifolds and lower bounds in real enumerative geometry.* Invent. Math. 162 (2005), no. 1, 195–234.

40. J.Y. Welschinger, *Spinor states of real rational curves in real algebraic convex 3-manifolds and enumerative invariants.* Duke Math. J. 127 (2005), no. 1, 89–121.
41. J.Y. Welschinger, *Invariants of real symplectic four-manifolds out of reducible and cuspidal curves.* Bull. Soc. Math. France 134 (2006), no. 2, 287–325.
42. J.Y. Welschinger, *Towards relative invariants of real symplectic four-manifolds.* Geom. Funct. Anal. 16 (2006), no. 5, 1157–1182.

Weyl Modules and Opers without Monodromy

Edward Frenkel* and Dennis Gaitsgory†

* Department of Mathematics, University of California, Berkeley, CA 94720, USA
frenkel@math.berkeley.edu
† Department of Mathematics, Harvard University, Cambridge, MA 02138, USA
gaitsgde@math.harvard.edu

Summary. We prove that the algebra of endomorphisms of a Weyl module of critical level is isomorphic to the algebra of functions on the space of monodromy-free opers on the disc with regular singularity and residue determined by the highest weight of the Weyl module. This result may be used to test the local geometric Langlands correspondence proposed in our earlier work.

AMS Classification Codes (2000). 17B67; 11R39

1 Introduction

Let $\mathfrak{g}$ be a simple finite-dimensional Lie algebra. For an invariant inner product κ on $\mathfrak{g}$ (which is unique up to a scalar) define the central extension $\widehat{\mathfrak{g}}_\kappa$ of the formal loop algebra $\mathfrak{g} \otimes \mathbb{C}((t))$, which fits into the short exact sequence

$$0 \to \mathbb{C}\mathbf{1} \to \widehat{\mathfrak{g}}_\kappa \to \mathfrak{g} \otimes \mathbb{C}((t)) \to 0.$$

This sequence is split as a vector space, and the commutation relations read

$$[x \otimes f(t), y \otimes g(t)] = [x, y] \otimes f(t)g(t) - (\kappa(x, y) \operatorname{Res} f dg)\mathbf{1}, \tag{1}$$

and $\mathbf{1}$ is a central element. The Lie algebra $\widehat{\mathfrak{g}}_\kappa$ is the *affine Kac–Moody algebra* associated to κ. We will denote by $\widehat{\mathfrak{g}}_\kappa$-mod the category of *discrete* representations of $\widehat{\mathfrak{g}}_\kappa$ (i.e., such that any vector is annihilated by $\mathfrak{g} \otimes t^n\mathbb{C}[[t]]$ for sufficiently large n), on which $\mathbf{1}$ acts as the identity.

Let $U_\kappa(\widehat{\mathfrak{g}})$ be the quotient of the universal enveloping algebra $U(\widehat{\mathfrak{g}}_\kappa)$ of $\widehat{\mathfrak{g}}_\kappa$ by the ideal generated by $(\mathbf{1} - 1)$. Define its completion $\widetilde{U}_\kappa(\widehat{\mathfrak{g}})$ as follows:

$$\widetilde{U}_\kappa(\widehat{\mathfrak{g}}) = \varprojlim U_\kappa(\widehat{\mathfrak{g}})/U_\kappa(\widehat{\mathfrak{g}}) \cdot (\mathfrak{g} \otimes t^n\mathbb{C}[[t]]).$$

It is clear that $\widetilde{U}_\kappa(\widehat{\mathfrak{g}})$ is a topological algebra, whose discrete continuous representations are the same as objects of $\widehat{\mathfrak{g}}_\kappa$-mod.

Ö. Ceyhan et al. (eds.), *Arithmetic and Geometry Around Quantization*,
Progress in Mathematics 279, DOI: 10.1007/978-0-8176-4831-2_5,

Let κ_{crit} be the *critical* inner product on $\mathfrak{g}$ defined by the formula

$$\kappa_{\text{crit}}(x, y) = -\frac{1}{2}\operatorname{Tr}(\operatorname{ad}(x) \circ \operatorname{ad}(y)).$$

In what follows we will use the subscript "crit" instead of κ_{crit}.

Let $\check{G}$ be the group of adjoint type whose Lie algebra $\check{\mathfrak{g}}$ is Langlands dual to $\mathfrak{g}$ (i.e., the Cartan matrix of $\check{\mathfrak{g}}$ is the transpose of that of $\mathfrak{g}$).

Let $\mathfrak{Z}_{\mathfrak{g}}$ be the center of $\widetilde{U}_{\text{crit}}(\widehat{\mathfrak{g}})$. According to a theorem of [FF, F2], $\mathfrak{Z}_{\mathfrak{g}}$ is isomorphic to the algebra $\operatorname{Fun}\operatorname{Op}_{\check{\mathfrak{g}}}(\mathcal{D}^{\times})$ of functions on the space of $\check{\mathfrak{g}}$-opers on the punctured disc $\mathcal{D}^{\times} = \operatorname{Spec}(\mathbb{C}((t)))$ (see [BD1, FG2] and Sect. 2 for the definition of opers).

It is interesting to understand how $\mathfrak{Z}_{\mathfrak{g}}$ acts on various $\widehat{\mathfrak{g}}_{\text{crit}}$-modules. The standard modules are the Verma modules and the Weyl modules. They are obtained by applying the induction functor

$$\begin{aligned} \operatorname{Ind} : \mathfrak{g}\text{-mod} &\to \widehat{\mathfrak{g}}_{\text{crit}}\text{-mod}, \\ M &\mapsto U(\widehat{\mathfrak{g}}_{\text{crit}}) \underset{U(\mathfrak{g}[[t]] \oplus \mathbf{1})}{\otimes} M, \end{aligned}$$

where $\mathfrak{g}[[t]]$ acts on M via the projection $\mathfrak{g}[[t]] \to \mathfrak{g}$ and $\mathbf{1}$ acts as the identity.

For $\lambda \in \mathfrak{h}^*$ let M_λ be the Verma module over $\mathfrak{g}$ with highest weight λ. The corresponding $\widehat{\mathfrak{g}}_{\text{crit}}$-module $\mathbb{M}_\lambda = \operatorname{Ind}(M_\lambda)$ is the *Verma module* of critical level with highest weight λ.

For a dominant integral weight λ let V_λ be the irreducible finite-dimensional $\mathfrak{g}$-module with highest weight λ. The corresponding $\widehat{\mathfrak{g}}_{\text{crit}}$-module $\mathbb{V}_\lambda = \operatorname{Ind}(V_\lambda)$ is the *Weyl module* of critical level with highest weight λ. The module $\mathbb{V}_0 = \operatorname{Ind}(\mathbb{C}_0)$ is also called the vacuum module.

It was proved in [FF, F2] that the algebra of $\widehat{\mathfrak{g}}_{\text{crit}}$-endomorphisms of $\mathbb{V}_0$ is isomorphic to the algebra $\operatorname{Fun}\operatorname{Op}^{\text{reg}}_{\check{\mathfrak{g}}}$ of functions on the space $\operatorname{Op}^{\text{reg}}_{\check{\mathfrak{g}}}$ of $\check{\mathfrak{g}}$-opers on the disc $\mathcal{D} = \operatorname{Spec}(\mathbb{C}[[t]])$. Moreover, there is a commutative diagram

$$\begin{array}{ccc} \mathfrak{Z}_{\mathfrak{g}} & \xrightarrow{\sim} & \operatorname{Fun}\operatorname{Op}_{\check{\mathfrak{g}}}(\mathcal{D}^{\times}) \\ \downarrow & & \downarrow \\ \operatorname{End}_{\widehat{\mathfrak{g}}_{\text{crit}}}(\mathbb{V}_0) & \xrightarrow{\sim} & \operatorname{Fun}\operatorname{Op}^{\text{reg}}_{\check{\mathfrak{g}}} \end{array}$$

We have shown in [FG2], Corollary 13.3.2, that a similar result holds for the Verma modules as well: the algebra of $\widehat{\mathfrak{g}}_{\text{crit}}$-endomorphisms of $\mathbb{M}_\lambda$ is isomorphic to $\operatorname{Fun}\operatorname{Op}^{\text{RS},\varpi(-\lambda-\rho)}_{\check{\mathfrak{g}}}$, where $\operatorname{Op}^{\text{RS},\varpi(-\lambda-\rho)}_{\check{\mathfrak{g}}}$ is the space of $\check{\mathfrak{g}}$-opers on $\mathcal{D}^{\times}$ with regular singularity and residue $\varpi(-\lambda-\rho)$, ϖ being the natural projection $\mathfrak{h}^* \to \operatorname{Spec}\operatorname{Fun}(\mathfrak{h}^*)^W$ (see [FG2], Sect. 2.4, for a precise definition). In addition, there is an analogue of the above commutative diagram for Verma modules.

In this paper we consider the Weyl modules $\mathbb{V}_\lambda$. In [FG2], Sect. 2.9, we defined the subspace $\operatorname{Op}^{\lambda,\text{reg}}_{\check{\mathfrak{g}}} \subset \operatorname{Op}^{\text{RS},\varpi(-\lambda-\rho)}_{\check{\mathfrak{g}}}$ of λ-*regular opers* (we recall this definition below). Its points are those opers in $\mathrm{Op}_{\check{\mathfrak{g}}}^{\mathrm{RS},\varpi(-\lambda-\rho)}$ which have trivial monodromy and are therefore $\check{G}(\hat{\mathcal{K}})$ gauge equivalent to the trivial local system on $\mathcal{D}^\times$. In particular, $\mathrm{Op}_{\check{\mathfrak{g}}}^{0,\mathrm{reg}} = \mathrm{Op}_{\check{\mathfrak{g}}}^{\mathrm{reg}}$. According to Lemma 1 below, the disjoint union of $\mathrm{Op}_{\check{\mathfrak{g}}}^{\lambda,\mathrm{reg}}$, where λ runs over the set P^+ of dominant integral weights of $\mathfrak{g}$, is precisely the locus of $\check{\mathfrak{g}}$-opers on $\mathcal{D}^\times$ with trivial monodromy. The main result of this paper is the following theorem, which generalizes the description of $\mathrm{End}_{\widehat{\mathfrak{g}}_{\mathrm{crit}}} \mathbb{V}_0$ from [FF, F2] to the case of an arbitrary dominant integral weight λ.

Theorem 1 *For any dominant integral weight λ the center $\mathfrak{Z}_{\mathfrak{g}}$ maps surjectively onto $\mathrm{End}_{\widehat{\mathfrak{g}}_{\mathrm{crit}}} \mathbb{V}_\lambda$, and we have the following commutative diagram*

$$\begin{array}{ccc} \mathfrak{Z}_{\mathfrak{g}} & \xrightarrow{\sim} & \mathrm{Fun}\,\mathrm{Op}_{\check{G}}(D^\times) \\ \downarrow & & \downarrow \\ \mathrm{End}_{\widehat{\mathfrak{g}}_{\mathrm{crit}}} \mathbb{V}_\lambda & \xrightarrow{\sim} & \mathrm{Fun}\,\mathrm{Op}_{\check{\mathfrak{g}}}^{\lambda,\mathrm{reg}} \end{array} \tag{2}$$

For $\mathfrak{g} = \mathfrak{sl}_2$ this follows from Prop. 1 of [F1]. This statement was also independently conjectured by A. Beilinson and V. Drinfeld (unpublished).

In addition, we prove that $\mathbb{V}_\lambda$ is a free module over $\mathrm{End}_{\widehat{\mathfrak{g}}_{\mathrm{crit}}} \mathbb{V}_\lambda \simeq \mathrm{Fun}\,\mathrm{Op}_{\check{\mathfrak{g}}}^{\lambda,\mathrm{reg}}$.

Theorem 1 has important consequences for the local geometric Langlands correspondence proposed in [FG2]. According to our proposal, to each "local Langlands parameter" σ, which is a $\check{G}$–local system on the punctured disc $\mathcal{D}^\times$ (or equivalently, a $\check{G}$-bundle with a connection on $\mathcal{D}^\times$), there should correspond a category $\mathcal{C}_\sigma$ equipped with an action of $G((t))$.

Now let χ be a fixed $\check{\mathfrak{g}}$-oper on $\mathcal{D}^\times$, which we regard as a character of the center $\mathfrak{Z}_{\mathfrak{g}}$. Consider the full subcategory $\widehat{\mathfrak{g}}_{\mathrm{crit}}\text{-mod}_\chi$ of the category $\widehat{\mathfrak{g}}_{\mathrm{crit}}\text{-mod}$ whose objects are $\widehat{\mathfrak{g}}_{\mathrm{crit}}$-modules, on which $\mathfrak{Z}_{\mathfrak{g}}$ acts according to this character. This category carries a canonical action of the ind-group $G((t))$ via its adjoint action on $\widehat{\mathfrak{g}}_{\mathrm{crit}}$. We proposed in [FG2] that $\widehat{\mathfrak{g}}_{\mathrm{crit}}\text{-mod}_\chi$ should be equivalent to the sought-after category $\mathcal{C}_\sigma$, where σ is the $\check{G}$-local system underlying the oper χ. This entails a far-reaching corollary that the categories $\widehat{\mathfrak{g}}_{\mathrm{crit}}\text{-mod}_{\chi_1}$ and $\widehat{\mathfrak{g}}_{\mathrm{crit}}\text{-mod}_{\chi_2}$ for two different opers χ_1 and χ_2 are equivalent if the underlying local systems of χ_1 and χ_2 are isomorphic to each other.

In particular, consider the simplest case when σ is the trivial local system. Then, by Lemma 1, χ must be a point of $\mathrm{Op}_{\check{\mathfrak{g}}}^{\lambda,\mathrm{reg}}$ for some $\lambda \in P^+$. Theorem 1 implies that the quotient $\mathbb{V}_\lambda(\chi)$ of the Weyl module $\mathbb{V}_\lambda$ by the central character corresponding to χ is non-zero. Therefore $\mathbb{V}_\lambda(\chi)$ is a non-trivial object of $\widehat{\mathfrak{g}}_{\mathrm{crit}}\text{-mod}_\chi$ and also of the corresponding $G[[t]]$-equivariant category $\widehat{\mathfrak{g}}_{\mathrm{crit}}\text{-mod}_\chi^{G[[t]]}$. In the case when $\lambda = 0$ we have proved in [FG1] that this is a unique irreducible object of $\widehat{\mathfrak{g}}_{\mathrm{crit}}\text{-mod}_\chi^{G[[t]]}$ and that this category is in fact equivalent to the category of vector spaces. Therefore we expect the same to be true for all other values of λ. This will be proved in a follow-up paper.

The chapter is organized as follows. In Sect. 2 we recall the relevant notions of opers, Cartan connections, and Miura transformation. In Sect. 3 we explain the strategy of the proof of the main result, Theorem 1, and reduce it to two statements, Theorem 2 and Proposition 1. We then prove Proposition 1 assuming Theorem 2 in Sect. 4. Our argument is based on the exactness of the functor of quantum Drinfeld–Sokolov reduction, which we derive from [Ar]. In Sect. 5 we compute the characters of the algebra of functions on $\mathrm{Op}_{\check{\mathfrak{g}}}^{\lambda,\mathrm{reg}}$ and of the semi-infinite cohomology of $\mathbb{V}_\lambda$ (they turn out to be the same). We then give two different proofs of Theorem 2 in Sect. 6. This completes the proof of the main result. We also show that the natural map from the Weyl module $\mathbb{V}_\lambda$ to the corresponding Wakimoto module is injective and that $\mathbb{V}_\lambda$ is a free module over its endomorphism algebra.

The research of E.F. was supported by the DARPA Program "Focus Areas in Fundamental Mathematics."

2 Some results on opers

In this section we recall the relevant notions of opers, Cartan connections, and Miura transformation, following [BD1, F2, FG2], where we refer the reader for more details.

Let $\mathfrak{g}$ be a simple Lie algebra and G the corresponding algebraic group of adjoint type. Let B be its Borel subgroup and $N = [B, B]$ its unipotent radical, with the corresponding Lie algebras $\mathfrak{n} \subset \mathfrak{b} \subset \mathfrak{g}$.

Let X be a smooth curve, or the disc $\mathcal{D} = \mathrm{Spec}(\hat{\mathcal{O}})$, where $\hat{\mathcal{O}}$ is a one-dimensional smooth complete local ring, or the punctured disc $\mathcal{D}^\times = \mathrm{Spec}(\hat{\mathcal{K}})$, where $\hat{\mathcal{K}}$ is the field of fractions of $\hat{\mathcal{O}}$.

Following Beilinson and Drinfeld (see [BD1], Sect. 3.1, and [BD2]), one defines a $\mathfrak{g}$-*oper* on X to be a triple $(\mathcal{F}_G, \nabla, \mathcal{F}_B)$, where $\mathcal{F}_G$ is a principal G-bundle $\mathcal{F}_G$ on X, ∇ is a connection on $\mathcal{F}_G$, and $\mathcal{F}_B$ is a B-reduction of $\mathcal{F}_G$ which is transversal to ∇, in the sense explained in the above references and in [FG2], Sect. 1.1. We note that the transversality condition allows us to identify canonically the B-bundles $\mathcal{F}_B$ underlying all opers.

More concretely, opers on the punctured disc $\mathcal{D}^\times$ may be described as follows. Let us choose a trivialization of $\mathcal{F}_B$ and a coordinate t on the disc $\mathcal{D}$ such that $\hat{\mathcal{O}} = \mathbb{C}[[t]]$ and $\hat{\mathcal{K}} = \mathbb{C}((t))$. Let us choose a nilpotent subalgebra $\mathfrak{n}_-$, which is in generic position with $\mathfrak{b}$ and a set of simple root generators $f_\imath, \imath \in I$, of $\mathfrak{n}_-$. Then a $\mathfrak{g}$-oper on $\mathcal{D}^\times$ is, by definition, an equivalence class of operators of the form

$$\nabla = \partial_t + \sum_{\imath \in I} f_\imath + \mathbf{v}, \qquad \mathbf{v} \in \mathfrak{b}(\hat{\mathcal{K}}), \tag{1}$$

with respect to the action of the group $N(\hat{\mathcal{K}})$ by gauge transformations. It is known that this action is free and the resulting set of equivalence classes is in bijection with $\hat{\mathcal{K}}^{\oplus \ell}$, where $\ell = \mathrm{rank}(\mathfrak{g})$. Opers may be defined in this way over any base, and this allows us to define the ind-affine scheme $\mathrm{Op}_{\mathfrak{g}}(\mathcal{D}^\times)$ of $\mathfrak{g}$-opers on $\mathcal{D}^\times$ (it is isomorphic to an inductive limit of affine spaces).

Let $\check{P}^+$ be the set of dominant integral coweights of $\mathfrak{g}$. For $\check{\lambda} \in \check{P}^+$ we define a $\mathfrak{g}$-*oper with* $\check{\lambda}$-*nilpotent singularity* as an equivalence class of operators

$$\nabla = \partial_t + \sum_{\imath \in I} t^{\langle \alpha_\imath, \check{\lambda} \rangle} \cdot f_\imath + \mathbf{v}(t) + \frac{v}{t}, \qquad \mathbf{v}(t) \in \mathfrak{b}(\hat{\mathcal{O}}), v \in \mathfrak{n}, \tag{2}$$

with respect to the action of the group $N(\hat{\mathcal{O}})$ by gauge transformations (see [FG2], Sect. 2.9). The corresponding scheme is denoted by $\mathrm{Op}_{\mathfrak{g}}^{\check{\lambda},\mathrm{nilp}}$. According to Theorem 2.9.1 of [FG2], the natural map $\mathrm{Op}_{\mathfrak{g}}^{\check{\lambda},\mathrm{nilp}} \to \mathrm{Op}_{\mathfrak{g}}(\mathcal{D}^\times)$ is injective, and its image is equal to the space of $\mathfrak{g}$-opers with regular singularity and residue $\varpi(-\check{\lambda} - \check{\rho})$ (here $\check{\rho}$ is the half-sum of positive coroots of $\mathfrak{g}$ and ϖ is the projection $\mathfrak{h} \to \mathrm{Spec}\,\mathrm{Fun}(\mathfrak{h})^W$).

Now we define the space $\mathrm{Op}_{\mathfrak{g}}^{\check{\lambda},\mathrm{reg}}$ of $\check{\lambda}$-*regular opers* as the subscheme of $\mathrm{Op}_{\mathfrak{g}}^{\check{\lambda},\mathrm{nilp}}$ corresponding to those operators (2) which satisfy $v = 0$ (so that ∇ is regular at $t = 0$). In particular, if $\check{\lambda} = 0$, then $\mathrm{Op}_{\mathfrak{g}}^{0,\mathrm{reg}}$ is just the space of regular $\check{\mathfrak{g}}$-opers on the disc $\mathcal{D}$. The geometric significance of $\check{\lambda}$-opers is explained by the following

Lemma 1 *Suppose that a* $\mathfrak{g}$*-oper* $\chi = (\mathcal{F}_G, \nabla, \mathcal{F}_B)$ *on* $\mathcal{D}^\times$ *is such that the corresponding* G*-local system is trivial (in other words, the corresponding operator* (1) *is in the* $G(\hat{\mathcal{K}})$ *gauge equivalence class of* $\nabla_0 = \partial_t$*). Then* $\chi \in \mathrm{Op}_{\mathfrak{g}}^{\check{\lambda},\mathrm{reg}}$ *for some* $\check{\lambda} \in \check{P}^+$.

Proof. It is clear from the definition that any oper in $\mathrm{Op}_{\mathfrak{g}}^{\check{\lambda},\mathrm{reg}}$ is regular on the disc $\mathcal{D}$ and is therefore $G(\hat{\mathcal{K}})$ gauge equivalent to the trivial connection $\nabla_0 = \partial_t$.

Now suppose that we have an oper $\chi = (\mathcal{F}_G, \nabla, \mathcal{F}_B)$ on $\mathcal{D}^\times$ such that the corresponding G-local system is trivial. Then ∇ is $G(\hat{\mathcal{K}})$ gauge equivalent to a regular connection on $\mathcal{D}$. We have the decomposition $G(\hat{\mathcal{K}}) = G(\hat{\mathcal{O}})B(\hat{\mathcal{K}})$. The gauge action of $G(\hat{\mathcal{O}})$ preserves the space of regular connections (in fact, it acts transitively on it). Therefore if an oper connection ∇ is gauge equivalent to a regular connection under the action of $G(\hat{\mathcal{K}})$, then its $B(\hat{\mathcal{K}})$ gauge equivalence class must contain a regular connection. The oper condition then implies that this gauge class contains a connection operator of the form (2) with $\mathbf{v}(0) = 0$, for some dominant integral coweight $\check{\lambda}$. Therefore $\chi \in \mathrm{Op}_{\mathfrak{g}}^{\check{\lambda},\mathrm{reg}}$.

Let us choose a coordinate t on $\mathcal{D}$. The vector field $L_0 = -t\partial_t$ then acts naturally on $\mathrm{Op}_{\mathfrak{g}}^{\check{\lambda},\mathrm{reg}}$ and defines a $\mathbb{Z}$-grading on the algebra of functions on it. In Sect. 5 we will compute the character of the algebra $\mathrm{Fun}\,\mathrm{Op}_{\mathfrak{g}}^{\check{\lambda},\mathrm{reg}}$ of functions on $\mathrm{Op}_{\mathfrak{g}}^{\check{\lambda},\mathrm{reg}}$ with respect to this grading.

Next, we introduce the space of H-connections and the Miura transformation.

Let X be as above. Denote by ω_X the $\mathbb{C}^\times$-torsor corresponding to the canonical line bundle on X. Let $\omega_X^{\check{\rho}}$ be the push-forward of ω_X to an H-torsor via the homomorphism $\check{\rho} : \mathbb{C}^\times \to H$. We denote by $\mathrm{Conn}_H(\omega_X^{\check{\rho}})$ the affine space of all connections on $\omega_X^{\check{\rho}}$. In particular, $\mathrm{Conn}_H(\omega_{\mathcal{D}^\times}^{\check{\rho}})$ is an inductive limit of affine spaces $\mathrm{Conn}_H(\omega_{\mathcal{D}}^{\check{\rho}})^{\mathrm{ord}_k}$ of connections with pole of order $\leq k$. We will use the notation $\mathrm{Conn}_H(\omega_{\mathcal{D}}^{\check{\rho}})^{\mathrm{RS}}$ for $\mathrm{Conn}_H(\omega_{\mathcal{D}}^{\check{\rho}})^{\mathrm{ord}_1}$. A connection $\overline{\nabla} \in \mathrm{Conn}_H(\omega_{\mathcal{D}}^{\check{\rho}})^{\mathrm{RS}}$ has a well-defined residue, which is an element of $\mathfrak{h}$. For $\check{\mu} \in \mathfrak{h}$ we denote by $\mathrm{Conn}_H(\omega_{\mathcal{D}}^{\check{\rho}})^{\mathrm{RS},\check{\mu}}$ the subspace of $\mathrm{Conn}_H(\omega_{\mathcal{D}}^{\check{\rho}})^{\mathrm{RS}}$ consisting of connections with residue $\check{\mu}$.

The *Miura transformation* is a morphism

$$\mathrm{MT} : \mathrm{Conn}_H(\omega_{\mathcal{D}^\times}^{\check{\rho}}) \to \mathrm{Op}_{\mathfrak{g}}(\mathcal{D}^\times), \tag{3}$$

introduced in [DS] (see also [F2] and [FG2], Sect. 3.3). It can be described as follows. If we choose a coordinate t on $\mathcal{D}$, then we trivialize $\omega_{\mathcal{D}}$ and hence $\omega_{\mathcal{D}}^{\check{\rho}}$. A point of $\mathrm{Conn}_H(\omega_{\mathcal{D}^\times}^{\check{\rho}})$ is then represented by an operator

$$\overline{\nabla} = \partial_t + \mathbf{u}(t), \qquad \mathbf{u}(t) \in \mathfrak{h}(\hat{\mathcal{K}}).$$

We associate to $\overline{\nabla}$ the $\mathfrak{g}$-oper which is the $N(\hat{\mathcal{K}})$ gauge equivalence class of the operator

$$\nabla = \partial_t + \sum_{\imath \in I} f_\imath + \mathbf{u}(t).$$

The following result is a corollary of Proposition 3.5.4 of [FG2]. Let $\mathcal{F}_{B,0}$ be the fiber at $0 \in \mathcal{D}$ of the B-bundle $\mathcal{F}_B$ underlying all $\mathfrak{g}$-opers. We denote by $N_{\mathcal{F}_{B,0}}$ the $\mathcal{F}_{B,0}$-twist of N.

Lemma 2 *Let $\check{\lambda}$ be a dominant integral coweight of $\mathfrak{g}$. The image of $\mathrm{Conn}_H(\omega_{\mathcal{D}}^{\check{\rho}})^{\mathrm{RS},-\check{\lambda}}$ in $\mathrm{Op}_{\mathfrak{g}}(\mathcal{D}^\times)$ under the Miura transformation is equal to $\mathrm{Op}_{\mathfrak{g}}^{\check{\lambda},\mathrm{reg}}$. Moreover, the map $\mathrm{Conn}_H(\omega_{\mathcal{D}}^{\check{\rho}})^{\mathrm{RS},-\check{\lambda}} \to \mathrm{Op}_{\mathfrak{g}}^{\check{\lambda},\mathrm{reg}}$ is a principal $N_{\mathcal{F}_{B,0}}$-bundle over $\mathrm{Op}_{\mathfrak{g}}^{\check{\lambda},\mathrm{reg}}$.*

In particular, this implies that the scheme $\mathrm{Op}_{\mathfrak{g}}^{\check{\lambda},\mathrm{reg}}$ is smooth and in fact isomorphic to an infinite-dimensional (pro)affine space.

3 Proof of the main theorem

Our strategy of the proof of Theorem 1 will be as follows: we will first construct natural maps

$$\mathrm{End}_{\widehat{\mathfrak{g}}_{\mathrm{crit}}} \mathbb{M}_\lambda \to \mathrm{End}_{\widehat{\mathfrak{g}}_{\mathrm{crit}}} \mathbb{V}_\lambda \to H^{\frac{\infty}{2}}(\mathfrak{n}_+((t)), \mathfrak{n}_+[[t]], \mathbb{V}_\lambda \otimes \Psi_0). \tag{1}$$

We already know from [FG2] that

$$\mathrm{End}_{\widehat{\mathfrak{g}}_{\mathrm{crit}}} \mathbb{M}_\lambda \simeq \mathrm{Fun}\, \mathrm{Op}_{\check{\mathfrak{g}}}^{\lambda,\mathrm{nilp}}.$$

We will show that the corresponding composition

$$\mathrm{Fun}\, \mathrm{Op}_{\check{\mathfrak{g}}}^{\lambda,\mathrm{nilp}} \to H^{\frac{\infty}{2}}(\mathfrak{n}_+((t)), \mathfrak{n}_+[[t]], \mathbb{V}_\lambda \otimes \Psi_0)$$

factors as follows:

$$\mathrm{Fun}\, \mathrm{Op}_{\check{\mathfrak{g}}}^{\lambda,\mathrm{nilp}} \twoheadrightarrow \mathrm{Fun}\, \mathrm{Op}_{\check{\mathfrak{g}}}^{\lambda,\mathrm{reg}} \xrightarrow{\sim} H^{\frac{\infty}{2}}(\mathfrak{n}_+((t)), \mathfrak{n}_+[[t]], \mathbb{V}_\lambda \otimes \Psi_0),$$

and that the map

$$\mathbb{V}_\lambda \to H^{\frac{\infty}{2}}(\mathfrak{n}_+((t)), \mathfrak{n}_+[[t]], \mathbb{V}_\lambda \otimes \Psi_0)$$

is injective. This will imply Theorem 1.

As a byproduct, we will obtain an isomorphism

$$H^{\frac{\infty}{2}}(\mathfrak{n}_+((t)), \mathfrak{n}_+[[t]], \mathbb{V}_\lambda \otimes \Psi_0) \simeq \mathrm{End}_{\widehat{\mathfrak{g}}_{\mathrm{crit}}} \mathbb{V}_\lambda$$

and find that the first map in (1) is surjective.

3.1 Homomorphisms of $\widehat{\mathfrak{g}}_{\mathrm{crit}}$-modules

Let us now proceed with the proof and construct the maps (1).

Note that a $\widehat{\mathfrak{g}}_{\mathrm{crit}}$-endomorphism of $\mathbb{M}_\lambda$ is uniquely determined by the image of the highest weight vector, which must be a vector in $\mathbb{M}_\lambda$ of weight λ annihilated by the Lie subalgebra

$$\widehat{\mathfrak{n}}_+ = (\mathfrak{n}_+ \otimes 1) \oplus (\mathfrak{g} \otimes t\mathbb{C}[[t]]).$$

This is the Lie algebra of the prounipotent proalgebraic group $I^0 = [I, I]$, where I is the Iwahori subgroup of $G((t))$. For a $\widehat{\mathfrak{g}}_{\mathrm{crit}}$-module M we denote the space of such vectors by $M_\lambda^{\widehat{\mathfrak{n}}_+}$.

According to Corollary 13.3.2 of [FG2], we have

$$\mathrm{End}_{\widehat{\mathfrak{g}}_{\mathrm{crit}}} \mathbb{M}_\lambda = (\mathbb{M}_\lambda)_\lambda^{\widehat{\mathfrak{n}}_+} \simeq \mathrm{Fun}\, \mathrm{Op}_{\check{\mathfrak{g}}}^{\lambda,\mathrm{nilp}} = \mathrm{Fun}\, \mathrm{Op}_{\check{G}}^{\mathrm{RS},\varpi(-\lambda-\rho)}$$

(see Sect. 2).

Likewise, any endomorphism of $\mathbb{V}_\lambda$ is uniquely determined by the image of the generating subspace V_λ. This subspace therefore defines a $\mathfrak{g}[[t]]$-invariant vector in $(\mathbb{V}_\lambda \otimes V_\lambda^*)^{\mathfrak{g}[[t]]}$. Note that for any $\mathfrak{g}$-integrable module M we have an isomorphism

$$(M \otimes V_\lambda^*)^{\mathfrak{g}[[t]]} \simeq M_\lambda^{\widehat{\mathfrak{n}}_+}.$$

Therefore we have

$$\mathrm{End}_{\widehat{\mathfrak{g}}_{\mathrm{crit}}} \mathbb{V}_\lambda = (\mathbb{V}_\lambda \otimes V_\lambda^*)^{\mathfrak{g}[[t]]} = (\mathbb{V}_\lambda)_\lambda^{\widehat{\mathfrak{n}}_+}. \tag{2}$$

The canonical surjective homomorphism

$$\mathbb{M}_\lambda \twoheadrightarrow \mathbb{V}_\lambda$$

of $\widehat{\mathfrak{g}}_{\mathrm{crit}}$-modules gives rise to a map $(\mathbb{M}_\lambda)_\lambda^{\widehat{\mathfrak{n}}_+} \to (\mathbb{V}_\lambda)_\lambda^{\widehat{\mathfrak{n}}_+}$. We obtain the following commutative diagram:

$$\begin{array}{ccccc} \mathfrak{Z}_{\mathfrak{g}} & \longrightarrow & \mathrm{End}_{\widehat{\mathfrak{g}}_{\mathrm{crit}}} \mathbb{M}_\lambda & \longrightarrow & \mathrm{End}_{\widehat{\mathfrak{g}}_{\mathrm{crit}}} \mathbb{V}_\lambda \\ \downarrow & & \downarrow\sim & & \downarrow\sim \\ \mathrm{Fun}\,\mathrm{Op}_{\check{\mathfrak{g}}}(\mathcal{D}^\times) & \longrightarrow & \mathrm{Fun}\,\mathrm{Op}_{\check{\mathfrak{g}}}^{\lambda,\mathrm{nilp}} & \longrightarrow & ? \end{array} \tag{3}$$

3.2 The functor of semi-infinite cohomology

Define the character

$$\Psi_0 : \mathfrak{n}_+((t)) \to \mathbb{C} \tag{4}$$

by the formula

$$\Psi_0(e_{\alpha,n}) = \begin{cases} 1, & \text{if } \alpha = \alpha_\imath, n = -1, \\ 0, & \text{otherwise}. \end{cases}$$

We have the functor of semi-infinite cohomology (the + quantum Drinfeld–Sokolov reduction) from the category of $\widehat{\mathfrak{g}}_{\mathrm{crit}}$-modules to the category of graded vector spaces,

$$M \mapsto H^{\frac{\infty}{2}+\bullet}(\mathfrak{n}_+((t)), \mathfrak{n}_+[[t]], M \otimes \Psi_0), \tag{5}$$

introduced in [FF, FKW] (see also [FB], Ch. 15, and [FG2], Sect. 18; we follow the notation of the latter).

Let M be a $\widehat{\mathfrak{g}}_{\mathrm{crit}}$-module. Consider the space $M_\lambda^{\widehat{\mathfrak{n}}_+}$ of $\widehat{\mathfrak{n}}_+$-invariant vectors in M of highest weight λ.

Lemma 3 *We have functorial maps*

$$M_\lambda^{\widehat{\mathfrak{n}}_+} \to H^{\frac{\infty}{2}}(\mathfrak{n}_+((t)), \mathfrak{n}_+[[t]], M \otimes \Psi_0). \tag{6}$$

Proof. Consider the complex $C^\bullet(M)$ computing the above semi-infinite cohomology (see, e.g., [FB], Sect. 15.2). It follows from the definition that the standard Chevalley complex computing the cohomology of the Lie algebra $\mathfrak{n}_+[[t]]$ with coefficients in M embeds into $C^\bullet(M)$. Therefore we obtain functorial maps

$$M_\lambda^{\widehat{\mathfrak{n}}_+} \to M^{\mathfrak{n}_+[[t]]} \to H^{\frac{\infty}{2}}(\mathfrak{n}_+((t)), \mathfrak{n}_+[[t]], \mathbb{V}_\lambda \otimes \Psi_0).$$

Introduce the notation

$$\mathfrak{z}^{\lambda,\mathrm{nilp}} = \operatorname{Fun} \operatorname{Op}_{\check{\mathfrak{g}}}^{\lambda,\mathrm{nilp}},$$
$$\mathfrak{z}^{\lambda,\mathrm{reg}} = \operatorname{Fun} \operatorname{Op}_{\check{\mathfrak{g}}}^{\lambda,\mathrm{reg}}$$

Our goal is to prove that

$$(\mathbb{V}_\lambda)_\lambda^{\widehat{\mathfrak{n}}_+} \simeq \mathfrak{z}^{\lambda,\mathrm{reg}}.$$

The proof will be based on analyzing the composition

$$\mathfrak{z}^{\lambda,\mathrm{nilp}} \to (\mathbb{V}_\lambda)_\lambda^{\widehat{\mathfrak{n}}_+} \to H^{\frac{\infty}{2}}(\mathfrak{n}_+((t)), \mathfrak{n}_+[[t]], \mathbb{V}_\lambda \otimes \Psi_0), \tag{7}$$

where the first map is obtained from the diagram (3), and the second map from Lemma 3. We will use the following two results.

Theorem 2 *The composition* (7) *factors as*

$$\mathfrak{z}^{\lambda,\mathrm{nilp}} \twoheadrightarrow \mathfrak{z}^{\lambda,\mathrm{reg}} \simeq H^{\frac{\infty}{2}}(\mathfrak{n}_+((t)), \mathfrak{n}_+[[t]], \mathbb{V}_\lambda \otimes \Psi_0). \tag{8}$$

Proposition 1 *The map*

$$(\mathbb{V}_\lambda)_\lambda^{\widehat{\mathfrak{n}}_+} \to H^{\frac{\infty}{2}}(\mathfrak{n}_+((t)), \mathfrak{n}_+[[t]], \mathbb{V}_\lambda \otimes \Psi_0) \tag{9}$$

is injective.

Assuming these two assertions, we can now prove our main result.

Proof of Theorem 1. By Theorem 2, we have the following commutative diagram:

$$\begin{array}{ccc}
\mathfrak{z}^{\mathrm{nilp},\lambda} & \longrightarrow & (\mathbb{V}^\lambda)_\lambda^{\mathfrak{n}_+} \\
\downarrow & & \downarrow \\
\mathfrak{z}^{\mathrm{reg},\lambda} & \xrightarrow{\sim} & H^{\frac{\infty}{2}}(\mathfrak{n}_+((t)), \mathfrak{n}_+[[t]], \mathbb{V}_\lambda \otimes \Psi_0)
\end{array}$$

The left vertical arrow is surjective, and the right vertical arrow is injective by Proposition 1. This readily implies that we have an isomorphism

$$(\mathbb{V}^\lambda)_\lambda^{\mathfrak{n}_+} \simeq \mathfrak{z}^{\mathrm{reg},\lambda}.$$

The assertion of Theorem 1 follows from this, the isomorphism (2), and the commutative diagram (3). □

The rest of this paper is devoted to the proof of Theorem 2 and Proposition 1.

4 Exactness

In this section we prove Proposition 1 assuming Theorem 2. The proof will rely on some properties of the semi-infinite cohomology functors.

Let $\widehat{\mathfrak{g}}_{\mathrm{crit}}\text{-mod}^{I,\mathbb{Z}}$ be the category of $\widehat{\mathfrak{g}}_{\mathrm{crit}}$-modules which are equivariant with respect to the Iwahori subgroup $I \subset G((t))$ and equipped with a $\mathbb{Z}$-grading with respect to the operator $L_0 = -t\partial_t$ which commutes with $\widehat{\mathfrak{g}}_{\mathrm{crit}}$ in the natural way. Since $I = H \ltimes I^0$, where $I^0 = [I, I]$, the first condition means that $\widehat{\mathfrak{n}}_+ = \mathrm{Lie}(I^0)$ acts locally nilpotently, and the constant Cartan subalgebra $\mathrm{Lie}(H) = \mathfrak{h} \otimes 1 \subset \mathfrak{g} \otimes 1 \subset \widehat{\mathfrak{g}}_{\mathrm{crit}}$ acts semi-simply with eigenvalues corresponding to integral weights. The second condition means that we have an action of the extended affine Kac–Moody algebra $\widehat{\mathfrak{g}}'_\kappa = \mathbb{C}L_0 \ltimes \widehat{\mathfrak{g}}_\kappa$. The category $\widehat{\mathfrak{g}}_{\mathrm{crit}}\text{-mod}^{I,\mathbb{Z}}$ is therefore the product of the blocks of the usual category $\mathcal{O}_{-h^\vee}$ of modules over the extended affine Kac–Moody algebra at the critical level $k = -h^\vee$, corresponding to the (finite) Weyl group orbits in the set of integral weights.

Let $\widehat{\mathfrak{g}}_{\mathrm{crit}}\text{-mod}^{G[[t]],\mathbb{Z}}$ be the category of $\widehat{\mathfrak{g}}_{\mathrm{crit}}$-modules which are equivariant with respect to the subgroup $G[[t]]$ and equipped with a $\mathbb{Z}$-grading with respect to the operator $L_0 = -t\partial_t$. This is the full subcategory of the category $\widehat{\mathfrak{g}}_{\mathrm{crit}}\text{-mod}^{I,\mathbb{Z}}$, whose objects are modules integrable with respect to the constant subalgebra $\mathfrak{g} = \mathfrak{g} \otimes 1 \subset \widehat{\mathfrak{g}}_{\mathrm{crit}}$.

We define $\mathbb{Z}$-gradings on the modules $\mathbb{M}_\lambda$ and $\mathbb{V}_\lambda$ in the standard way, by setting the degrees of the generating vectors to be equal to 0 and using the commutation relations of L_0 and $\widehat{\mathfrak{g}}_{\mathrm{crit}}$ to define the grading on the entire modules. Thus, $\mathbb{M}_\lambda$ and $\mathbb{V}_\lambda$ become objects of the category $\widehat{\mathfrak{g}}_{\mathrm{crit}}\text{-mod}^{I,\mathbb{Z}}$, and $\mathbb{V}_\lambda$ also an object of the category $\widehat{\mathfrak{g}}_{\mathrm{crit}}\text{-mod}^{G[[t]],\mathbb{Z}}$. Moreover, the homomorphism $\mathbb{M}_\lambda \to \mathbb{V}_\lambda$, and therefore the map $(\mathbb{M}_\lambda)^{\widehat{\mathfrak{n}}_+}_\lambda \to (\mathbb{V}_\lambda)^{\widehat{\mathfrak{n}}_+}_\lambda$, preserve these gradings.

We will now derive Proposition 1 from Theorem 2 and the following statement.

Proposition 2 *The functor* (5) *is right exact on the category* $\widehat{\mathfrak{g}}_{\mathrm{crit}}\text{-mod}^{I,\mathbb{Z}}$ *and is exact on the category* $\widehat{\mathfrak{g}}_{\mathrm{crit}}\text{-mod}^{G[[t]],\mathbb{Z}}$.

Introduce a $\mathbb{Z}$-grading operator on the standard complex of semi-infinite cohomology $H^{\frac{\infty}{2}+\bullet}(\mathfrak{n}_+((t)), \mathfrak{n}_+[[t]], M \otimes \Psi_0)$, where $M = \mathbb{M}_\lambda$ or $\mathbb{V}_\lambda$, by the formula

$$L_0 - \check{\rho} \otimes 1 + \langle \lambda, \check{\rho} \rangle.$$

Here L_0 is the natural grading operator and $\check{\rho} \in \mathfrak{h}$ is such that $\langle \alpha_\imath, \check{\rho} \rangle = 1$ for all $\imath \in I$. In the same way as in [FB], Sect. 15.1.8, we check that this $\mathbb{Z}$-grading operator commutes with the differential of the complex and hence induces a $\mathbb{Z}$-grading operator on the cohomology. Observe that $\check{\rho} \otimes 1$ acts by multiplication by $\langle \lambda, \check{\rho} \rangle$ on any element in $(\mathbb{V}_\lambda)^{\widehat{\mathfrak{n}}_+}_\lambda$. Therefore the map (9) preserves $\mathbb{Z}$-gradings.

Proof of Proposition 1. Let $A \in (\mathbb{V}_\lambda)^{\widehat{\mathfrak{n}}_+}_\lambda$ be an element in the kernel of the map (9). Since this map preserves $\mathbb{Z}$-gradings, without loss of generality we may, and will, assume that A is homogeneous. Under the identification

$$(\mathbb{V}_\lambda)_\lambda^{\widehat{\mathfrak{n}}_+} \simeq \operatorname{End}_{\widehat{\mathfrak{g}}_{\mathrm{crit}}} \mathbb{V}_\lambda$$

it gives rise to a homogeneous $\widehat{\mathfrak{g}}_{\mathrm{crit}}$-endomorphism E of $\mathbb{V}_\lambda$. Then the induced map $H(E)$ on $H^{\frac{\infty}{2}}(\mathfrak{n}_+((t)),\mathfrak{n}_+[[t]],\mathbb{V}_\lambda\otimes\Psi_0)$ is identically zero. Indeed, the image of the generating vector of $(\mathbb{V}_\lambda)_\lambda^{\widehat{\mathfrak{n}}_+}$ under the map (9) is identified with the element

$$1\in\mathfrak{z}^{\lambda,\mathrm{reg}}\simeq H^{\frac{\infty}{2}}(\mathfrak{n}_+((t)),\mathfrak{n}_+[[t]],\mathbb{V}_\lambda\otimes\Psi_0),$$

where we use the isomorphism of Theorem 2. Therefore the image of this element 1 of $H^{\frac{\infty}{2}}(\mathfrak{n}_+((t)),\mathfrak{n}_+[[t]],\mathbb{V}_\lambda\otimes\Psi_0)$ under $H(E)$ is equal to the image of A in $H^{\frac{\infty}{2}}(\mathfrak{n}_+((t)),\mathfrak{n}_+[[t]],\mathbb{V}_\lambda\otimes\Psi_0)$, which is 0. By Theorem 2, $H^{\frac{\infty}{2}}(\mathfrak{n}_+((t)),\mathfrak{n}_+[[t]],\mathbb{V}_\lambda\otimes\Psi_0)$ is a free $\mathfrak{z}^{\lambda,\mathrm{reg}}$-module generated by the element 1. Therefore we find that $H(E)\equiv 0$.

Now let M and N be the kernel and cokernel of E,

$$0\to M\to\mathbb{V}_\lambda\xrightarrow{E}\mathbb{V}_\lambda\to N\to 0.$$

Note that both M and N, as well as $\mathbb{V}_\lambda$, are objects of the category $\widehat{\mathfrak{g}}_{\mathrm{crit}}\text{-mod}^{G[[t]],\mathbb{Z}}$. By Proposition 2, the functor of semi-infinite cohomology is exact on this category. Therefore we obtain an exact sequence

$$0\to H^{\frac{\infty}{2}}(\mathfrak{n}_+((t)),\mathfrak{n}_+[[t]],M\otimes\Psi_0)\to H^{\frac{\infty}{2}}(\mathfrak{n}_+((t)),\mathfrak{n}_+[[t]],\mathbb{V}_\lambda\otimes\Psi_0)\xrightarrow{H(E)}$$
$$H^{\frac{\infty}{2}}(\mathfrak{n}_+((t)),\mathfrak{n}_+[[t]],\mathbb{V}_\lambda\otimes\Psi_0)\to H^{\frac{\infty}{2}}(\mathfrak{n}_+((t)),\mathfrak{n}_+[[t]],N\otimes\Psi_0)\to 0,$$

where the middle map $H(E)$ is equal to zero. If $A\neq 0$, then M is a proper submodule of $\mathbb{V}_\lambda$ which does not contain the generating vector of $\mathbb{V}_\lambda$. We obtain that the values of the $\mathbb{Z}$-grading on M are strictly greater than those on $\mathbb{V}_\lambda$. Therefore the above sequence cannot be exact. Hence $A=0$ and we obtain the assertion of the proposition.

In the rest of this section we prove Proposition 2. Introduce the second semi-infinite cohomology functor (the – quantum Drinfeld–Sokolov reduction of [FKW])

$$M\mapsto H^{\frac{\infty}{2}+\bullet}(\mathfrak{n}_-((t)),t\mathfrak{n}_-[[t]],M\otimes\Psi_{-\check{\rho}}),\tag{1}$$

where

$$\Psi^-_{-\check{\rho}}:\mathfrak{n}_-((t))\to\mathbb{C}\tag{2}$$

is given by the formula

$$\Psi^-_{-\check{\rho}}(f_{\alpha,n})=\begin{cases}1, & \text{if } \alpha=\alpha_\imath, n=0,\\ 0, & \text{otherwise}.\end{cases}$$

Then we have the following important result due to Arakawa [Ar], Main Theorem 1,(1) (note that the functor (1) is the functor $H^\bullet_-$ in the notation [Ar]):

Theorem 3 *The functor* (1) *is exact on the category* $\widehat{\mathfrak{g}}_{\text{crit}}\text{-mod}^{I,\mathbb{Z}}$.

We now derive Proposition 2 from Theorem 3.

Proof of Proposition 2. Recall that we have the convolution functors

$$M \mapsto \mathcal{F} \star M$$

on the category of I-equivariant $\widehat{\mathfrak{g}}_{\text{crit}}$ modules, for each I-equivariant right D-module $\mathcal{F}$ on $G((t))/I$ (see [FG2], Sect. 22, for the precise definition).

According to Proposition 18.1.1 of [FG2], we have

$$H^{\frac{\infty}{2}+\bullet}(\mathfrak{n}_+((t)), \mathfrak{n}_+[[t]], M \otimes \Psi_0) \simeq H^{\frac{\infty}{2}+\bullet}(\mathfrak{n}_-((t)), t\mathfrak{n}_-[[t]], j_{w_0\check{\rho},*} \star M \otimes \Psi_{-\check{\rho}}) \quad (3)$$

for any I-equivariant $\widehat{\mathfrak{g}}_{\text{crit}}$-module M. We recall that the D-module $j_{w_0\check{\rho},*}$ is defined as the $*$-extension of the "constant" D-module on the I-orbit in the affine flag scheme $G((t))/I$ corresponding to the element $w_0\check{\rho}$ of the affine Weyl group. Hence the functor

$$M \mapsto j_{w_0\check{\rho},*} \star M$$

is right exact. Combining this with Theorem 3, we obtain that the functor

$$M \mapsto H^{\frac{\infty}{2}+\bullet}(\mathfrak{n}_-((t)), t\mathfrak{n}_-[[t]], j_{w_0\check{\rho},*} \star M \otimes \Psi_{-\check{\rho}})$$

is right exact on the category $\widehat{\mathfrak{g}}_{\text{crit}}\text{-mod}^{I,\mathbb{Z}}$ (note that the convolution with $j_{w_0\check{\rho},*}$ sends $\mathbb{Z}$-graded modules to $\mathbb{Z}$-graded modules). The isomorphism (3) then implies that the functor (5) is right exact on the category $\widehat{\mathfrak{g}}_{\text{crit}}\text{-mod}^{I,\mathbb{Z}}$.

On the other hand, let $j_{w_0\check{\rho},!}$ be the !-extension of the "constant" D-module on the same I-orbit. The convolution functor with $j_{w_0\check{\rho},!}$ is both left and right adjoint to the convolution with $j_{w_0\check{\rho},*}$. Therefore we find that the functor

$$M \mapsto j_{w_0\check{\rho},!} \star M$$

is left exact. Combining this with Theorem 3, we obtain that the functor

$$M \mapsto H^{\frac{\infty}{2}+\bullet}(\mathfrak{n}_-((t)), t\mathfrak{n}_-[[t]], j_{w_0\check{\rho},!} \star M \otimes \Psi_{-\check{\rho}})$$

is left exact on the category $\widehat{\mathfrak{g}}_{\text{crit}}\text{-mod}^{I,\mathbb{Z}}$ (again, note that the convolution with $j_{w_0\check{\rho},!}$ sends $\mathbb{Z}$-graded modules to $\mathbb{Z}$-graded modules).

Now consider the homomorphism

$$j_{w_0\check{\rho},!} \star M \to j_{w_0\check{\rho},*} \star M \quad (4)$$

of $\widehat{\mathfrak{g}}_{\text{crit}}$-modules induced by the morphism

$$j_{w_0\check{\rho},!} \to j_{w_0\check{\rho},*}$$

of D-modules on $G((t))/I$. Suppose in addition that M is $G[[t]]$-equivariant.

Then we have

$$j_{w_0\check{\rho},!} \star M \simeq j_{w_0\cdot\check{\rho},!} \star \delta_{1_{Gr_G}} \underset{G[[t]]}{\star} M, \qquad j_{w_0\check{\rho},*} \star M \simeq j_{w_0\cdot\check{\rho},*} \star \delta_{1_{Gr_G}} \underset{G[[t]]}{\star} M,$$

where $\delta_{1_{Gr_G}}$ is the "delta-function" D-module on the affine Grassmannian $Gr_G = G((t))/G[[t]]$ supported at the identity coset. It then follows from Lemma 15.1.2 of [FG2] that the kernel and the cokernel of the map (4) are partially integrable $\widehat{\mathfrak{g}}_{\mathrm{crit}}$-modules. We recall from [FG2], Sect. 6.3, that a $\widehat{\mathfrak{g}}_{\mathrm{crit}}$-module is called partially integrable if it admits a filtration such that each successive quotient is equivariant with respect to the parahoric Lie subalgebra $\mathfrak{p}^\iota = \mathrm{Lie}(I) + \mathfrak{sl}_2^\iota$ for some vertex of the Dynkin diagram of $\mathfrak{g}$, $\iota \in I$.

But, according to Lemma 18.1.2 of [FG2], if M is a partially integrable $\widehat{\mathfrak{g}}_{\mathrm{crit}}$-module, then

$$H^{\frac{\infty}{2}+\bullet}(\mathfrak{n}_-((t)), t\mathfrak{n}_-[[t]], M \otimes \Psi_{-\check{\rho}}) = 0$$

for all $i \in \mathbb{Z}$. Therefore we obtain that the map

$$H^{\frac{\infty}{2}+\bullet}(\mathfrak{n}_-((t)), t\mathfrak{n}_-[[t]], j_{w_0\check{\rho},!} \star M \otimes \Psi_{-\check{\rho}}) \to H^{\frac{\infty}{2}+\bullet}(\mathfrak{n}_-((t)), t\mathfrak{n}_-[[t]], j_{w_0\check{\rho},*} \star M \otimes \Psi_{-\check{\rho}})$$

induced by (4) is an isomorphism for any $G[[t]]$-equivariant $\widehat{\mathfrak{g}}_{\mathrm{crit}}$-module M. Since the former is right exact and the latter is left exact on the category $\widehat{\mathfrak{g}}_{\mathrm{crit}}\text{-mod}^{G[[t]],\mathbb{Z}}$, we obtain that both functors are exact on this category. Combining this with the isomorphism (3), we find that the functor

$$M \mapsto H^{\frac{\infty}{2}+\bullet}(\mathfrak{n}_+((t)), \mathfrak{n}_+[[t]], M \otimes \Psi_0)$$

is exact on the category $\widehat{\mathfrak{g}}_{\mathrm{crit}}\text{-mod}^{G[[t]],\mathbb{Z}}$.

This completes the proof of Proposition 2. □

Remark 1. There are obvious analogues of the categories $\widehat{\mathfrak{g}}_{\mathrm{crit}}\text{-mod}^{I,\mathbb{Z}}$ and $\widehat{\mathfrak{g}}_{\mathrm{crit}}\text{-mod}^{G[[t]],\mathbb{Z}}$ for an arbitrary level κ. The same proof as above works for any κ, so Proposition 2 actually holds for an arbitrary level. □

It remains to prove Theorem 2. We will give two proofs: one relies on the results of [FG3], and the other uses the Wakimoto modules. Both proofs use the computation of the characters of $\mathfrak{z}^{\lambda,\mathrm{reg}}$ and $H^{\frac{\infty}{2}+\bullet}(\mathfrak{n}_+((t)), \mathfrak{n}_+[[t]], \mathbb{V}_\lambda \otimes \Psi_0)$ which is performed in the next section.

5 Computation of characters

5.1 Character of $\mathfrak{z}^{\lambda,\mathrm{reg}}$

Let us compute the character of the algebra

$$\mathfrak{z}^{\lambda,\mathrm{reg}} = \mathrm{Fun}\,\mathrm{Op}_{\check{\mathfrak{g}}}^{\lambda,\mathrm{reg}}$$

of functions on $\mathrm{Op}_{\check{\mathfrak{g}}}^{\lambda,\mathrm{reg}}$ with respect to the $\mathbb{Z}$-grading by the operator L_0. This space was defined in Sect. 2, but note that we now switch to the Langlands dual Lie algebra $\check{\mathfrak{g}}$.

We give another, more convenient, realization of the space $\mathrm{Op}_{\check{\mathfrak{g}}}^{\lambda,\mathrm{reg}}$. Suppose that we are given an operator of the form

$$\nabla = \partial_t + \sum_{\imath\in I} t^{\langle\check{\alpha}_\imath,\lambda\rangle} f_\imath + \mathbf{v}(t), \qquad \mathbf{v}(t)\in\check{\mathfrak{b}}(\hat{\mathcal{O}}). \tag{1}$$

Applying gauge transformation with $(\lambda+\rho)(t)$, we obtain an operator of the form

$$\nabla' = \partial_t + \frac{1}{t}\left(\sum_{\imath\in I} f_\imath - (\lambda+\rho)\right) + \mathbf{v}(t), \qquad \mathbf{v}(t)\in(\lambda+\rho)(t)\check{\mathfrak{b}}(\hat{\mathcal{O}})(\lambda+\rho)(t)^{-1}. \tag{2}$$

The space $\mathrm{Op}_{\check{\mathfrak{g}}}^{\lambda,\mathrm{reg}}$ is defined as the space of $\check{N}(\hat{\mathcal{O}})$-equivalence classes of operators (1). Equivalently, this is the space of $(\lambda+\rho)(t)\check{N}(\hat{\mathcal{O}})(\lambda+\rho)(t)^{-1}$-equivalence classes of operators (2). It follows from Theorem 2.21 of [FG2] that the action of the group $(\lambda+\rho)(t)\check{N}(\hat{\mathcal{O}})(\lambda+\rho)(t)^{-1}$ on this space is free. Therefore the character of $\mathrm{Fun}\,\mathrm{Op}_{\check{\mathfrak{g}}}^{\lambda,\mathrm{reg}}$ is equal to the the character of the algebra of functions on the space of operators (2) divided by the character of the algebra $\mathrm{Fun}((\lambda+\rho)(t)\check{N}(\hat{\mathcal{O}})(\lambda+\rho)(t)^{-1})$.

By definition, a character of a $\mathbb{Z}_+$-graded vector space $V = \bigoplus_{n\in\mathbb{Z}_+} V_n$, where each V_n is finite-dimensional, is the formal power series

$$\mathrm{ch}\, V = \sum_{n\geq 0} \dim V_n \cdot q^n.$$

Applying the dilation $t\mapsto at$ to (2), we obtain the action $\mathbf{v}(t)\mapsto a\mathbf{v}(at)$. Therefore the character of the algebra of functions on the space of operators (2) is equal to

$$\prod_{n>0}(1-q^n)^{-\ell}\cdot\prod_{\check{\alpha}\in\check{\Delta}_+}\prod_{n>0}(1-q^{n+\langle\check{\alpha},\lambda+\rho\rangle})^{-1}.$$

On the other hand, the character of $\mathrm{Fun}((\lambda+\rho)(t)\check{N}(\hat{\mathcal{O}})(\lambda+\rho)(t)^{-1})$ is equal to

$$\prod_{\check{\alpha}\in\check{\Delta}_+}\prod_{n\geq 0}(1-q^{n+\langle\check{\alpha},\lambda+\rho\rangle})^{-1}.$$

Therefore the character of $\mathfrak{z}^{\lambda,\mathrm{reg}} = \mathrm{Fun}\,\mathrm{Op}_{\check{\mathfrak{g}}}^{\lambda,\mathrm{reg}}$ is equal to

$$\prod_{n>0}(1-q^n)^{-\ell}\cdot\prod_{\check{\alpha}\in\check{\Delta}_+}(1-q^{\langle\alpha,\lambda+\rho\rangle}).$$

We rewrite this in the form

$$\operatorname{ch}\mathfrak{z}^{\lambda,\mathrm{reg}} = \prod_{\check{\alpha}\in\check{\Delta}_+} \frac{1-q^{\langle\check{\alpha},\lambda+\rho\rangle}}{1-q^{\langle\check{\alpha},\rho\rangle}} \prod_{i=1}^{\ell} \prod_{n_i\geq d_i+1} (1-q^{n_i})^{-1}, \tag{3}$$

using the identify

$$\prod_{\check{\alpha}\in\check{\Delta}_+} (1-q^{\langle\check{\alpha},\rho\rangle}) = \prod_{i=1}^{\ell}\prod_{m_i=1}^{d_i} (1-q^{m_i}),$$

where $d_1,\ldots,d_\ell$ are the exponents of $\mathfrak{g}$.

5.2 Computation of semi-infinite cohomology

Let us now compute the semi-infinite cohomology of $\mathbb{V}_\lambda$. The complex $C^\bullet(\mathbb{V}_\lambda)$ computing this cohomology is described in [FB], Ch. 15. In particular, as explained in Sect. 4, it carries a $\mathbb{Z}$-grading operator which commutes with the differential and hence gives rise to a grading operator on the cohomology. We will compute the character with respect to this grading operator.

Theorem 4 *We have*

$$\operatorname{ch} H^{\frac{\infty}{2}}(\mathfrak{n}_+((t)),\mathfrak{n}_+[[t]],\mathbb{V}_\lambda\otimes\Psi_0) = \operatorname{ch}\mathfrak{z}^{\lambda,\mathrm{reg}}$$

given by formula (3), *and*

$$H^{\frac{\infty}{2}+i}(\mathfrak{n}_+((t)),\mathfrak{n}_+[[t]],\mathbb{V}_\lambda\otimes\Psi_0) = 0, \qquad i\neq 0.$$

Proof. The vanishing of the ith cohomology for $i\neq 0$ follows from Proposition 2. We will give an alternative proof of this, as well as the computation of the character of the 0th cohomology, using the argument of [FB], Sect. 15.2.

Consider the complex $C^\bullet(\mathbb{V}_\lambda)$ computing our semi-infinite cohomology. This complex was studied in detail in [FB], Sect. 15.2, in the case when $\lambda=0$. We decompose $C^\bullet(\mathbb{V}_\lambda)$ into the tensor product of two subcomplexes as in [FB], Sect. 15.2.1,

$$C^\bullet(\mathbb{V}_\lambda) = C^\bullet(\mathbb{V}_\lambda)_0\otimes C^\bullet(\mathbb{V}_\lambda)'$$

defined in the same way as the subcomplexes $C^\bullet_{-h^\vee}(\mathfrak{g})_0$ and $C^\bullet_{-h^\vee}(\mathfrak{g})'$, respectively. In fact, $C^\bullet(\mathbb{V}_\lambda)_0$ is equal to $C^\bullet_{-h^\vee}(\mathfrak{g})_0$, and

$$C^\bullet(\mathbb{V}_\lambda)' \simeq V_\lambda\otimes U(t^{-1}\mathfrak{b}_-[t^{-1}])\otimes{\bigwedge}^\bullet(\mathfrak{n}_+[[t]]^*).$$

In particular, its cohomological grading takes only non-negative values on $C^\bullet(\mathbb{V}_\lambda)'$.

We show, in the same way as in [FB], Lemma 15.2.5, that the cohomology of $C^\bullet(\mathbb{V}_\lambda)$ is isomorphic to the tensor product of the cohomologies of the subcomplexes $C^\bullet(\mathbb{V}_\lambda)_0$ and $C^\bullet(\mathbb{V}_\lambda)'$. The former is one-dimensional, according to [FB], Lemma 15.2.7, and hence we find that our semi-infinite cohomology is isomorphic to the cohomology of the subcomplex $C^\bullet(\mathbb{V}_\lambda)'$.

Following verbatim the computation in [FB], Sect. 15.2.9, in the case when $\lambda = 0$, we find that the 0th cohomology of $C^\bullet(\mathbb{V}_\lambda)'$ is isomorphic to

$$H^{\frac{\infty}{2}}(\mathfrak{n}_+((t)), \mathfrak{n}_+[[t]], \mathbb{V}_\lambda \otimes \Psi_0) \simeq V_\lambda \otimes V(\mathfrak{a}_-)$$

(where $V(\mathfrak{a}_-)$ is defined in [FB], Sect. 15.2.9), and all other cohomologies vanish.

In particular, we find that the character of $H^{\frac{\infty}{2}}(\mathfrak{n}_+((t)), \mathfrak{n}_+[[t]], \mathbb{V}_\lambda \otimes \Psi_0)$ is equal to

$$\operatorname{ch} V_\lambda \cdot \operatorname{ch} V(\mathfrak{a}_-),$$

where $\operatorname{ch} V_\lambda$ is the character of V_λ with respect to the principal grading. By [FB], Sect. 15.2.9, we have

$$\operatorname{ch} V(\mathfrak{a}_-) = \prod_{i=1}^{\ell} \prod_{m_i \geq d_i+1} (1 - q^{m_i})^{-1}.$$

According to formula (10.9.4) of [Kac], $\operatorname{ch} V_\lambda$ is equal to

$$\operatorname{ch} V_\lambda = \prod_{\check{\alpha} \in \check{\Delta}_+} \frac{1 - q^{\langle \check{\alpha}, \lambda + \rho \rangle}}{1 - q^{\langle \check{\alpha}, \rho \rangle}}. \tag{4}$$

Therefore the character

$$\operatorname{ch} H^{\frac{\infty}{2}}(\mathfrak{n}_+((t)), \mathfrak{n}_+[[t]], \mathbb{V}_\lambda \otimes \Psi_0)$$

is given by formula (3), which coincides with the character of $\mathfrak{z}^{\lambda,\mathrm{reg}}$.

6 Proof of Theorem 2

6.1 First proof

The following result is proved in [FG3], Lemma 1.7.

Proposition 3 *The action of the center $\mathfrak{Z}_\mathfrak{g}$ on $\mathbb{V}_\lambda$ factors through $\mathfrak{z}^{\lambda,\mathrm{reg}}$.*

Let I_λ be the ideal of $\mathrm{Op}_{\check{\mathfrak{g}}}^{\mathrm{reg},\lambda} = \operatorname{Spec} \mathfrak{z}^{\mathrm{reg},\lambda}$ in the center $\mathfrak{Z}_\mathfrak{g} = \mathrm{Fun}(\mathrm{Op}_{\check{\mathfrak{g}}}(\mathcal{D}^\times))$. As explained in [FG2], Sect. 4.6 (see [BD1], Sect. 3.6, in the case when $\lambda = 0$), the Poisson structure on $\mathfrak{Z}_\mathfrak{g}$ gives rise to the structure of a Lie algebroid on the quotient $I_\lambda/(I_\lambda)^2$, which we denote by $N^*_{\mathrm{Op}_{\check{\mathfrak{g}}}^{\lambda,\mathrm{reg}}/\mathrm{Op}_{\check{\mathfrak{g}}}(\mathcal{D}^\times)}$. Recall from [FF, F2] that the Poisson structure on $\mathfrak{Z}_\mathfrak{g}$ is obtained by deforming the completed enveloping algebra of $\widehat{\mathfrak{g}}$ to non-critical levels. By Proposition 3, I_λ annihilates the module $\mathbb{V}_\lambda$. Since this module may be deformed to the Weyl modules at non-critical levels, we obtain that the Lie algebroid $N^*_{\mathrm{Op}_{\check{\mathfrak{g}}}^{\lambda,\mathrm{reg}} / \mathrm{Op}_{\check{\mathfrak{g}}}(\mathcal{D}^\times)}$ naturally acts on $\mathbb{V}_\lambda$ (see [BD1], Sect. 5.6, in the case when $\lambda = 0$) and on its semi-infinite cohomology. This action is compatible with the action of $\mathrm{Fun}(\mathrm{Op}_{\check{\mathfrak{g}}}(\mathcal{D}^\times))/I_\lambda = \mathfrak{z}^{\mathrm{reg},\lambda}$.

Using the commutative diagram (3) and Proposition 3, we obtain that the composition (7) factors through a map

$$\mathfrak{z}^{\lambda,\mathrm{reg}} \to H^{\frac{\infty}{2}}(\mathfrak{n}_+((t)), \mathfrak{n}_+[[t]], \mathbb{V}_\lambda \otimes \Psi_0).$$

By applying the same argument as in the proof of Proposition 18.3.2 of [FG2], we obtain that the above map is a homomorphism of modules over the Lie algebroid $N^*_{\mathrm{Op}_{\check{\mathfrak{g}}}^{\lambda,\mathrm{reg}} / \mathrm{Op}_{\check{\mathfrak{g}}}(\mathcal{D}^\times)}$. This homomorphism is clearly non-zero, because we can identify the image of the generator $1 \in \mathfrak{z}^{\lambda,\mathrm{reg}}$ with the cohomology class represented by the highest weight vector of $\mathbb{V}_\lambda$. Since $\mathfrak{z}^{\mathrm{reg},\lambda}$ is irreducible as a module over $N^*_{\mathrm{Op}_{\check{\mathfrak{g}}}^{\lambda,\mathrm{reg}} / \mathrm{Op}_{\check{\mathfrak{g}}}(\mathcal{D}^\times)}$, this homomorphism is injective. Moreover, it is clear that this map preserves the natural $\mathbb{Z}$-gradings on both modules. Therefore the equality of the two characters established in Sect. 5 shows that it is an isomorphism. □

6.2 Second proof

Let us recall some results about Wakimoto modules of critical level from [F2] (see also [FG2]). Specifically, we will consider the module which was denoted by W_{λ,κ_c} in [F2] and by $\mathbb{W}^{w_0}_{\mathrm{crit},\lambda}$ in [FG2]. Here we will denote it simply by W_λ. As a vector space, it is isomorphic to the tensor product

$$W_\lambda = M_{\mathfrak{g}} \otimes \mathfrak{H}^\lambda, \tag{1}$$

where we use the notation

$$\mathfrak{H}^\lambda = \mathrm{Fun}\,\mathrm{Conn}_{\check{H}}(\omega^\rho_{\mathcal{D}})^{\mathrm{RS},-\lambda}$$

(see Sect. 2), and $M_{\mathfrak{g}}$ is the Fock representation of a Weyl algebra with generators $a_{\alpha,n}, a^*_{\alpha,n}, \alpha \in \Delta_+, n \in \mathbb{Z}$.

We now construct a map $\mathbb{V}_\lambda \to W_\lambda$. Let us observe that the action of the constant subalgebra $\mathfrak{g} \subset \widehat{\mathfrak{g}}_\kappa$ on the subspace

$$W^0_\lambda = \mathbb{C}[a^*_{\alpha,0}]_{\alpha \in \Delta_+}|0\rangle \simeq \mathrm{Fun}\,N \subset M_{\mathfrak{g}}$$

coincides with the natural action of $\mathfrak{g}$ on the contragredient Verma module M^*_λ realized as $\mathrm{Fun}\,N$. In addition, the Lie subalgebra $\mathfrak{g} \otimes t \otimes \mathbb{C}[[t]] \subset \widehat{\mathfrak{g}}_{\mathrm{crit}}$ acts by zero on the subspace W^0_λ, and $\mathbf{1}$ acts as the identity.

Therefore the injective $\mathfrak{g}$-homomorphism $V_\lambda \hookrightarrow M^*_\lambda$ gives rise to a non-zero $\widehat{\mathfrak{g}}_{\mathrm{crit}}$-homomorphism

$$\imath_\lambda : \mathbb{V}_\lambda \to W_\lambda$$

sending the generating subspace $V_\lambda \subset \mathbb{V}_\lambda$ to the image of V_λ in $W^0_\lambda \simeq M^*_\lambda$.

Now we obtain a sequence of maps

$$\mathbb{M}_\lambda \twoheadrightarrow \mathbb{V}_\lambda \to W_\lambda. \tag{2}$$

Recall from Sect. 3.1 that

$$(\mathbb{M}_\lambda)_\lambda^{\widehat{\mathfrak{n}}_+} \simeq \mathfrak{z}^{\lambda,\mathrm{nilp}}.$$

We also prove, by using the argument of Lemma 6.5 of [F2] that

$$(W_\lambda)_\lambda^{\widehat{\mathfrak{n}}_+} = \mathfrak{H}^\lambda,$$

where $\mathfrak{H}^\lambda$ is identified with the second factor of the decomposition (1).

We obtain the following commutative diagram, in which all maps preserve $\mathbb{Z}$-gradings:

$$\begin{array}{ccc} \mathfrak{z}^{\lambda,\mathrm{nilp}} & \xrightarrow{\ b\ } & \mathfrak{H}^\lambda \\ a\big\downarrow & & c\big\downarrow \\ H^{\frac{\infty}{2}}(\mathfrak{n}_+((t)), \mathfrak{n}_+[[t]], \mathbb{V}_\lambda \otimes \Psi_0) & \xrightarrow{\ d\ } & H^{\frac{\infty}{2}}(\mathfrak{n}_+((t)), \mathfrak{n}_+[[t]], W_\lambda \otimes \Psi_0) \end{array} \tag{3}$$

Here the map a is obtained as the composition

$$\mathfrak{z}^{\lambda,\mathrm{nilp}} \simeq (\mathbb{M}_\lambda)_\lambda^{\widehat{\mathfrak{n}}_+} \to (\mathbb{V}_\lambda)_\lambda^{\widehat{\mathfrak{n}}_+} \to H^{\frac{\infty}{2}}(\mathfrak{n}_+((t)), \mathfrak{n}_+[[t]], \mathbb{V}_\lambda \otimes \Psi_0),$$

the map c as the composition

$$\mathfrak{H}^\lambda \simeq (W_\lambda)_\lambda^{\widehat{\mathfrak{n}}_+} \to H^{\frac{\infty}{2}}(\mathfrak{n}_+((t)), \mathfrak{n}_+[[t]], W_\lambda \otimes \Psi_0)$$

(see Lemma 3), and the map b as the composition

$$\mathfrak{z}^{\lambda,\mathrm{nilp}} \simeq (\mathbb{M}_\lambda)_\lambda^{\widehat{\mathfrak{n}}_+} \to (W_\lambda)_\lambda^{\widehat{\mathfrak{n}}_+} \simeq \mathfrak{H}^\lambda.$$

Using Theorem 12.5 of [F2], we identify the map b with the homomorphism of the algebras of functions corresponding to the map

$$\mathrm{Conn}_{\check{H}}(\omega_{\mathcal{D}}^{\rho})^{\mathrm{RS},-\lambda} \to \mathrm{Op}_{\check{\mathfrak{g}}}^{\lambda,\mathrm{nilp}}$$

obtained by restriction from the Miura transformation (3). Therefore we obtain from Lemma 2 that the image of this map is precisely $\mathrm{Op}_{\check{\mathfrak{g}}}^{\lambda,\mathrm{reg}}$, and so the image of the homomorphism b is equal to

$$\mathfrak{z}^{\lambda,\mathrm{reg}} = \mathrm{Fun}\,\mathrm{Op}_{\check{\mathfrak{g}}}^{\lambda,\mathrm{reg}}.$$

On the other hand, we have the decomposition (1) and $\mathfrak{n}_+((t))$ acts along the first factor $M_{\mathfrak{g}}$. It follows from the definition of $M_{\mathfrak{g}}$ that there is a canonical identification

$$H^{\frac{\infty}{2}}(\mathfrak{n}_+((t)), \mathfrak{n}_+[[t]], M_{\mathfrak{g}} \otimes \Psi_0) \simeq \mathbb{C}|0\rangle.$$

Therefore we obtain that

$$H^{\frac{\infty}{2}}(\mathfrak{n}_+((t)), \mathfrak{n}_+[[t]], W_\lambda \otimes \Psi_0) \simeq \mathfrak{H}^\lambda$$

and so the map c is an isomorphism.

This implies that the image of the composition cob is $\mathfrak{z}^{\lambda,\mathrm{reg}} \subset \mathfrak{H}^\lambda$. Therefore d factors as follows:

$$H^{\frac{\infty}{2}}(\mathfrak{n}_+((t)), \mathfrak{n}_+[[t]], \mathbb{V}_\lambda \otimes \Psi_0) \twoheadrightarrow \mathfrak{z}^{\lambda,\mathrm{reg}} \hookrightarrow \mathfrak{H}^\lambda.$$

But the characters of the first two spaces coincide, according to Theorem 4. Hence

$$H^{\frac{\infty}{2}}(\mathfrak{n}_+((t)), \mathfrak{n}_+[[t]], \mathbb{V}_\lambda \otimes \Psi_0) \simeq \mathfrak{z}^{\lambda,\mathrm{reg}}.$$

This proves Theorem 2. □

We obtain the following corollary, which for $\lambda = 0$ was proved in [F2], Prop. 5.2.

Corollary 1 *The map $\imath_\lambda : \mathbb{V}_\lambda \to W_\lambda$ is injective for any dominant integral weight λ.*

Proof. Let us extend the action of $\widehat{\mathfrak{g}}_\kappa$ to an action of $\widehat{\mathfrak{g}}'_\kappa = \mathbb{C}L_0 \ltimes \widehat{\mathfrak{g}}_\kappa$ in the same way as above. Denote by K_λ the kernel of the map $\imath_\lambda$. Suppose that $K_\lambda \neq 0$. Since both $\mathbb{V}_\lambda$ and $\mathbb{M}_\lambda$ become graded with respect to the extended Cartan subalgebra $\widetilde{\mathfrak{h}} = (\mathfrak{h} \otimes 1) \oplus \mathbb{C}L_0$, we find that the $\widehat{\mathfrak{g}}_{\mathrm{crit}}$-module K_λ contains a non-zero highest weight vector annihilated by the Lie subalgebra $\widehat{\mathfrak{n}}_+$. Let $\widetilde{\mu}$ be its weight. Since $\mathbb{V}_\lambda$ is $\mathfrak{g}$-integrable, so is K_λ, and therefore the restriction of $\widetilde{\mu}$ to $\mathfrak{h} \subset \widetilde{\mathfrak{h}}$ is a dominant integral weight μ.

Now, since $\mathbb{V}_\lambda$ is a quotient of the Verma module $\mathbb{M}_\lambda$, and the action of the center $\mathfrak{Z}_\mathfrak{g}$ on $\mathbb{M}_\lambda$ factors through $\operatorname{Fun}\operatorname{Op}_{\check{\mathfrak{g}}}^{\lambda,\mathrm{nilp}}$ (see the diagram (3)), we find that the same is true for $\mathbb{V}_\lambda$. (Actually, we have already determined in Proposition 3 that the support of $\mathbb{V}_\lambda$ is contained in $\operatorname{Op}_{\check{\mathfrak{g}}}^{\lambda,\mathrm{reg}}$, but it is not necessary to use this result here.) According to [F2], Prop. 12.8, the degree 0 part of $\operatorname{Fun}\operatorname{Op}_{\check{\mathfrak{g}}}^{\lambda,\mathrm{nilp}}$ is isomorphic to $(\operatorname{Fun}\mathfrak{h}^*)^W$, and it acts on any highest weight vector v through the quotient by the maximal ideal in $(\operatorname{Fun}\mathfrak{h}^*)^W$ corresponding to $-\lambda - \rho$. Moreover, its action coincides with the action of the center $Z(\mathfrak{g})$ of $U(\mathfrak{g})$ on the vector v under the Harish-Chandra isomorphism $Z(\mathfrak{g}) \simeq (\operatorname{Fun}\mathfrak{h}^*)^W$ followed by the sign isomorphism $x \mapsto -x$. Therefore if v has $\mathfrak{h}$-weight μ, the action of $(\operatorname{Fun}\mathfrak{h}^*)^W$ factors through the quotient by the maximal ideal in $(\operatorname{Fun}\mathfrak{h}^*)^W$ corresponding to $-\mu - \rho$. Since μ is dominant integral, this implies that $\mu = \lambda$.

Thus, K_λ contains a non-zero highest weight vector of $\mathfrak{h}$-weight λ. This vector then lies in $(\mathbb{V}_\lambda)_\lambda^{\widehat{\mathfrak{n}}_+}$. But Theorem 1 and the diagram (3) imply that the map

$$(\mathbb{V}_\lambda)_\lambda^{\widehat{\mathfrak{n}}_+} \to W_\lambda$$

is injective. Indeed, the image of the composition

$$(\mathbb{M}_\lambda)^{\widehat{\mathfrak{n}}_+}_\lambda \to (\mathbb{V}_\lambda)^{\widehat{\mathfrak{n}}_+}_\lambda \to W_\lambda$$

is equal to $\mathfrak{z}^{\lambda,\mathrm{reg}} \subset \mathfrak{H}^\lambda$, which is isomorphic to $(\mathbb{V}_\lambda)^{\widehat{\mathfrak{n}}_+}_\lambda$.

This leads to a contradiction and hence proves the desired assertion.

The following result is also useful in applications. Note that for $\lambda = 0$ it follows from the corresponding statement for the associated graded module proved in [EF] or from the results of [BD1], Sect. 6.2.

Theorem 5 *For any dominant integral weight λ the Weyl module $\mathbb{V}_\lambda$ is free as a* $\mathrm{Fun}\,\mathrm{Op}^{\lambda,\mathrm{reg}}_{\check{\mathfrak{g}}}$*-module.*

Proof. Recall from Sect. 6.1 that $\mathbb{V}_\lambda$ carries an action of the Lie algebroid $N^*_{\mathrm{Op}^{\lambda,\mathrm{reg}}_{\check{\mathfrak{g}}} / \mathrm{Op}_{\check{\mathfrak{g}}}(\mathcal{D}^\times)}$ compatible with the action of $\mathfrak{z}^{\mathrm{reg},\lambda} = \mathrm{Fun}(\mathrm{Op}^{\lambda,\mathrm{reg}}_{\check{\mathfrak{g}}})$. According to the results of [FG2], Sect. 4.6, this Lie algebroid is nothing but the Atiyah algebroid of the universal $\check{G}$-bundle on $\mathrm{Op}^{\lambda,\mathrm{reg}}_{\check{\mathfrak{g}}}$ whose fiber at $\chi \in \mathrm{Op}^{\lambda,\mathrm{reg}}_{\check{\mathfrak{g}}}$ is the fiber of the $\check{G}$-torsor on $\mathcal{D}$ underlying χ at $0 \in \mathcal{D}$. This bundle is isomorphic (non-canonically) to the trivial $\check{G}$-bundle on $\mathrm{Op}^{\lambda,\mathrm{reg}}_{\check{\mathfrak{g}}}$. Let us choose such an isomorphism. Then $N^*_{\mathrm{Op}^{\lambda,\mathrm{reg}}_{\check{\mathfrak{g}}} / \mathrm{Op}_{\check{\mathfrak{g}}}(\mathcal{D}^\times)}$ splits as a direct sum of the Lie algebra $\mathrm{Vect}(\mathrm{Op}^{\lambda,\mathrm{reg}}_{\check{\mathfrak{g}}})$ of vector fields on $\mathrm{Op}^{\lambda,\mathrm{reg}}_{\check{\mathfrak{g}}}$ and the Lie algebra $\check{\mathfrak{g}} \otimes \mathrm{Fun}(\mathrm{Op}^{\lambda,\mathrm{reg}}_{\check{\mathfrak{g}}})$.

Thus, we obtain an action of $\mathrm{Vect}(\mathrm{Op}^{\lambda,\mathrm{reg}}_{\check{\mathfrak{g}}})$ on $\mathbb{V}_\lambda$ compatible with the action of $\mathrm{Fun}(\mathrm{Op}^{\lambda,\mathrm{reg}}_{\check{\mathfrak{g}}})$. Note that the algebra $\mathfrak{z}^{\mathrm{reg},\lambda}$ and the Lie algebra $\mathrm{Vect}(\mathrm{Op}^{\lambda,\mathrm{reg}}_{\check{\mathfrak{g}}})$ are $\mathbb{Z}$-graded by the operator $L_0 = -t\partial_t$. According to Lemma 2, $\mathrm{Op}^{\lambda,\mathrm{reg}}_{\check{\mathfrak{g}}}$ is an infinite-dimensional affine space, and there exists a system of coordinates $x_i, i = 1, 2, \ldots$, on it such that these coordinates are homogeneous with respect to L_0. The character formula (3) shows that the degrees of all of these generators are strictly positive.

Thus, we find that the action on $\mathbb{V}_\lambda$ of the polynomial algebra $\mathrm{Fun}(\mathrm{Op}^{\lambda,\mathrm{reg}}_{\check{\mathfrak{g}}})$ generated by the x_i's extends to an action of the Weyl algebra generated by the x_i's and the $\partial/\partial x_i$'s. Recall that the $\mathbb{Z}$-grading on $\mathbb{V}_\lambda$ with respect to the operator L_0 takes non-negative integer values. Repeating the argument of Lemma 6.2.2 of [BD1] (see Lemma 9.13 of [Kac]), we obtain that $\mathbb{V}_\lambda$ is a free module over $\mathfrak{z}^{\mathrm{reg},\lambda} = \mathrm{Fun}(\mathrm{Op}^{\lambda,\mathrm{reg}}_{\check{\mathfrak{g}}})$.

References

[Ar] T. Arakawa, *Representation theory of $\mathcal{W}$-algebras*, Preprint arXiv: math/0506056.

[BD1] A.Beilinson and V.Drinfeld, *Quantization of Hitchin's integrable system and Hecke eigensheaves*, available at http://www.math.uchicago.edu/~arinkin/langlands/

[BD2] A. Beilinson and V. Drinfeld, *Opers*, Preprint arXiv:math.AG/0501398.

[DS] V. Drinfeld and V. Sokolov, *Lie algebras and KdV type equations*, J. Sov. Math. **30** (1985) 1975–2036.

[EF] D. Eisenbud and E. Frenkel, Appendix to M. Mustaţă, *Jet schemes of locally complete intersection canonical singularities*, Inv. Math. **145** (2001) 397–424.

[FF] B. Feigin and E. Frenkel, *Affine Kac–Moody algebras at the critical level and Gelfand–Dikii algebras*, in *Infinite Analysis*, eds. A. Tsuchiya, T. Eguchi, M. Jimbo, Adv. Ser. in Math. Phys. **16**, 197–215, Singapore: World Scientific, 1992.

[F1] E. Frenkel, *Affine algebras, Langlands duality and Bethe Ansatz*, in Proceedings of the International Congress of Mathematical Physics, Paris, 1994, ed. D. Iagolnitzer, pp. 606–642, International Press, 1995; arXiv:q-alg/9506003.

[F2] E. Frenkel, *Wakimoto modules, opers and the center at the critical level*, Adv. Math. **195** (2005) 297–404; arXiv:math.QA/0210029.

[FB] E. Frenkel and D. Ben-Zvi, *Vertex Algebras and Algebraic Curves*, Mathematical Surveys and Monographs **88**, Second Edition, AMS, 2004.

[FG1] E. Frenkel and D. Gaitsgory, *D-modules on the affine Grassmannian and representations of affine Kac–Moody algebras*, Duke Math. J. **125** (2004) 279–327.

[FG2] E. Frenkel, D. Gaitsgory, *Local geometric Langlands correspondence and affine Kac–Moody algebras*, in *Algebraic Geometry and Number Theory*, Progress in Math. **253**, pp. 69–260, Birkhäuser Boston, 2006.

[FG3] E. Frenkel and D. Gaitsgory, *Fusion and convolution: applications to affine Kac–Moody algebras at the critical level*, Pure and Applied Math. Quart. **2**, (2006) 1255–1312.

[FKW] E. Frenkel, V. Kac, and M. Wakimoto, *Characters and fusion rules for $\mathcal{W}$-algebras via quantized Drinfeld-Sokolov reduction*, Comm. Math. Phys. **147** (1992) 295–328.

[Kac] V.G. Kac, *Infinite-dimensional Lie Algebras*, 3rd Edition, Cambridge University Press, 1990.

Differentiable Operads, the Kuranishi Correspondence, and Foundations of Topological Field Theories Based on Pseudo-Holomorphic Curves

Kenji Fukaya

Department of Mathematics, Kyoto University, Kitashirakawa, Sakyo-ku, Kyoto, Japan
fukaya@math.kyoto-u.ac.jp

Summary. In this article the author describes a general framework to establish foundation of various topological field theories. By taking the case of Lagrangian Floer theory as an example, we explain it in a way so that it is applicable to many similar situations including, for example, the case of 'symplectic field theory'. The results of this article is not really new in the sense that its proof was already written in [33], in detail. However several statements are formulated here, for the first time. Especially the relation to the theory of operad is clarified.

1 Introduction

The purpose of this article is to describe a general framework to construct topological field theory by using smooth correspondence (by various moduli spaces typically). We explain our general construction by taking the case of Lagrangian Floer theory [32,33,34] as an example. However we explain it in the way so that it is applicable to many similar situations, including, for example, the case of 'symplectic field theory' [18] (that is Gromov-Witten theory of symplectic manifold with cylindrical ends). This article is also useful for the readers who are interested in the general procedure which was established in [33] but are not familiar with the theory of pseudo-holomorphic curves (especially with its analytic detail).

In this article, we extract from [33] various results and techniques and formulate them in such a way so that its generalizations to other similar situations are apparent. We do so by clarifying its relation to the theory of operads. In this way, we may regard the analytic parts of the story as a 'black box', and separate geometric (topological) and algebraic constructions from analytic part of the story. The geometric and algebraic constructions include in particular the transversality and orientation issue, which are the heart of

Ö. Ceyhan et al. (eds.), *Arithmetic and Geometry Around Quantization*,
Progress in Mathematics 279, DOI: 10.1007/978-0-8176-4831-2_6,

the rigorous construction of topological field theories of various kinds. The analytic parts of the story, such as Fredholm theory, gluing, compactness, etc., are to be worked out for each individual case. In many (but not all) of the cases which are important for applications to topological field theory, the analytic part of the story, by now, is well-established or understood by experts in principle. For example, in the case of pseudo-holomorphic discs with boundary condition given by a Lagrangian submanifold, we carried it out in [33] especially in its §7.1. (This part of [33] is not discussed in this article, except for its conclusion.) The results of this article (and its cousin for other operads or PROPs) clarify the output of the analytic part of the story which is required for the foundations of topological field theory, in a way so that one can state them without looking at the proof of the analytic part. This seems to be useful for various researchers, since building the foundations of topological field theory is now becoming rather a massive work to carry out which requires many different kinds of mathematics and is becoming harder to be worked out by a single researcher.

It is possible to formulate the axioms under which the framework of [33] and of this article are applicable. Those axioms are to be formulated in terms of a 'differential topology analogue' of operads (or PROPs) (See for example [1,49] for a review of operads, PROPs, etc. and Definition 2 for its 'differential topology analogue') and correspondence by spaces with Kuranishi structure (see [30]) parametrized by such an operad (or PROP). In other words, the output of the analytic part is to be formulated as the existence of spaces with Kuranishi structure with appropriate compatibility conditions. To formulate the compatibility conditions in a precise way is the main part of this article. In this article we give a precise formulation in the case of the A_∞ operad. The author is planning to discuss it in a more general situation elsewhere.

The main theorem of this article is as follows. We define the notion of G-gapped filtered Kuranishi A_∞ correspondence in §10. There we define the notion of morphisms between them and also homotopy between morphisms. Thus we have a homotopy category of G-gapped filtered Kuranishi A_∞ correspondences, which we denote by $\mathfrak{HAKCorr}_G$. We also have a homotopy category of G-gapped filtered A_∞ algebras (with $\mathbb{Q}$ coefficients). This notion is defined in [33] Chapter 4. See also §7 and §9 of this chapter. We denote this category by $\mathfrak{HAlg}_G^{\mathbb{Q}}$.

Theorem 12. *There exists a functor* $\mathfrak{HAKCorr}_G \to \mathfrak{HAlg}_G^{\mathbb{Q}}$.

This theorem is in §10. Roughly speaking, Theorem 12 says that we can associate an A_∞ algebra in a canonical way to a Kuranishi correspondence. Thus it reduces the construction of A_∞ algebras to the construction of Kuranishi correspondences.

Actually once the statement is given, we can extract the proof of Theorem 12 from [33]. So the main new point of this article is the statement itself. In other words, it is the idea to formulate the construction using operads and the correspondence by Kuranishi structures.

The contents of each section are as follows. §2 is a review of the general ideas of topological field theory and smooth correspondences. We emphasize the important role of chain level intersection theory in it. In §3, we exhibit our construction in the simplest case, that is, the Bott-Morse theory on a finite dimensional manifold. §4 and §5 are reviews of A_∞ spaces and A_∞ algebras, respectively. Thus, up to §5, this article is a review and there is nothing new there. We start discussing our main theorem in §6. In §6, we study the case of a correspondence by a manifold which is parametrized by an A_∞ operad. In §7 we study morphisms between such correspondences and in §8 we study homotopy between morphisms. We generalize it to its filtered version in §9. Such a generalization is essential in order to apply it to various topological field theories. Then in §10 we introduce the notion of Kuranishi correspondence and Theorem 12. §11 is again a review and explains how Theorem 12 is used in Lagrangian Floer theory. As we mentioned already the heart of the proof of Theorem 12 is the study of transversality and orientation. They are discussed in detail in [33] §7.2 and Chapter 8 respectively. The argument there can be directly applied to prove Theorem 12. In §12 we give the transversality part of the proof of Theorem 12 with $\mathbb{R}$ coefficients, in a way different from [33]. See the beginning of §12, where we discuss various known techniques to handle transversality. In §13 we discuss orientation. There we explain the way to translate the argument on orientation in [33] Chapter 8 to our abstract situation.

Part of this chapter is a survey. But the main result Theorem 12 is new and its proof is completed in this chapter (using the results quoted from [33]).

The author would like to thank Y.-G.Oh, H.Ohta, K.Ono with whom most of the work presented in this chapter was done. He would also like to thank the organizers of the conference "Arithmetic and Geometry Around Quantization" Istanbul 2006, especially Ozgur Ceyhan, for giving him an opportunity to write this chapter. It was developed from the lecture delivered there by the author.

2 Smooth correspondence and chain level intersection theory

Let us begin with explaining the notion of smooth correspondence.

A typical example of smooth correspondence is given by the following diagram:

$$M \xleftarrow{\pi_1} \mathfrak{M} \xrightarrow{\pi_2} N \tag{1}$$

of oriented closed manifolds of dimension $m, \mathfrak{r}, n$, respectively. It induces a homomorphism

$$\mathrm{Corr}_{\mathfrak{M}} : H_d(M) \to H_{d+\mathfrak{r}-m}(N) \tag{2}$$

by

$$\mathrm{Corr}_{\mathfrak{M}}([c]) = (\pi_{2*} \circ PD \circ \pi_1^* \circ PD)([c]) \tag{3}$$

where PD is Poincaré duality. More explicitly, we can define this homomorphism by using singular homology as follows. Let

$$c = \sum \sigma_i, \quad \sigma_i : \Delta^d \to M$$

be a singular chain representing the homology class $[c]$. We assume that the map σ_i are smooth and transversal to π_1. Then we take a simplicial decomposition

$$\Delta^d{}_{\sigma_i} \times_{\pi_1} \mathfrak{M} = \sum_j \Delta_{i,j}^{d+\mathfrak{r}-m} \tag{4}$$

of the fiber product. The map π_2 induces $\sigma_{ij} : \Delta_{i,j}^{d+\mathfrak{r}-m} \to N$. We thus obtain a singular chain on N by

$$\mathrm{Corr}_{\mathfrak{M}}(c) = \sum_{i,j} \left(\Delta_{i,j}^{d+\mathfrak{r}-m}, \sigma_{ij} \right). \tag{5}$$

An immediate generalization of it is

$$\underbrace{M \times \cdots \times M}_{k} \xleftarrow{\pi_1} \mathfrak{M} \xrightarrow{\pi_2} N \tag{6}$$

which defines a multi-linear map

$$\mathrm{Corr}_{\mathfrak{M}} : (H(M)^{\otimes k})_d \to H_{d+\mathfrak{r}-m}(N), \tag{7}$$

or, in other words, a family of operations on homology groups.

An example is given by the diagram

$$M \times M \xleftarrow{\pi_1} M \xrightarrow{\pi_2} M \tag{8}$$

where

$$\pi_1(p) = (p, p), \qquad \pi_2(p) = p.$$

The homomorphism (7) in this case is nothing but the intersection pairing. We can apply a similar idea to the case when $\mathfrak{M}$ is a moduli space of various kinds.

The notion of correspondence is extensively used in algebraic geometry. (In a situation closely related to topological field theory, correspondences were used by H. Nakajima [53] to construct various algebraic structures. His concept of 'generating space' ([54]) is somewhat similar to the notion of Kuranishi correspondence.) In complex algebraic geometry, it is, in principle, possible to include the case when M, $\mathfrak{M}$, N are singular spaces, since the singularity occurs in real codimension two.

However in the case of a real C^∞ manifold, if we include the manifold $\mathfrak{M}$ which is not necessarily closed (that is, a manifold which may have boundary and/or corners), we will be in trouble. This is because Poincaré duality does not hold in the way appearing in formula (3). We can still define operations *in the chain level* by the formula (5). However, the operation does not then induce a map between homology groups, directly.

This problem can also be reformulated as follows. Let $f_i : P_i \to M$ and $f : Q \to N$ be maps from smooth oriented manifolds (without boundary), which represent cycles on M or N, respectively. Under appropriate transversality conditions, we count (with sign) the cardinality of the set

$$\{(x, p_1, \dots, p_k, q) \in \mathfrak{M} \times M^k \times N \mid \pi_1(x) = f_i(p_i), \pi_2(x) = f(q)\}, \quad (9)$$

in the case when its (virtual) dimension is zero. The cardinality (counted with sign) of (9) is an 'invariant' of various kinds in the case $\mathfrak{M}$ is a moduli space. For example the Donaldson invariants of a 4-manifold and Gromov–Witten invariants of a symplectic manifold both can be regarded as invariants of this kind[1]. When the boundary of $\mathfrak{M}$ is not empty, the cardinality (counted with sign) of (9) is not an invariant of the homology classes but depends on the chains P_i, Q which represent the homology classes. So we need to perform our construction at the chain level. The first example where one needs such a chain level construction is the theory of Floer homology. In that case the 'invariant' obtained by counting (with sign) of something similar to (9) depends on various choices involved. What is invariant in Floer's case is the homology group of the chain complex, of which the matrix coefficient of the boundary operator is obtained by such counting.

In various important cases, the boundary of the moduli space $\mathfrak{M}$ is described as a *fiber* product of other moduli spaces. To be slightly more precise, we consider the following situation. (See §9 and also [33] §7.2.2 for a more detailed exposition.)

Let $(\mathcal{O}, \leq)$ be a partially ordered set. We suppose that, for each $\alpha \in \mathcal{O}$, we have a space $\mathfrak{M}_\alpha$ together with a diagram

$$\underbrace{M \times \cdots \times M}_{k_\alpha} \xleftarrow{\pi_1} \mathfrak{M}_\alpha \xrightarrow{\pi_2} M \quad (10)$$

Here $\mathfrak{M}_\alpha$ is a manifold with boundary. (In the more general case, $\mathfrak{M}_\alpha$ is a space with Kuranishi structure with boundary (See [30] and §10).) Moreover, the boundary of $\mathfrak{M}_\alpha$ is described as a *fiber product* of various $\mathfrak{M}_\beta$ with $\beta \in \mathcal{O}$, $\beta < \alpha$.

Such a situation occurs in the case when the moduli space has Uhlenbeck type bubbles and noncompactness of the moduli space occurs only by this bubbling phenomenon. In those cases, the partially ordered set A encodes the energy of elements of our moduli space together with data describing the complexity of the combinatorial structure of the singular objects appearing in the compactification.

In such a situation, we are going to define an algebraic system on the set of chains on M so that the numbers defined by counting the order of a set like (9)

[1] We need to include the case when M or N is of infinite dimension in the case of Donaldson invariants. Namely, they are the spaces of connections in that case.

are its structure constants. The structure constant itself is *not* well-defined. Namely, it depends on various choices.

One of such choices is the choice of perturbation (or multisection of our Kuranishi structure) required to achieve the relevant transversality condition.

Another choice is an extra geometric structure on our manifolds which we need to determine a partial differential equation defining the moduli space $\mathfrak{M}_{\alpha}$. In the case of self-dual Yang–Mills gauge theory, it is a conformal structure on 4-manifolds. In the case of Gromov–Witten theory, it is an almost complex structure which is compatible with a given symplectic structure.

This extra structure we impose later on plays a very different role from the structure we start with. In Gromov–Witten theory, for example, we start with a symplectic structure and add an almost complex structure later on. The invariant we finally obtain is independent of the almost complex structure but may depend on the symplectic structure. We remark that symplectic structure is of 'topological' nature. Namely, its moduli space is of finite dimension by Moser's theorem. On the other hand, the moduli space of almost complex structures is of infinite dimension. The word 'topological' in 'topological field theory' in the title of this article, means that it depends only on 'topological' structure such as symplectic structure but is independent of the 'geometric' structure such as an almost complex structure. In this sense the word 'topological field theory' in the title of this article is slightly different from those axiomatized by Atiyah [2]. Our terminology coincides with Witten [73].

In order to establish the independence of our 'topological field theory' from perturbation and 'geometric' structures, it is important to define an appropriate notion of *homotopy equivalence* among algebraic systems which appear. We then prove that the algebraic system we obtain by smooth correspondence is independent of the choices *up to homotopy equivalence*. We remark that establishing the appropriate notion of equivalence is the most important part of the application of homological algebra to our story, since the main role of homological algebra is to overcome the difficulty of dependence of the cardinality of the set (9) on various choices.

The story we outlined above is initiated by Donaldson and Floer in the 1980s, originally in gauge theory. Based on it, Witten [73] introduced the notion of topological field theory. Around the same time, Gromov and Floer studied the case of pseudo-holomorphic curves in symplectic geometry. In those days, the relevant algebraic systems were chain complexes mainly. In the early 1990s, more advanced homological algebra was introduced and several researchers started to use it more systematically. It appeared in Mathematics (for example [66, 20, 21, 43, 40]) and in physics (for example [72, 74]) independently.

For the development of the mathematical side of the story, one of the main obstacles to building topological field theory, at the level where advanced homological algebra is included, was the fact that we did not have an adequate general framework for the transversality issue at that time. The virtual fundamental cycle technique which was introduced by [30, 47, 59, 62] at the year

1996, resolved this problem in sufficiently general situation. Then this obstacle was removed, in the case when the group of automorphisms of objects involved is of finite order. Actually this was the main motivation for the author and K. Ono to introduce this technique and to write [30] Chapter 1 in a way so that it can be *directly* applied to other situations than those we need in [30]. Applying virtual fundamental chain/cycle technique in the chain level, sometimes requires more careful discussion, which was completed soon after and was written in detail in [33].

As we already mentioned, we need homological (or homotopical) algebra of various kinds, to develop topological field theory in our sense. The relevant algebraic structure sometimes had been known before. Especially, the notion of A_∞ algebra and L_∞ algebra were already known much earlier in algebraic topology. The importance of such structures in topological field theory was realized more and more by various researchers during 1990's. At the same time, homological algebra to handle those structures itself has been developed. Since the main motivation to use homological algebra in topological field theory is slightly different from those in homotopy theory, one needs to clean it up in a slightly different way. We need also to introduce several new algebraic structures in order to study various problems in topological field theory. Study of such homological algebra is still on the way of rapid progress by various researchers.

The general strategy we mentioned above (together with the basic general technology to realize it) was well established, as a principle, was known to experts around the end of 1990's, and was written in several articles. (See for example [21, 24, 57].) The main focus of the development then turned to rigorously establishing it in various important cases. Another main topic of the subject is a calculation and application of the structure obtained. Around the same time, the number of researchers working on topological field theory (in the sense we use in this article) increased much. Working out the above mentioned strategy in a considerable level of generality, is a heavy work. So its completion took lots of time after the establishment of the general strategy. In [33] we completed the case of Lagrangian Floer theory. Several other projects are now in progress by various authors in various situations.

In this article the author comes back to general frame work, and axiomatize it in a package, so that one can safely use it without repeating the proof.

3 Bott-Morse theory : a baby example

In this section, we discuss Bott-Morse theory as a simplest example of our story. Let X be a compact smooth manifold of finite dimension and $f : X \to \mathbb{R}$ be a smooth function on X. We put

$$\mathrm{Crit}(f) = \{x \in X \mid df(x) = 0\}. \tag{11}$$

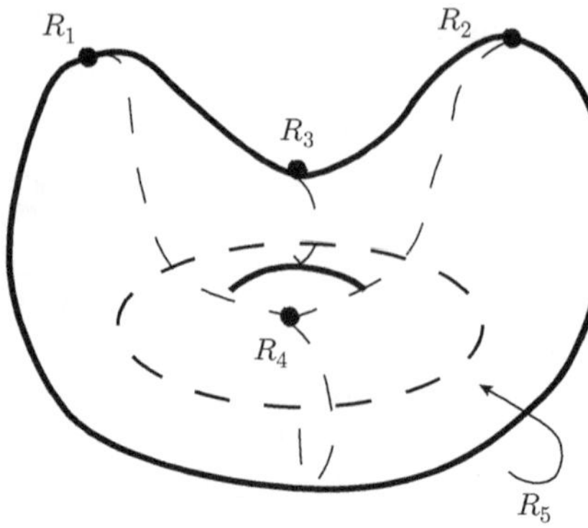

Fig. 1.

Definition 1. *f is said to be a Bott-Morse function if each connected component R_i of $\mathrm{Crit}(f)$ is a smooth submanifold and the restriction of Hessian $\mathrm{Hess}_x f$ to the normal bundle $N_{R_i}X$ is non- degenerate.*

We define the Morse index of f at R_i to be the sum of the multiplicities of the negative eigenvalues of $\mathrm{Hess}_x f$ on $N_{R_i}X$. (Here $x \in R_i$.) We denote it by $\mu(R_i)$

An example of Bott-Morse function is as in the Figure 1 above.

In this example the critical point set is a union of 4 points R_1, R_2, R_3, R_4 and a circle R_5. The Morse indices of them are 2,2,1,1,0 respectively.

The main result of Bott-Morse theory is the following result due to Bott. We enumerate critical submanifolds R_i such that $f(R_i) \ge f(R_j)$ for $i < j$.

Let $N^-_{R_i}X$ be the subbundle of the normal bundle generated by the negative eigenspaces of $\mathrm{Hess}_x f$. Let $\Theta^-_{R_i}$ be the local system associated with the determinant real line bundle of $N^-_{R_i}X$. (It corresponds to a homomorphism $\pi_1(R_i) \to \{\pm 1\} = \mathrm{Aut}\mathbb{Z}$.)

Theorem 1 (Bott [8]). *There exists a spectral sequence E^*_{**} such that*

$$E^2_{i,j} = \bigoplus_i H_{j-\mu(R_i)}(R_i; \Theta^-_{R_i}) \tag{12}$$

and such that it converges to $H(X;\mathbb{Z})$.

Classical proof is based on the stratification of the space X to the union of stable manifolds of critical submanifolds. This approach is not suitable for its generalization to some of its infinite dimensional version, especially to the situation where Morse index is infinite. (This is the situation of Floer homology.) We need to use Floer's approach [19] to Morse theory in such cases. We explain Bott-Morse version of Floer's approach here following [22]. (See [4] for related results. The restriction on Bott-Morse function which was assumed in [4] can by now be removed as we explained in [33] §7.2.2 Remark 7.2.20.)

We take a Riemannian metric g on X. We then have a gradient vector field grad$f$ of f. Let $\widetilde{\mathcal{M}}(R_i, R_j)$ be the set of all maps $\ell : \mathbb{R} \to X$ such that

$$\frac{d\ell}{dt}(t) = -\mathrm{grad}_{\ell(t)} f \tag{13}$$

$$\lim_{t\to-\infty} \ell(t) \in R_i, \qquad \lim_{t\to+\infty} \ell(t) \in R_j. \tag{14}$$

The group $\mathbb{R}$ acts on $\widetilde{\mathcal{M}}(R_i, R_j)$ by $(s \cdot \ell) = \ell(t+s)$. Let $\mathcal{M}(R_i, R_j)$ be the quotient space. We define the maps π_i by

$$\pi_1(\ell) = \lim_{t\to-\infty} \ell(t), \qquad \pi_2(\ell) = \lim_{t\to+\infty} \ell(t). \tag{15}$$

They define a diagram

$$R_i \xleftarrow{\ \pi_1\ } \mathcal{M}(R_i, R_j) \xrightarrow{\ \pi_2\ } R_j. \tag{16}$$

Now we have

Lemma 1. *By perturbing f on a set away from* Crit$f$, *we may choose f so that $\mathcal{M}(R_i, R_j)$ is a smooth manifold with boundary and corners. Moreover we have*

$$\partial\mathcal{M}(R_i, R_j) = \bigcup_{i<k<j} \mathcal{M}(R_i, R_k) \times_{R_k} \mathcal{M}(R_k, R_j). \tag{17}$$

Let us exhibit the lemma in the case of the example of the Morse function in Figure 1. In this case we have the following:

$$\begin{aligned} \mathcal{M}(R_1, R_3) = \mathcal{M}(R_2, R_3) = \mathcal{M}(R_1, R_4) = \mathcal{M}(R_2, R_4) &= \text{one point.} \\ \mathcal{M}(R_3, R_5) = \mathcal{M}(R_4, R_5) &= \text{two points.} \\ \mathcal{M}(R_1, R_5) = \mathcal{M}(R_2, R_5) &= \text{union of two arcs.} \end{aligned}$$

We then have

$$\begin{aligned} \partial(\mathcal{M}(R_1, R_5)) &= (\mathcal{M}(R_1, R_3) \times \mathcal{M}(R_3, R_5)) \cup (\mathcal{M}(R_1, R_4) \times \mathcal{M}(R_4, R_5)) \\ &= 4 \text{ points.} \end{aligned}$$

Now we have the following:

Lemma 2. *There exists a subcomplex $C\left(R_i; \Theta^-_{R_i}\right) \subset S\left(R_i; \Theta^-_{R_i}\right)$ of the singular chain complex $S\left(R_i; \Theta^-_{R_i}\right)$ of R_i with $\Theta^-_{R_i}$ coefficients, such that the inclusion induces an isomorphism of homologies and such that the following holds.*

If $c \in C\left(R_i; \Theta^-_{R_i}\right)$ then $\mathrm{Corr}_{\mathcal{M}(R_i,R_j)}(c)$ *is well-defined by (5) and is in $C\left(R_j; \Theta^-_{R_j}\right)$. Moreover we have*

$$[\partial, \mathrm{Corr}_{\mathcal{M}}] + \mathrm{Corr}_{\mathcal{M}} \circ \mathrm{Corr}_{\mathcal{M}} = 0. \tag{18}$$

Here in (18) we write $\mathrm{Corr}_{\mathcal{M}}$ in place of $\mathrm{Corr}_{\mathcal{M}(R_i,R_j)}$ for various i, j.

Lemmas 1 and 2 are in [22]. (See also [33] §7.2.2.)

We remark that (18) is a version of the Maurer–Cartan equation. We define

$$\partial_{\mathcal{M}} = \partial + \mathrm{Corr}_{\mathcal{M}} : \bigoplus_i C\left(R_i; \Theta^-_{R_i}\right) \to \bigoplus_i C\left(R_i; \Theta^-_{R_i}\right). \tag{19}$$

(18) implies

$$\partial_{\mathcal{M}} \circ \partial_{\mathcal{M}} = 0. \tag{20}$$

Lemma 3. $\mathrm{Ker}\,\partial_{\mathcal{M}}/\mathrm{Im}\,\partial_{\mathcal{M}}$ *is isomorphic to the homology group of* M.

We can prove Lemma 3 as follows. First we prove that $\mathrm{Ker}\,\partial_{\mathcal{M}}/\mathrm{Im}\,\partial_{\mathcal{M}}$ is independent of the choice of the Bott–Morse function f. (An infinite dimensional version of this fact (whose proof is similar to and is harder than Lemma 3) is proved in [22].) Moreover, in the case when $f \equiv 0$ the lemma is obvious.

By construction

$$F_j = \bigoplus_{i \geq j} C\left(R_i; \Theta^-_{R_i}\right)$$

is a filtration of $\left(\bigoplus_i C\left(R_i; \Theta^-_{R_i}\right), \partial_{\mathcal{M}}\right)$. The spectral sequence associated to this filtration is the one required in Theorem 1.

We remark that in the case $j > i + 1$ the correspondence $\mathrm{Corr}_{\mathcal{M}(R_i,R_j)}$ may not define a homomorphism $H\left(R_i; \Theta^-_{R_i}\right) \to H\left(R_j; \Theta^-_{R_j}\right)$ at the homology level, because $\partial\mathcal{M}(R_i, R_j) \neq \emptyset$ in general. This point is taken care of by the spectral sequence. Namely the third differential d_3 of the spectral sequence is defined only partially and has an ambiguity, which is controlled by the second differential d_2. This is a prototype of the phenomenon which appears in the situation where we need more advanced homological algebra. (For example the Massey product has similar properties.) In the situation of Theorem 1, we study the chain complex without any additional structure on it. In the situation we will discuss later, we consider the chain complex together with various multiplicative structures. We remark that including multiplicative structure (the cup and Massey products) in Theorem 1 is rather a delicate issue, if we prove it in the way described above. See [33] §7.2.2 on this point. Actually there are many errors and confusions in various references on this point. If we prove it by stratifying the space X, it is easy to prove that the spectral sequence in Theorem 1 is multiplicative.

4 A_∞ spaces.

In the rest of this chapter we describe the construction outlined in §1 in more detail. To be specific, we concentrate to the case of A_∞ structure. We first recall the notion of A_∞ structure introduced by Stasheff [65]. (See also Sugawara [68].) (There are excellent books [1, 7, 49] etc., on related topics.)

The original motivation for Stasheff to introduce A_∞ space is to study loop space. Let us recall it briefly. Let (X, p) be a topological space with base point. We put

$$\Omega(X,p) = \{\ell : [0,1] \to X \mid \ell(0) = \ell(1) = p\}. \tag{21}$$

We define $\mathsf{m}_2 : \Omega(X,p) \times \Omega(X,p) \to \Omega(X,p)$ by

$$\mathsf{m}_2(\ell_1,\ell_2)(t) = \begin{cases} \ell_1(2t) & \text{if } t \le 1/2 \\ \ell_2(2t-1) & \text{if } t \ge 1/2 \end{cases} \tag{22}$$

$\mathfrak{m}_2$ is associative only modulo parametrization. In fact, for $t \le 1/4$, we have

$$(\mathsf{m}_2(\ell_1, \mathsf{m}_2(\ell_2,\ell_3)))(t) = \ell_1(2t) \neq$$
$$(\mathsf{m}_2(\mathsf{m}_2(\ell_1,\ell_2),\ell_3))(t) = \ell_1(4t).$$

On the other hand, there is a canonical homotopy between $\mathsf{m}_2(\ell_1, \mathsf{m}_2(\ell_2,\ell_3))$ and $\mathsf{m}_2(\mathsf{m}_2(\ell_1,\ell_2),\ell_3)$. Actually some stronger statements than the existence of homotopy hold. The A_∞ structure is the way to state it precisely.

To define A_∞ structure we need a series of contractible spaces $\mathcal{M}_{k+1}$ together with continuous maps

$$\circ_i : \mathcal{M}_{k+1} \times \mathcal{M}_{l+1} \to \mathcal{M}_{k+l}, \tag{23}$$

for $i = 1, \cdots, l$, such that the following holds for $a \in \mathcal{M}_{k+1}$, $b \in \mathcal{M}_{l+1}$, $c \in \mathcal{M}_{m+1}$.

$$(a \circ_j b) \circ_i c = a \circ_j (b \circ_{i-j+1} c), \tag{24}$$

for $i \ge j$ (see Figure 2) and

$$(a \circ_j b) \circ_i c = (a \circ_i c) \circ_{j+k_2-1} b \tag{25}$$

for $i < j$, $a \in \mathcal{M}_{k_1+1}$, $b \in \mathcal{M}_{k_2+1}$, $c \in \mathcal{M}_{k_3+1}$ (see Figure 3).

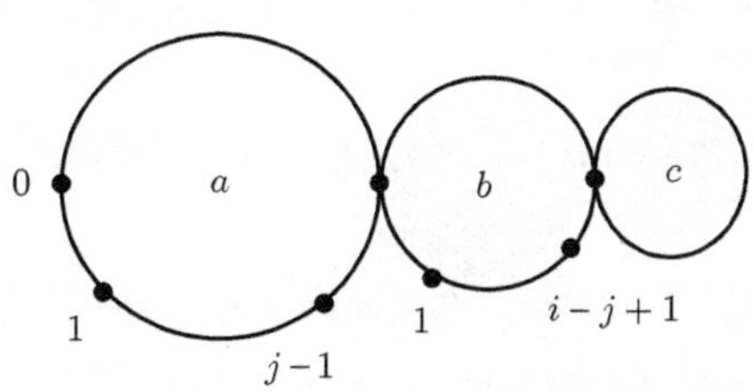

Fig. 2.

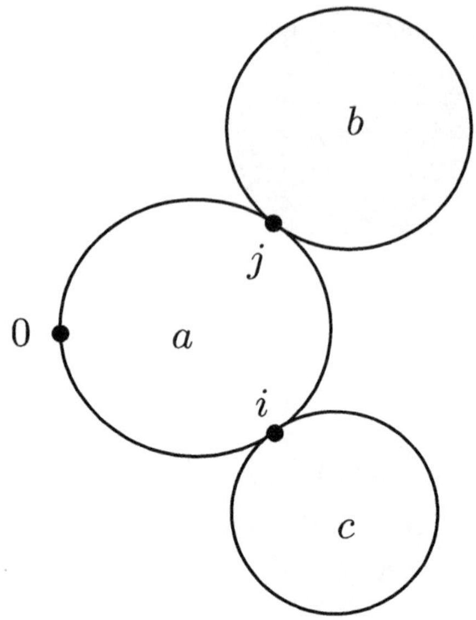

Fig. 3.

A topological space X is an A_∞ space if there is a sequence of continuous maps

$$\mathcal{M}_{k+1} \times X^k \to X; (a; x_1, \cdots, x_k) \mapsto \mathsf{m}(a; x_1, \cdots, x_k) \tag{26}$$

such that

$$\begin{aligned} &\mathsf{m}(a \circ_i b; x_1, \cdots, x_{k+\ell-1}) \\ &= \mathsf{m}(a; x_1, \cdots, x_{i-1}, \mathsf{m}(b; x_i, \cdots, x_{i+\ell-1}), \cdots, x_{k+\ell-1}). \end{aligned} \tag{27}$$

We remark that (24), (25) are compatible with (27).

Suppose that X is an A_∞ space. Then, by taking an (arbitrary but fixed) element $a_0 \in \mathcal{M}_{2+1}$ we define

$$\mathsf{m}_2 = \mathsf{m}(a_0; \cdot, \cdot) : M^2 \to M.$$

Since $\mathcal{M}_{3+1}$ is contractible there exists a path joing $a_0 \circ_1 a_0$ to $a_0 \circ_2 a_0$. Using it we have a homotopy between

$$\mathsf{m}_2(\mathsf{m}_2(x, y), z) = \mathsf{m}(a_0; \mathsf{m}_2(a_0; x, y), z) = \mathsf{m}(a_0 \circ_1 a_0; x, y, z)$$

and

$$\mathsf{m}_2(x, \mathsf{m}_2(y, z)) = \mathsf{m}(a_0; x, \mathsf{m}_2(a_0; y, z)) = \mathsf{m}(a_0 \circ_2 a_0; x, y, z).$$

Namely m_2 is homotopy associative. The condition (27) is more involved than homotopy associativity.

An important result by Stasheff is that an H-space is an A_∞ space if and only if it is homotopy equivalent to a loop space $\Omega(X, p)$ for some (X, p). More precisely loop space corresponds to an A_∞ space with unit for this theorem. We do not discuss unit here. (See §14.1.)

In the rest of this section, we give two examples of $\mathcal{M}_{k+1}$ which satisfy (24). We call such system of $\mathcal{M}_{k+1}$ an A_∞ *operad.*

Remark 1. We remark that the notion of operad which was introduced by May [50] is similar to but is slightly different from above. It is a family of spaces $\mathcal{P}(k)$ together with operations

$$\mathcal{P}(l) \times (\mathcal{P}(k_1) \times \cdots \times \mathcal{P}(k_l)) \to \mathcal{P}(k_1 + \cdots + k_l). \tag{28}$$

Its axiom contains an associativity of the operation (28) and also symmetry for exchanging $\mathcal{P}_{k_i}$'s. In our case, the structure map (23) is slightly different from (28) and is closer to something called non Σ-operad. One important difference is that we do not require any kinds of the commutativity to our operations $\mathsf{m}(a; x_1, \cdots, x_k)$.

There are several other variants of operad or prop. (The difference between operad and prop is as follows. An operad has several inputs but has only one output. A prop has several inputs and several outputs.) See [49] for those variants and history of its development. We can discuss correspondence parametrized by them in a way similar to the case of A_∞ operad which we are discussing in this paper.

The first example of A_∞ operad is classical and due to Boardman-Vogt [7] Definition 1.19. (See [33] §3.4.) Let us consider the planer tree $|T|$ (that is a tree embedded in $\mathbb{R}^2$). We divide the set $C^0(|T|)$ of the set of vertices of $|T|$ into a disjoint union

$$C^0(|T|) = C^0_{\text{int}}(|T|) \cup C^0_{\text{ext}}(|T|)$$

where every vertex in $C^0_{\text{int}}(|T|)$ has at least 3 edges and all the vertices in $C^0_{\text{ext}}(|T|)$ have exactly one edge. We assume that there is no vertex with two edges. Elements of $C^0_{\text{int}}(|T|)$, $C^0_{\text{ext}}(|T|)$ are said to be interior edges and exterior edges, respectively. $C^1(|T|)$ denotes the set of all edges. We say an edge to be exterior if it contains an exterior vertex. Otherwise the edge is said to be interior.

We consider such $|T|$ together with a function $l : C^1(|T|) \to (0, \infty]$ which assigns the length $l(e)$ to each edge e. We assume $l(e) = \infty$ if e is an exterior edge.

We consider $(|T|, l)$ as above such that the tree has exactly $k+1$ exterior vertices. We fix one exterior vertex v_0 and consider the set of all the isomorphism classes of such $(|T|, l, v_0)$. We denote it by Gr_{k+1} and call its element *a rooted metric ribbon tree with* $k+1$ *exterior vertices.* We enumerate the exterior vertices as $\{v_0, v_1, \cdots, v_k\}$ so that it respects the counter-clockwise orientation of $\mathbb{R}^2$.

We can prove (see [31] for example) that Gr_{k+1} is homeomorphic to D^{k-2} and hence is contractible.

We define

$$\circ_i : Gr_{k+1} \times Gr_{l+1} \to Gr_{k+l}$$

as follows. Let $T = (|T|, l, v_0) \in Gr_{k+1}$, $T' = (|T'|, l', v'_0) \in Gr_{l+1}$. Let $v'_0, \cdots, v'_l$ be the exterior vertices of Gr_{l+1} enumerated according to the counter-clockwise orientation. We identify $v_i \in |T|$ and $v'_0 \in |T'|$ to obtain $|T| \circ_i |T'|$. The length of the edges of it is the same as one for $|T|$ or $|T'|$ except the new edge, whose length is defined to be infinity. We thus obtain an element $T \circ_i T'$ of Gr_{k+l}.

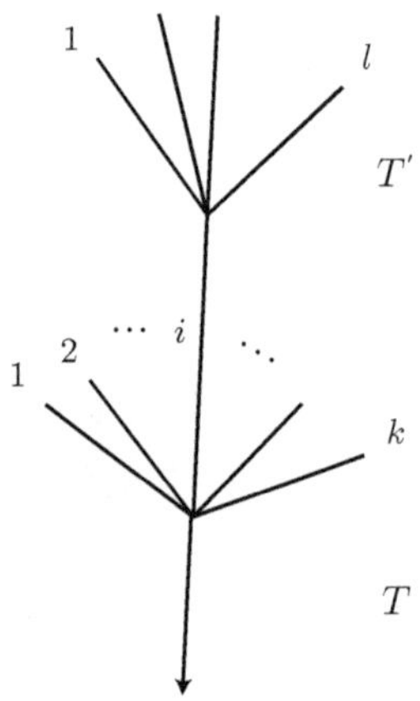

Fig. 4.

(24), (25) can be checked easily. Thus, by putting $\mathcal{M}_{k+1} = Gr_{k+1}$, we obtain an example of A_∞-operad.

We next discuss another example of A_∞-operad, which is closely related to Lagrangian Floer theory.

We consider $(D^2; z_0, \cdots, z_k)$ where D^2 is the unit disc in $\mathbb{C}$ centered at origin. z_i, $i = 0, \cdots, k$ are pair-wise distinct points of ∂D^2. We assume that $z_0, \cdots, z_k$ respects counter- clockwise cyclic order of ∂D^2. Let $PSL(2;\mathbb{R})$ be the group of all biholomorphic maps $D^2 \to D^2$. For $u \in PSL(2;\mathbb{R})$ we put

$$u \cdot (D^2; z_0, \cdots, z_k) = (D^2; u(z_0), \cdots, u(z_k)) \tag{29}$$

We denote by $\overset{\circ}{\mathcal{M}}_{k+1}$ the set of all the equivalence classes of such $(D^2; z_0, \cdots, z_k)$ with respect to the relation (29).

It is easy to see that $\overset{\circ}{\mathcal{M}}_{k+1}$ is diffeomorphic to $\mathbb{R}^{k-2}$. We can compactify $\overset{\circ}{\mathcal{M}}_{k+1}$ to obtain $\mathcal{M}_{k+1}$. An idea to do so is to take double and use Deligne-Mumford compactification of the moduli space of Riemann surface. (See [32] §3, where its generalization to higher genus is also discussed.)

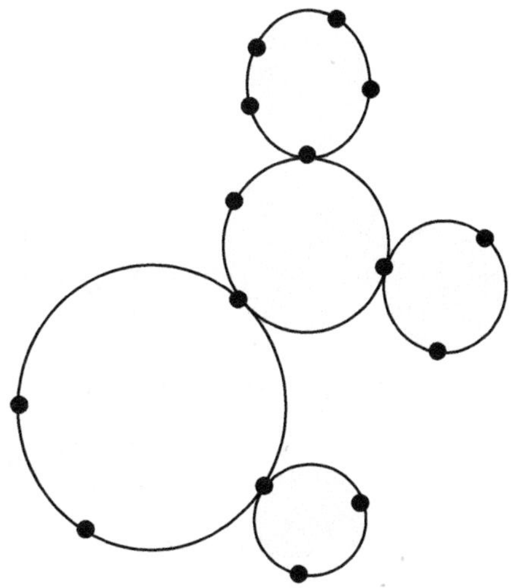

Fig. 5.

An element of $\mathcal{M}_{k+1}$ is regarded as $(\Sigma; z_0, \cdots, z_k)$ which satisfies the following conditions. (See Figure 5.)

We consider a Hausdorff topological space Σ which is a union of discs $D_1^2, \cdots, D_k^2$. We call D_i^2 a *components* of Σ. We assume that, for each $i \neq j$, the intersection $D_i^2 \cap D_j^2$ is either empty or consists of one point which lies on the boundaries of D_i^2 and of D_j^2. We assume moreover that intersection of three components are empty. Furthermore Σ is assumed to be connected and simply connected.

The set of all points on Σ which belongs to more than 2 components are called *singular.*

We put

$$\partial \Sigma = \bigcup \partial D_i^2.$$

We assume that $z_i \in \partial \Sigma$. We also assume that z_i is not singular. We embed Σ to $\mathbb{C}$ so that it is biholomorphic on each D_i^2. We require (the image of) $z_0, \cdots, z_k$ respects the counter- clockwise cyclic orientation induced by the orientation of $\mathbb{C}$.

Finally we assume the following stability condition. We say

$$\varphi : \Sigma \to \Sigma'$$

is biholomorphic if it is a homeomorphism and if its restriction to D_i^2 induces a biholomorphic maps $D_i^2 \to D_j^2$ for some j. Here D_i^2 and D_j^2 are components of Σ and Σ', respectively. Let $\mathrm{Aut}(\Sigma; z_0, \cdots, z_k)$ be the group of all biholomorphic maps $\varphi : \Sigma \to \Sigma$ such that $\varphi(z_i) = z_i$. We say that $(\Sigma; z_0, \cdots, z_k)$ is *stable* if $\mathrm{Aut}(\Sigma; z_0, \cdots, z_k)$ is of finite order.

Two such $(\Sigma; z_0, \cdots, z_k)$, $(\Sigma'; z'_0, \cdots, z'_k)$ are said to be *biholomorphic* to each other if there exists a biholomorphic map $\varphi : \Sigma \to \Sigma'$ such that $\varphi(z_i) = z'_i$.

$\mathcal{M}_{k+1}$ is the set of all the biholomorphic equivalence classes of $(\Sigma; z_0, \cdots, z_k)$ which is stable.

We can show that $(\Sigma; z_0, \cdots, z_k)$ is stable if and only if each components contain at least 3 marked or singular points. We remark that in our case of genus zero, $\mathrm{Aut}(\Sigma; z_0, \cdots, z_k)$ is trivial if $(\Sigma; z_0, \cdots, z_k)$ is stable.

We define

$$\circ_i : \mathcal{M}_{k+1} \times \mathcal{M}_{l+1} \to \mathcal{M}_{k+l}$$

as follows. Let $(\Sigma; z_0, \cdots, z_k) \in \mathcal{M}_{k+1}$ and $(\Sigma'; z'_0, \cdots, z'_l) \in \mathcal{M}_{l+1}$. We identify $z_i \in \Sigma$ and $z'_0 \in \Sigma'$ to obtain Σ''. We put

$$(z''_0, \cdots, z''_{k+l-1}) = (z_0, \cdots, z_{i-1}, z'_1, \cdots, z'_l, z_{i+1}, \cdots, z_k).$$

We now define

$$(\Sigma; z_0, \cdots, z_k) \circ_i (\Sigma'; z'_0, \cdots, z'_l) = (\Sigma''; z''_0, \cdots, z''_{k+l-1})$$

which represents an element of $\mathcal{M}_{k+\ell}$. We can easily check (24), (25).

Actually the two A_∞-operads we described above are isomorphic to each other. In fact the following theorem is proved in [31]. (This is a result along the line of theory of the quadratic differential by Strebel [67] etc.)

Theorem 2. *There exists a homeomorphism $Gr_{k+1} \cong \mathcal{M}_{k+1}$ which is compatible with $\circ_i$.*

We remark that $\mathcal{M}_{k+1} \cong D^{k-2}$. Moreover we have

$$\partial \mathcal{M}_{k+1} = \sum_{1 \le i < j \le k} \mathcal{M}_{j-i+1} \circ_i \mathcal{M}_{k-j+i} \tag{30}$$

where the images of the right hand sides intersect each other only at their boundaries. We remark that (30) can be regarded as a Maurer-Cartan equation.

Using (30) inductively we obtain a cell decomposition of the cell D^{k-2}. A famous example is the case $k = 5$. In that case we obtain a cell decomposition of D^3 which is called Stasheff cell.

In the rest of this article, we use $\mathcal{M}_{k+1}$ which is the moduli space of bordered Rieman surface of genus 0 with $k + 1$ marked points at boundary, as the A_∞ operad, unless otherwise mentioned.

For the purpose of this paper, the structure of differentiable manifold on $\mathcal{M}_{k+1}$ is important. So we make the following definition.

Definition 2. *A differentiable A_∞ operad is an A_∞ operad $\mathcal{M}_{k+1}$ such that $\mathcal{M}_{k+1}$ is a compact and oriented smooth manifold (with boundary or corner) and the structure map* (23) *is an smooth embedding. Moreover we assume* (30).

It is straightforward to extend this definition to the case of variants of operad or PROP.

5 A_∞ algebra.

A_∞ algebra is an algebraic analogue of A_∞ space and is a generalization of differential graded algebra. It is defined as follows.

Hereafter R is a commutative ring with unit. Let C be a graded R module. We assume it is free as R module. We define its suspension $C[1]$ (the degree shift) by $C[1]^k = C^{k+1}$. Hereafter we denote by $\deg'$ the degree after shifted and by deg the degree before shifted.

We define the *Bar complex* $BC[1]$ by

$$B_kC[1] = \underbrace{C[1] \otimes \cdots \otimes C[1]}_{k \text{ times}}, \qquad BC[1] = \bigoplus_{k=0}^{\infty} B_kC[1].$$

We define a coalgebra structure Δ on $BC[1]$ by

$$\Delta(x_1 \otimes \cdots \otimes x_k) = \sum_{i=0}^{k} (x_1 \otimes \cdots \otimes x_i) \otimes (x_{i+1} \otimes \cdots \otimes x_k)$$

here in case $i = 0$ the right hand side is $1 \otimes (x_1 \otimes \cdots \otimes x_k)$.

We consider a series of operations

$$\mathfrak{m}_k : B_kC[1] \to C[1] \tag{31}$$

of degree 1 for each $k \geq 1$. It is extended uniquely to a coderivation

$$\hat{\mathfrak{m}}_k : BC[1] \to BC[1]$$

whose $Hom(B_kC[1], B_1C[1]) = Hom(B_kC[1], C[1])$ component is $\mathfrak{m}_k$.

We recall $\varphi : BC[1] \to BC[1]$ is said to be a *coderivation* if and only if

$$\Delta \circ \varphi = (\varphi \otimes 1 + 1 \otimes \varphi) \circ \Delta. \tag{32}$$

(Note that the right hand side is defined by $(1 \otimes \varphi)(x \otimes y) = (-1)^{\deg' x \deg \varphi} x \otimes \varphi(y)$.)

We put

$$\hat{d} = \sum_k \hat{\mathfrak{m}}_k : BC[1] \to BC[1].$$

Definition 3. *$(C, \mathfrak{m}_*)$ is an A_∞ algebra if $\hat{d} \circ \hat{d} = 0$.*

We can rewrite the condition $\hat{d} \circ \hat{d} = 0$ to the following relation, which is called the A_∞ relation.

$$\sum_{1\le i<j\le k} (-1)^* \mathfrak{m}_{k-j+i}(x_1,\cdots,\mathfrak{m}_{j-i+1}(x_i,\cdots,x_j),\cdots,x_k) = 0 \tag{33}$$

where

$$* = \deg' x_1 + \cdots \deg' x_{i-1} = \deg x_1 + \cdots + \deg x_{i-1} + i - 1.$$

Our sign convention is slightly different from Stasheff's [65].

We remark that (33) implies $\mathfrak{m}_1 \circ \mathfrak{m}_1 = 0$. Namely $(C, \mathfrak{m}_1)$ is a chain complex.

Example 1. If $(C, d, \wedge)$ is a differential graded algebra, we may regard it as an A_∞ algebra by putting

$$\mathfrak{m}_1(x) = (-1)^{\deg x} dx, \qquad \mathfrak{m}_2(x,y) = (-1)^{\deg x(\deg y+1)} x \wedge y. \tag{34}$$

An alternative choice of sign $\mathfrak{m}_1(x) = dx$, $\mathfrak{m}_2(x,y) = (-1)^{\deg x} x \wedge y$, also works. Here we follow the convention of [33].

The following result is classical and is certainly known to Stasheff.

Theorem 3. *A structure of A_∞ space on X induces a structure of A_∞ algebra on its singular chain complex.*

Sketch of the proof : By using (30), we may take a simplicial decomposition of $\mathcal{M}_{k+1}$ so that $\circ_i$ are all simplicial embedding.

We use the 'cohomological' notation for the degree of singular chain complex $S(X)$. Namely we put $S^{-d}(X) = S_d(X)$. (We remark that we do *not* assume X is a manifold. So we can not use Poincaré duality to identify chain with cochain.)

Let $\sigma_i : \Delta^{d_i} \to X$, $i = 1, \cdots, k$ be singular chains. We take the standard simplicial decomposition

$$\mathcal{M}_{k+1} \times \Delta^{d_1} \times \cdots \times \Delta^{d_k} = \sum_j \Delta_j^d$$

induced by the simplicial decomposition of $\mathcal{M}_{k+1}$, where $d = \sum d_i + k - 2$. We have

$$\mathfrak{m}_k(\sigma_1,\cdots,\sigma_k) = \sum \pm(\Delta_j^d; \sigma)$$

where

$$\sigma : \mathcal{M}_{k+1} \times \Delta^{d_1} \times \cdots \times \Delta^{d_k} \to X$$

is defined by

$$\sigma(a; p_1, \cdots, p_k) = \mathfrak{m}_k(a; \sigma_1(p_1), \cdots, \sigma_k(p_k)).$$

Since $-d-1 = \sum(-d_i - 1) + 1$ the degree is as required. We do not discuss sign here. (33) can be checked by using (30). □

6 A_∞ correspondence.

Theorem 3 we discussed at the end of the last section is a result on algebraic topology where we never use manifold structure etc.

Contrary to Theorem 3 we use manifold structure and Poincaré duality in the next theorem. To state it we need some notation.

We consider the following diagram of smooth maps :

$$\begin{array}{ccccc} & & \mathcal{M}_{k+1} & & \\ & & \uparrow{\scriptstyle \pi_0} & & \\ M^k & \xleftarrow[\pi_2]{} & \mathfrak{M}_{k+1} & \xrightarrow[\pi_1]{} & M \end{array} \tag{35}$$

where M is a closed oriented manifold and $\mathfrak{M}_{k+1}$ is a compact oriented manifold which may have boundary or corner. We assume

$$\dim \mathfrak{M}_{k+1} = \dim M + \dim \mathcal{M}_{k+1} = \dim M + k - 2. \tag{36}$$

We define $ev_i : \mathfrak{M}_{k+1} \to M$ by :

$$\pi_2 = (ev_1, \cdots, ev_k), \qquad \pi_1 = ev_0.$$

Definition 4. *We say that $\mathfrak{M}_{k+1}$, $k = 1, 2, \cdots$ is an A_∞ correspondence on M, if there exists a family of smooth maps*

$$\circ_{\mathrm{m},i} : \mathfrak{M}_{k+1\ ev_i} \times_{ev_0} \mathfrak{M}_{l+1} \to \mathfrak{M}_{k+l} \tag{37}$$

for $i = 1, \cdots, l$ with the following properties.

(1) (operad axiom) *The following diagram commutes.*

$$\begin{array}{ccc} \mathcal{M}_{k+1} \times \mathcal{M}_{l+1} & \xrightarrow[\circ_i]{} & \mathcal{M}_{k+l} \\ {\scriptstyle \pi_0 \times \pi_0}\uparrow & & {\scriptstyle \pi_0}\uparrow \\ \mathfrak{M}_{k+1\ ev_i} \times_{ev_0} \mathfrak{M}_{l+1} & \xrightarrow[\circ_{\mathrm{m},i}]{} & \mathfrak{M}_{k+l} \end{array} \tag{38}$$

(2) (cartesian axiom) $\mathfrak{M}_{k+1\ ev_i} \times_{ev_0} \mathfrak{M}_{l+1}$ *coincides with the fiber product*

$$(\mathcal{M}_{k+1} \times \mathcal{M}_{l+1}) \times_{\mathcal{M}_{k+l}} \mathfrak{M}_{k+l}$$

obtained by Diagram (38).

(3) (associativity axiom) *We have*

$$(\mathbf{a} \circ_{\mathrm{m},j} \mathbf{b}) \circ_{\mathrm{m},i} \mathbf{c} = \mathbf{a} \circ_{\mathrm{m},j} (\mathbf{b} \circ_{\mathrm{m},i-j+1} \mathbf{c}), \tag{39}$$

for $i < j$ and

$$(\mathbf{a} \circ_{\mathrm{m},j} \mathbf{b}) \circ_{\mathrm{m},i} \mathbf{c} = (\mathbf{a} \circ_{\mathrm{m},i} \mathbf{c}) \circ_{\mathrm{m},j+k_2-1} \mathbf{b} \tag{40}$$

for $i > j$, $\mathbf{c} \in \mathfrak{M}_{k_2+1}$.
(4) (evaluation map axiom) *The following diagram commutes.*

$$\begin{array}{ccccc} M^{k+l} & \xleftarrow[\pi_2]{} & \mathfrak{M}_{k+l} & \xrightarrow[\pi_1]{} & M \\ \| & & \uparrow \circ_{\mathrm{m},i} & & \| \\ M^{k+l} & \longleftarrow & \mathfrak{M}_{k+1\ ev_i} \times_{ev_0} \mathfrak{M}_{l+1} & \xrightarrow[\pi_1 \circ \mathrm{pr}_1]{} & M \end{array} \tag{41}$$

Here the first arrow in the second line is

$$(ev_1 \circ \mathrm{pr}_1, \cdots, ev_{i-1} \circ \mathrm{pr}_1, ev_1 \circ \mathrm{pr}_2, \cdots, ev_l \circ \mathrm{pr}_2, ev_{i+1} \circ \mathrm{pr}_1, \cdots, ev_l \circ \mathrm{pr}_1)$$

where $\mathrm{pr}_1 : \mathfrak{M}_{k+1\ ev_i} \times_{ev_0} \mathfrak{M}_{l+1} \to \mathfrak{M}_{k+1}$ *is the projection to the first factor and* $\mathrm{pr}_2 : \mathfrak{M}_{k+1\ ev_i} \times_{ev_0} \mathfrak{M}_{l+1} \to \mathfrak{M}_{l+1}$ *is the projection to the second factor.*
(5) (Maurer-Cartan axiom) *We have*

$$\partial \mathfrak{M}_{n+1} = \bigcup_{\substack{k+l=n+1 \\ 1 \le i \le k}} \circ_{\mathrm{m},i}(\mathfrak{M}_{k+1\ ev_i} \times_{ev_0} \mathfrak{M}_{l+1}). \tag{42}$$

The images of the fiber product of the right hand sides, intersect to each other only at their boundaries.
(6) (orientation axiom) *The isomorphism* (42) *preserves orientation.*

We need a sign for (6) which will be discussed as Definition 27. The property (3) is regarded as a Maurer-Cartan equation. We recall that the diagram (38) is said to be *cartesian* when the condition (2) above holds.

A typical example of A_∞ correspondence is as follows.

Example 2. $\mathfrak{M}_{k+1} = \mathcal{M}_k \times M^{k+1}$ and π_0, π_1, π_2 are obvious projections.

Example 3. Let M be a manifold which is an A_∞ space such that the structure map $\mathsf{m} : \mathcal{M}_{k+1} \times M^k \to M$ is smooth. We put $\mathfrak{M}_{k+1} = \mathcal{M}_{k+1} \times M^k$, $\pi_1 = \mathsf{m}$, (π_0, π_1) =identity. This gives another example of A_∞ correspondence.

Now the next results can be proved in the same way as [33] §7.2.

Theorem 4. *If there is an* A_∞ *correspondence on* M *then there exists a cochain complex* $C(M;\mathbb{Z})$ *whose homology group is* $H^*(M;\mathbb{Z})$ *and such that* $C(M;\mathbb{Z})$ *has a structure of* A_∞ *algebra.*

Remark 2. In our situation, we can prove Theorem 4 over $\mathbb{Z}_2$ coefficient without assuming none of the conditions on orientations in Definition 4.

Applying Theorem 4 to Example 2 we obtain the following :

Corollary 1. *For any oriented closed manifold* M, *there exists an* A_∞ *algebra whose cohomology group is* $H^*(M;\mathbb{Z})$.

Corollary 1 is [33] Theorem 3.4.8. It was also proved by McClure [51] and by Wilson [71].

Remark 3. Actually the statement of Corollary 1 itself is a consequence of a classical fact. In fact the singular cochain complex has a cup product which is associative at the chain level. What is important here is that the A_∞ operations are realized by the *chain level intersection theory* and by identifying chains with cochains by *Poincaré duality.* We emphasize that the chain level Poincaré duality is still a mysterious subject. We emphasize that using chains (instead of cochains) is more natural in our situation since the moduli space can be naturally regarded as a chain but can be regarded as a cochain only via Poincaré duality.

Sketch of the proof of Theorem 4: Let $f_i : P_i \to M$ be 'chains' of dimension $\dim M - d_i$. (Actually the precise choice of the chain complex to work with is the important part of the proof. See [33] Remark 1.32 and the beginning of §12.) We consider the fiber product

$$\mathfrak{M}_{k+1} \ {}_{\pi_2}\times_{f_1,\dots,f_k} (P_1 \times \cdots \times P_k) \tag{43}$$

over M^k. Assuming transversality, (43) is a smooth manifold with boundary or corners. $\pi_1 = ev_0 : \mathfrak{M}_{k+1} \to M$ induces a smooth map ev_0 from the manifold (43). We now put

$$\mathfrak{m}_k(P_1, \dots, P_k) = (ev_0)_* \left(\mathfrak{M}_{k+1} \ {}_{\pi_2}\times_{f_1,\dots,f_k} (P_1 \times \cdots \times P_k)\right). \tag{44}$$

(44) is a chain of dimension

$$\dim M + k - 2 + \sum(\dim M - d_i) - k \dim M = \dim M - \left(\sum d_i - (k-2)\right).$$

Using Poincaré duality, we identify a chain of degree $\dim M - d$ on M with a cochain of degree d. Then (44) induces a map of degree $k - 2$ on cochains. This is (after degree shift) a map of the required degree.

(42) implies (33) modulo transversality and sign. □

As we mentioned already, the main difficulty in proving Theorem 4 is the study of orientation and transversality. Transversality is discussed in detail in [33] §7.2 and orientation is discussed in detail in [33] Chapter 8. These points will be discussed also in §12 and §13 of this chapter.

We remark that if we replace 'manifold' by 'space with Kuranishi structure', we can still prove Theorem 4 with $\mathbb{Q}$ coefficient. See §10.

Remark 4. In this article we discuss A_∞ structure since it is the only case which is worked out and written down in detail, at the time of writing this article. However the argument of [33] can be generalized to show the analogy of Theorem 4 for various other (differentiable) operads or PROPs. In particular we can generalize it to the case of L_∞ structure, which appears in the loop space formulation of Floer theory ([29]) and involutive-bi-Lie infinity (or BV infinity) structure, which appears when we study symplectic manifolds with cylindrical end [16, 18] and also in the higher genus generalization of Lagrangian Floer theory. It appears also in string topology [12].

7 A_∞ homomorphism.

As we mentioned before, the main motivation for the author to study homotopy theory of A_∞ algebra etc. is to find a correct way to state the well-definedness of the algebraic system induced by the smooth correspondence by moduli space. Actually those algebraic systems are well-defined up to homotopy equivalence. To prove it is our main purpose. For this purpose, it is very important to define the notion of homotopy equivalence and derive its basic properties. In this section we define A_∞ homomorphism and describe a way to obtain it from smooth correspondence.

Let $(C, \mathfrak{m})$ and $(C', \mathfrak{m}')$ be A_∞ algebras. We consider a series of homomorphisms

$$\mathfrak{f}_k : B_k C[1] \to C'[1], \tag{45}$$

$k = 1, 2, \cdots$ of degree 0. We can extend it uniquely to a coalgebra homomorphism

$$\hat{\mathfrak{f}} : BC[1] \to BC[1]$$

whose $Hom(B_k C[1], C'[1])$ component is φ_k. Here $\hat{\varphi}$ is said to be a *coalgebra homomorphism* if $(\hat{\mathfrak{f}} \otimes \hat{\mathfrak{f}}) \circ \Delta = \Delta \circ \hat{\mathfrak{f}}$.

Definition 5. $\mathfrak{f}_k$ $(k = 1, 2, \cdots)$ *is said to be an* A_∞ *homomorphism if* $\hat{d}' \circ \hat{\mathfrak{f}} = \hat{\mathfrak{f}} \circ \hat{d}$. *An* A_∞ *homomorphism is said to be linear if* $\mathfrak{f}_k = 0$ *for* $k \neq 1$.

We can rewrite the condition $\hat{d}' \circ \hat{\mathfrak{f}} = \hat{\mathfrak{f}} \circ \hat{d}$ as follows.

$$\begin{aligned}\sum_l \sum_{k_1+\cdots+k_l=k} & \mathfrak{m}'_l(\mathfrak{f}_{k_1}(x_1, \cdots, x_{k_1}), \cdots, \mathfrak{f}_{k_l}(x_{k-k_l+1}, \cdots, x_k)) \\ &= \sum_{1 \le i < j \le k} (-1)^* \mathfrak{f}_{k-j+i}(x_1, \cdots, \mathfrak{m}_{j-i+1}(x_i, \cdots, x_j), \cdots, x_k)\end{aligned} \tag{46}$$

where

$$* = \deg' x_1 + \cdots + \deg' x_{i-1}.$$

We remark that (46) implies that $\mathfrak{f}_1 : (C, \mathfrak{m}_1) \to (C', \mathfrak{m}_1)$ is a chain map.

Definition 6. *Let* $\mathfrak{f} : C^{(1)} \to C^{(2)}$ *and* $\mathfrak{g} : C^{(2)} \to C^{(3)}$ *be* A_∞ *homomorphisms. The composition* $\mathfrak{f} \circ \mathfrak{g}$ *of them is defined by* $\widehat{\mathfrak{f} \circ \mathfrak{g}} = \hat{\mathfrak{f}} \circ \hat{\mathfrak{g}}$.

We next are going to discuss how we obtain A_∞ homomorphism by smooth correspondence. For this purpose we need to find spaces $\mathcal{F}_{k+1}$ whose relation to A_∞ homomorphism is the same as the relation of the spaces $\mathcal{M}_{k+1}$ to

A_∞ operations. We use the space $\mathcal{F}_{k+1}$ also to define the notion of A_∞ maps between two A_∞ spaces.

The following result, which is [33] Theorem 7.1.51, gives such spaces $\mathcal{F}_{k+1}$. (Note: in [33] we wrote $\mathcal{N}_{k+1}$ in place of $\mathcal{F}_{k+1}$.)

Theorem 5 (FOOO). *There exists a cell decomposition $\mathcal{F}_{k+1}$ of D^{k-1} such that the boundary $\partial\mathcal{F}_{k+1}$ is a union of the following two types of spaces, which intersect only at their boundaries.*

(1) *The spaces $\mathcal{M}_{l+1} \times \mathcal{F}_{k_1+1} \times \cdots \times \mathcal{F}_{k_l+1}$ where $\sum_{i=1}^{l} k_i = k$.*
(2) *The spaces $\mathcal{F}_{(k-j+i)+1} \times \mathcal{M}_{(j-i+1)+1}$ where $1 \le i < j \le k$.*

Sketch of the proof: We define $\mathcal{F}_{k+1}$ by modifying $\mathcal{M}_{k+1}$. Let $(\Sigma; z_0, \dots, z_k)$ be an element of $\mathcal{M}_{k+1}$ and $\Sigma = \bigcup_{a\in A} D^2_a$ be the decomposition into components. We assume that $z_0 \in D_{a_0}$. We say $a \le b$ if any path connecting D_a to D_{a_0} intersects D_b. The relation $\le$ defines a partial order on A.

We say a map $\rho : A \to [0,1]$ is a *time allocation* if $\rho(a) \le \rho(b)$ for all $a, b \in A$ with $a \le b$. Let $\mathcal{F}_{k+1}$ be the set of all isomorphism classes of $(\Sigma; z_0, \dots, z_k; \rho)$ where $(\Sigma; z_0, \dots, z_k) \in \mathcal{M}_{k+1}$ and ρ is a time allocation. We define a topology on it in an obvious way.

Lemma 4. *$\mathcal{F}_{k+1}$ is homeomorphic to D^{k-1}.*

We omit the proof. See [33] §7.1.5.

Let us consider the boundary of $\mathcal{F}_{k+1}$. Let $\left(\Sigma^{(i)}; z_0^{(i)}, \dots, z_k^{(i)}; \rho^{(i)}\right)$ be a sequence of elements of $\mathcal{F}_{k+1}$. There are several cases where it converges to a potential boundary point. These can be classified as follows.

(I) A component $D_a^{(i)}$ splits into two components in the limit $i \to \infty$.
(II) There exist a, b such that $D_a^{(i)} \cap D_b^{(i)} \neq \emptyset$ and such that $\lim_{i\to\infty} \rho^{(i)}(a) = \lim_{i\to\infty} \rho^{(i)}(b)$.
(III) $\lim_{i\to\infty} \rho^{(i)}(a_0) = 1$.
(IV) $\lim_{i\to\infty} \rho^{(i)}(a) = 0$, for some a.

We observe that the set of limits of type (I) is identified to the set of limits of type (II). Therefore they do not correspond to a boundary point of $\mathcal{F}_{k+1}$.

On the other hand, we can check that (III) corresponds to (1) of Theorem 5 and (IV) corresponds to (2) of Theorem 5. This implies the theorem. □

Let us explain how we use $\mathcal{F}_{k+1}$ to define an A_∞ map between two A_∞ spaces. We denote by

$$\circ_{\mathrm{mf}} : \mathcal{M}_{l+1} \times \mathcal{F}_{k_1+1} \times \cdots \times \mathcal{F}_{k_l+1} \to \mathcal{F}_{k_1+\cdots+k_l+1} \tag{47}$$

and

$$\circ_{\mathrm{fm},i} : \mathcal{F}_{(k-j+i)+1} \times \mathcal{M}_{(j-i+1)+1} \to \mathcal{F}_{k+1} \tag{48}$$

the inclusions obtained by Theorem 5 (1) and (2), respectively. They satisfy the following compatibility conditions (49), (50), (51), (52).

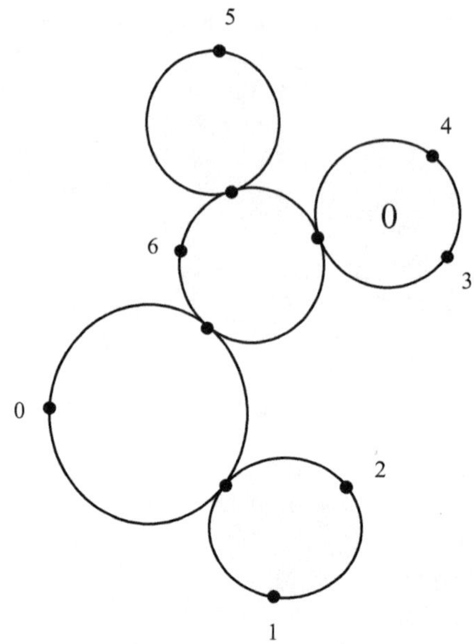

Fig. 6.

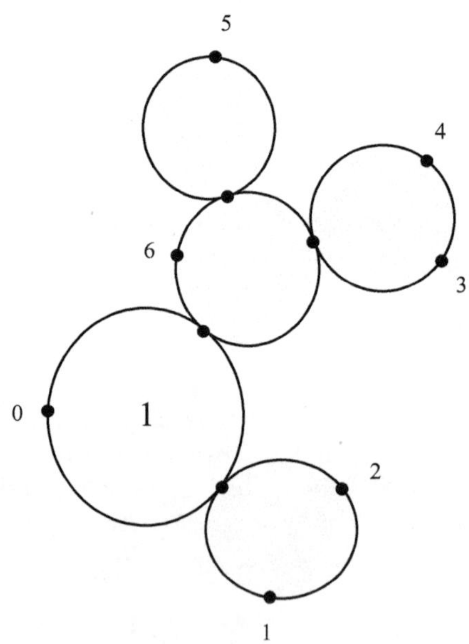

Fig. 7.

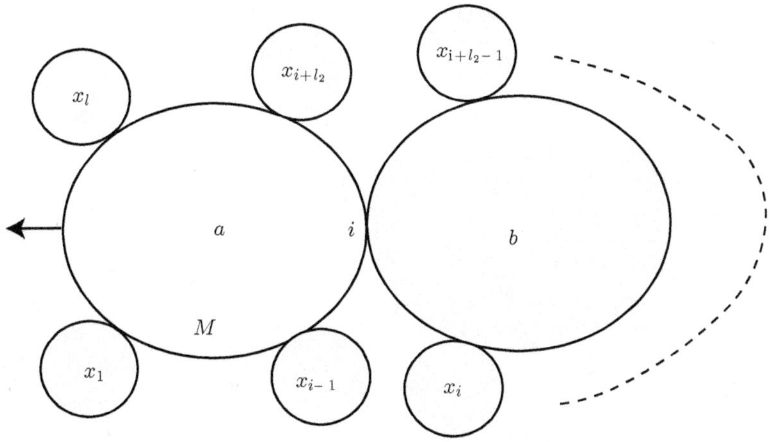

Fig. 8.

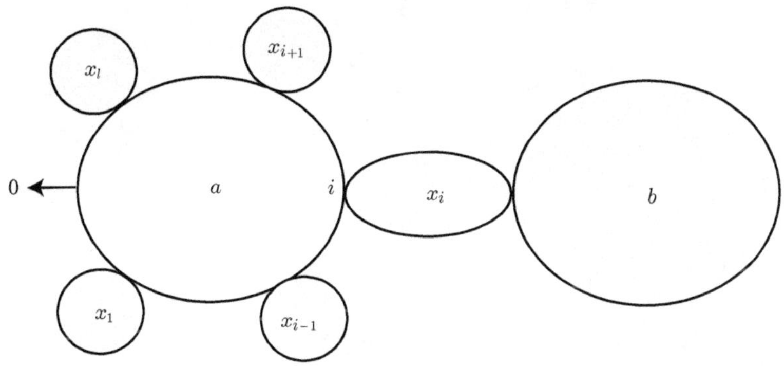

Fig. 9.

$$\begin{aligned}&(a \circ_i b) \circ_{\text{mf}} (x_1, \cdots, x_l)\\ &= a \circ_{\text{mf}} (x_1, \cdots, x_{i-1}, b \circ_{\text{mf}} (x_i, \cdots, x_{i+l_2-1}), \cdots, x_l)\end{aligned} \tag{49}$$

where $a \in \mathcal{M}_{l_1+1}$, $b \in \mathcal{M}_{l_2+1}$, $x_n \in \mathcal{F}_{k_n+1}$. See Figure 8.

$$\begin{aligned}&a \circ_{\text{mf}} (x_1, \cdots, x_{i-1}, x_i \circ_{\text{fm},j} b, x_{i+1}, \cdots, x_l)\\ &= (a \circ_{\text{mf}} (x_1, \cdots, x_l)) \circ_{\text{fm},j'} b,\end{aligned} \tag{50}$$

where $x_n \in \mathcal{F}_{k_n+1}$, $j' = j + k_1 + \cdots + k_{i-1}$. See Figure 9.

$$(x \circ_{\text{fm},i} a) \circ_{\text{fm},j+l_1-1} b = (x \circ_{\text{fm},j} b) \circ_{\text{fm},i} a \tag{51}$$

where $x \in \mathcal{F}_{k+1}$, $a \in \mathcal{M}_{l_1+1}$, $b \in \mathcal{M}_{l_2+1}$, $i < j$. See Figure 10.

$$(x \circ_{\text{fm},i} a) \circ_{\text{fm},i+j-1} b = x \circ_{\text{fm},i} (a \circ_j b). \tag{52}$$

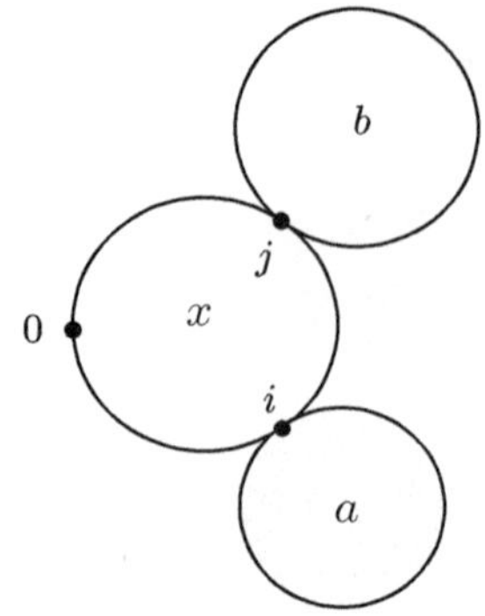

Fig. 10.

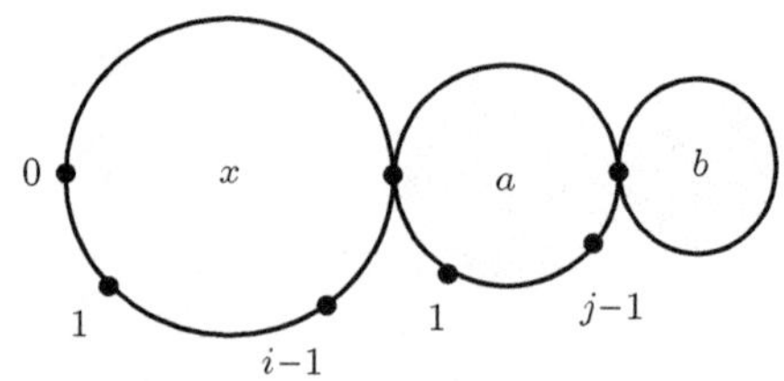

Fig. 11.

See Figure 11.

Let X, X' be A_∞ spaces with structure maps

$$\mathsf{m}_k : \mathcal{M}_{k+1} \times X^k \to X, \qquad \mathsf{m}'_k : \mathcal{M}_{k+1} \times X'^k \to X'$$

respectively.

Definition 7. *An A_∞ map from X to X' is a sequence of maps*

$$\mathsf{f}_k : \mathcal{F}_{k+1} \times X^k \to X'$$

with the following properties.

(1) *Let $k = k_1 + \cdots + k_l$, $x_{i,j} \in X$ for $1 \le i \le l, 1 \le j \le k_i$, and $a \in \mathcal{M}_{l+1}$, $c_i \in \mathcal{F}_{k_i+1}$. Then, we have*

$$\begin{aligned} &\mathsf{f}_k((a \circ_{\mathrm{mf}} (c_1, \cdots, c_l)), (x_{1,1}, \cdots, x_{l,k_l})) \\ &\quad = \mathsf{m}_l(a; \mathsf{f}_{k_1}(c_1; x_{1,1}, \cdots, x_{1,k_1}), \cdots, \mathsf{f}_{k_l}(c_l; x_{l,1}, \cdots, x_{l,k_l})) \end{aligned} \tag{53}$$

(2) *Let $0 \le i \le k$ and $a \in \mathcal{M}_{(j-i+1)+1}$, $c \in \mathcal{F}_{(k-j+i)+1}$, $x_1, \cdots, x_k \in X$. Then we have*

$$\begin{aligned} &\mathsf{f}_k(c \circ_{\mathrm{mf},i} a; x_1, \cdots, x_k) \\ &\quad = \mathsf{f}_{k-j+i}(c; x_1, \cdots, \mathsf{m}_{j-i+1}(a; x_i, \cdots, x_j), \cdots, x_k). \end{aligned} \tag{54}$$

We remark that in Stasheff [65] and in [49], the notion of A_∞ map is defined in the case X' is a monoid. (See [7].) Stasheff told the author that A_∞ map between A_∞ spaces is defined in his thesis. A construction of $\mathcal{F}_{k+1}$ can be found in [41]. It does not seem to be easy to see that the space in [41] is a cell. For the purpose of homotopy theory, this point is not important at all. However to apply it to the study of smooth correspondence, the fact that $\mathcal{F}_{k+1}$ is a smooth manifold is useful.

We mention the following which the author believes to be a classical result in homotopy theory.

Proposition 1. *An A_∞ map $X \to X'$ between two A_∞ spaces induces an A_∞ homomorphisms between A_∞ algebras in Theorem 3.*

The proof is similar to the proof of Theorem 3 and is omitted.

Now we use the spaces $\mathcal{F}_{k+1}$ to define the notion of a morphism between A_∞ correspondences. Let

$$\begin{array}{ccccc} & & \mathcal{M}_{k+1} & & \\ & & \uparrow{\scriptstyle \pi_0} & & \\ M^k & \xleftarrow[\pi_2]{} & \mathfrak{M}_{k+1} & \xrightarrow[\pi_1]{} & M \end{array} \tag{55}$$

and

$$\begin{array}{ccccc} & & \mathcal{M}_{k+1} & & \\ & & \uparrow{\scriptstyle \pi_0} & & \\ M'^k & \xleftarrow[\pi_2]{} & \mathfrak{M}'_{k+1} & \xrightarrow[\pi_1]{} & M' \end{array} \tag{56}$$

be A_∞ correspondences.

Definition 8. *A morphism between two A_∞ correspondences is the following diagram of smooth manifolds (with boundary or corners)*

$$\begin{array}{ccccc} & & \mathcal{F}_{k+1} & & \\ & & \uparrow{\scriptstyle \pi_0} & & \\ M^k & \xleftarrow[\pi_2]{} & \mathfrak{F}_{k+1} & \xrightarrow[\pi_1]{} & M' \end{array} \tag{57}$$

together with smooth maps

$$\circ_{\mathrm{mf}} : \mathfrak{M}'_{l+1} \ {}_{\pi_2}\times_{\pi_1,\cdots,\pi_1} (\mathfrak{F}_{k_1+1} \times \cdots \times \mathfrak{F}_{k_l+1}) \to \mathfrak{F}_{k_1+\cdots+k_l+1} \tag{58}$$

$$\circ_{\mathrm{fm},i} : \mathfrak{F}_{(k-j+i)+1} \ {}_{ev_i}\times_{ev_0} \mathfrak{M}_{(j-i+1)+1} \to \mathfrak{F}_{k+1} \tag{59}$$

with the following properties.

(1) *The following diagrams commute.*

$$\begin{array}{ccc}
\mathcal{M}_{l+1} \times \mathcal{F}_{k_1+1} \times \cdots \times \mathcal{F}_{k_l+1} & \xrightarrow[\circ_{\rm mf}]{} & \mathcal{F}_{k_1+\cdots+k_l+1} \\
{\scriptstyle \pi_0\times\cdots\times\pi_0}\uparrow & & {\scriptstyle \pi_0}\uparrow \\
\mathfrak{M}_{l+1} \ {}_{\pi_2}\times_{\pi_1,\cdots,\pi_1} (\mathfrak{F}_{k_1+1} \times \cdots \times \mathfrak{F}_{k_l+1}) & \xrightarrow[\circ_{\mathfrak{mf}}]{} & \mathfrak{F}_{k_1+\cdots+k_l+1}
\end{array} \tag{60}$$

$$\begin{array}{ccc}
\mathcal{F}_{(k-j+i)+1} \times \mathcal{M}_{(j-i+1)+1} & \xrightarrow[\circ_{{\rm fm},i}]{} & \mathcal{F}_{k+1} \\
{\scriptstyle \pi_0\times\pi_0}\uparrow & & {\scriptstyle \pi_0}\uparrow \\
\mathfrak{F}_{(k-j+i)+1} \ {}_{ev_i}\times_{ev_0} \mathfrak{M}_{(j-i+1)+1} & \xrightarrow[\circ_{\mathfrak{fm},i}]{} & \mathfrak{F}_{k+1}
\end{array} \tag{61}$$

(2) *Diagrams* (60), (61) *are cartesian.*
(3) *Formulae* (49), (50), (51), (52) *hold after replacing* $\circ_{{\rm fm},i}$, $\circ_{\rm mf}$ *by* $\circ_{\mathfrak{fm},i}$, $\circ_{\mathfrak{mf}}$.
(4) *The following diagrams commute. We put* $k = k_1 + \cdots + k_l$

$$\begin{array}{ccccc}
M^k & \xleftarrow[\pi_2]{} & \mathfrak{F}_{k+1} & \xrightarrow[\pi_1]{} & M' \\
\| & & \uparrow{\scriptstyle \circ_{\mathfrak{mf}}} & & \| \\
M^k & \longleftarrow & \mathfrak{M}'_{l+1} \ {}_{\pi_2}\times_{\pi_1,\cdots,\pi_1} (\mathfrak{F}_{k_1+1} \times \cdots \times \mathfrak{F}_{k_l+1}) & \xrightarrow[\pi_1\circ{\rm pr}_1]{} & M'
\end{array} \tag{62}$$

Here the first arrow in the second line is $(\pi_2 \cdots \pi_2)$ *on the factor* $\mathfrak{F}_{k_1+1} \times \cdots \times \mathfrak{F}_{k_l+1}$.

$$\begin{array}{ccccc}
M^k & \xleftarrow[\pi_2]{} & \mathfrak{F}_{k+1} & \xrightarrow[\pi_1]{} & M' \\
\| & & \uparrow{\scriptstyle \circ_{\mathfrak{fm},i}} & & \| \\
M^k & \longleftarrow & \mathfrak{F}_{(k-j+i)+1} \ {}_{ev_i}\times_{ev_0} \mathfrak{M}_{(j-i+1)+1} & \xrightarrow[\pi_1\circ{\rm pr}_1]{} & M'
\end{array} \tag{63}$$

Here the first arrow in the second line is

$$(ev_1\circ{\rm pr}_1, \cdots, ev_{i-1}\circ{\rm pr}_1, ev_1\circ{\rm pr}_2, \cdots, ev_{j-i+1}\circ{\rm pr}_2, ev_{i+1}\circ{\rm pr}_1, \cdots, ev_{k-j+i}\circ{\rm pr}_1).$$

(5) *The union of the images of* $\circ_{\mathfrak{mf}}$ *and of* $\circ_{\mathfrak{fm},i}$ *is the boundary of* $\mathfrak{F}_{k+1}$. *Those images intersect only at their boundaries.*
(6) *The identification of* (5) *preserves orientation.*

Now we have :

Theorem 6. *A morphism between two* A_∞ *correspondences induces an* A_∞ *homomorphism between* A_∞ *algebras in Theorem 4.*

The proof is similar to the proof of Theorem 6 and can be extracted from [33] §7.2.

Before closing this section, we remark that there is a map:

$$\mathrm{Comp}_{\mathbf{k},\mathbf{k}'} : \mathcal{F}_{k'+l+1} \times (\mathcal{F}_{k_1+1} \times \cdots \mathcal{F}_{k_l+1}) \to \mathcal{F}_{k+k'+1} \tag{64}$$

where $\mathbf{k} = (k_1, \cdots, k_l)$, $\mathbf{k}' = (k'_0, \cdots, k'_l)$, $k_1 + \cdots + k_l = k$ and $k'_0 + \cdots + k'_l = k'$. The map (64) describes the way how the A_∞ maps and A_∞ correspondences are composed. The map (64) is defined as follows.

Let $\mathcal{S} = (\Sigma; z_0, \cdots, z_{k'+l}, \rho) \in \mathcal{F}_{k'+l}$ and $\mathcal{S}_i = (\Sigma_i; z_0^{(i)}, \cdots, z_{k_i}^{(i)}, \rho_i) \in \mathcal{F}_{k_i+1}$ $(i = 1, \cdots, l)$.

We identify $z_{k'_0+\cdots+k'_{i-1}+i} \in \partial\Sigma$ with $z_0^{(i)} \in \partial\Sigma_i$ for each i and obtain Σ'.

Time allocations ρ and ρ_i induce a time allocation ρ' on Σ' as follows. When D_a is a component of Σ, we regard it as a component of Σ' and put $\rho'(a) = (1 + \rho(a))/2$. When D_a is a component of Σ_i, we regard it as a component of Σ' and put $\rho'(a) = \rho_i(a)/2$.

We define $(z'_0, z'_1, \cdots, z'_{k+k'})$ as

$$\begin{aligned}(z_0, z_1, \cdots, z_{k'_0}, z_1^{(1)}, \cdots, z_{k_1}^{(1)}, z_{k'_0+2}, \cdots \\ \cdots, z_{k'_1+\cdots+k'_{l-1}+l-1}, z_1^{(l)}, \cdots, z_{k_l}^{(l)}, z_{k'_1+\cdots+k'_{l-1}+l+1}, \cdots, z_{k+k'}).\end{aligned} \tag{65}$$

(See Figure 12.)

Then, we put

$$\mathrm{Comp}_{\mathbf{k},\mathbf{k}'}(\mathcal{S}, \mathcal{S}_1, \cdots, \mathcal{S}_l) = (\Sigma'; z'_0, z'_1, \cdots, z'_{k+k'}; \rho'). \tag{66}$$

Lemma 5. *$\mathcal{F}_{n+1}$ is a union of the images of $\mathrm{Comp}_{\mathbf{k},\mathbf{k}'}$ for various $\mathbf{k}, \mathbf{k}'$ with $n = k_1 + \cdots + k_l + k'_0 + \cdots + k'_l$. Those images intersect only at their boundaries.*

The proof is easy from the combinatorial description of the elements of $\mathcal{F}_{k+1}$ and is omitted. (See also [33] §4.6.3.)

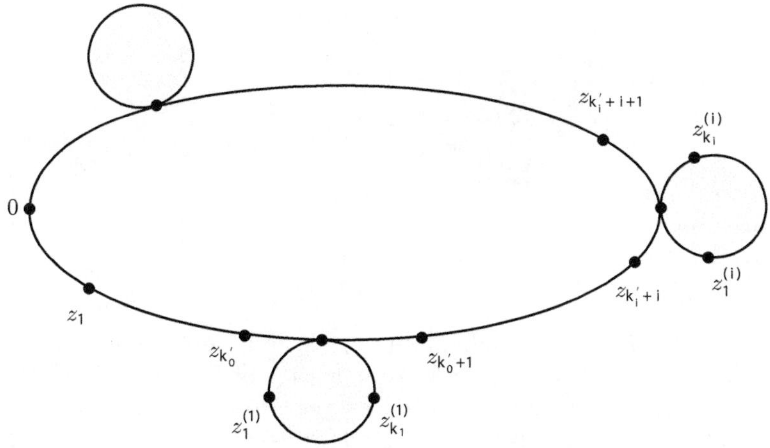

Fig. 12.

The map (64) is compatible with (47) and (48). We omit the precise description of the compatibility condition and leave it to interested readers. By using (64) and Lemma 5 we can define a composition of two A_∞ maps. We omit it since we do not use it. The map (64) is related to the composition of A_∞ correspondences. Since composition of A_∞ correspondences is more naturally defined in the case in which we include correspondence by spaces with Kuranishi structure, we will introduce it in §10.

8 A_∞ homotopy.

We begin this section with the algebraic side of the story. In this section we define the notion of homotopy between two A_∞ homomorphisms. As far as the author knows, there are two definitions of homotopy between two A_∞ homomorphisms in the literature. One of then can be found, for example, in [64, 46]. (In the case of graded commutative differential graded algebras, a similar formulation is due to Sullivan [69].) Another is an A_∞ version of the definition of homotopy which is written down in [35] in the case of differential graded algebras. (See [28] for the A_∞ version of this second definition.) When we were writing [33], we were trying to find a relation between these two definitions but were not able to find it in the literature. (The author believes that this equivalence was known to experts long ago.) So in [33] Chapter 4 we took an axiomatic approach and gave a definition which is equivalent to both of them (and hence proved the equivalence of those two definitions as a consequence). We will discuss this approach here.

Let $(C, \mathfrak{m})$ be an A_∞ algebra.

Definition 9. *An A_∞ algebra $(\mathfrak{C}, \mathfrak{m})$ together with the following diagram is said to be a model of* $[0,1] \times C$ *if the conditions* (1–4) *below hold.*

$$\begin{array}{ccccc} & & C & & \\ & & \big\downarrow{\scriptstyle \mathrm{Incl}} & & \\ C & \xleftarrow[\mathrm{Eval}_0]{} & \mathfrak{C} & \xrightarrow[\mathrm{Eval}_1]{} & C \end{array} \tag{67}$$

(1) Incl, Eval_0, Eval_1 *are linear A_∞ homomorphisms.*
(2) $\mathrm{Eval}_0 \circ \mathrm{Incl} = \mathrm{Eval}_1 \circ \mathrm{Incl} = \mathrm{identity}$.
(3) $\mathrm{Eval}_0 \oplus \mathrm{Eval}_1 : \mathfrak{C} \to C \oplus C$ *is surjective.*
(4) $\mathrm{Incl} : C \to \mathfrak{C}$ *is a chain homotopy equivalence.*

Example 4. Let M be a manifold and C be its de Rham complex regarded as an A_∞ algebra. Let $\mathfrak{C}$ be the de Rham complex of $\mathbb{R} \times M$. Let Incl, Eval_0, and Eval_1 be the linear maps induced by the projection $\mathbb{R} \times M \to M$, the inclusion $M \to \{0\} \times M \subset \mathbb{R} \times M$, and the inclusion $M \to \{1\} \times M \subset \mathbb{R} \times M$, respectively. It is easy to see that $\mathfrak{C}$ is a model of $[0,1] \times C$.

Proposition 2. *For any* $(C, \mathfrak{m})$ *there exists a model of* $[0,1] \times C$.

We can prove this either by an explicit construction or by using obstruction theory to show the existence. We omit the proof and refer to [33] §4.2.1.

The following is a kind of uniqueness theorem of the model of $[0,1] \times C$.

Theorem 7. *Let* $(C, \mathfrak{m})$, $(C', \mathfrak{m}')$ *be* A_∞ *algebras and* $\mathfrak{f} : C \to C'$ *be an* A_∞ *homomorphism. Let* $\mathfrak{C}$, $\mathfrak{C}'$ *be models of* $[0,1] \times C$, $[0,1] \times C'$, *respectively. Then there exists an* A_∞ *homomorphism* $\mathfrak{F} : \mathfrak{C} \to \mathfrak{C}'$ *such that*

$$\mathfrak{F} \circ \mathrm{Incl} = \mathrm{Incl} \circ \mathfrak{f}, \qquad \mathrm{Eval}_{s_0} \circ \mathfrak{F} = \mathfrak{f} \circ \mathrm{Eval}_{s_0}$$

where $s_0 = 0$ *or* 1.

This is [33] Theorem 4.2.34.

Definition 10. *Let* $(C, \mathfrak{m})$, $(C', \mathfrak{m}')$ *be* A_∞ *algebras and* $\mathfrak{f} : (C, \mathfrak{m}) \to (C', \mathfrak{m}')$ *and* $\mathfrak{g} : (C, \mathfrak{m}) \to (C', \mathfrak{m}')$ *be* A_∞ *homomorphisms. Let* $\mathfrak{C}'$ *be a model of* $[0,1] \times C'$.

We say that an A_∞ *homomorphism* $\mathfrak{H} : C \to \mathfrak{C}'$ *is a homotopy from* $\mathfrak{f}$ *to* $\mathfrak{g}$ *if*

$$\mathrm{Eval}_0 \circ \mathfrak{H} = \mathfrak{f}, \quad \mathrm{Eval}_1 \circ \mathfrak{H} = \mathfrak{g}.$$

We write $\mathfrak{f} \sim_{\mathfrak{C}'} \mathfrak{g}$ *if there exists a homotopy* $\mathfrak{H} : C \to \mathfrak{C}'$ *between them.*

Using Theorem 7, we can prove the following.

Proposition 3. (1) $\sim_{\mathfrak{C}'}$ *is independent of the choice of* $\mathfrak{C}'$. *(We write* $\sim$ *in place of* $\sim_{\mathfrak{C}'}$ *hereafter.)*
(2) $\sim$ *is an equivalence relation.*
(3) $\sim$ *is compatible with composition of* A_∞ *homomorphisms. Namely if* $\mathfrak{f} \sim \mathfrak{g}$ *then*

$$\mathfrak{f} \circ \mathfrak{h} \sim \mathfrak{g} \circ \mathfrak{h}, \qquad \mathfrak{h}' \circ \mathfrak{f} \sim \mathfrak{h}' \circ \mathfrak{g}.$$

Sketch of the proof: We assume $\mathfrak{f} \sim_{\mathfrak{C}'} \mathfrak{g}$. Let $\mathfrak{H} : C \to \mathfrak{C}'$ be the homotopy between them. Let $\mathfrak{C}''$ be another model of $[0,1] \times C'$. We apply Theorem 7 to the identity from C' to C' and obtain $\mathfrak{ID} : \mathfrak{C}' \to \mathfrak{C}''$. It is easy to see that $\mathfrak{ID} \circ \mathfrak{H}$ is a homotopy from $\mathfrak{f}$ to $\mathfrak{g}$. We have $\mathfrak{f} \sim_{\mathfrak{C}''} \mathfrak{g}$. (1) follows.

Let $\mathfrak{f} \sim \mathfrak{g}$ and $\mathfrak{g} \sim \mathfrak{h}$. Let $\mathfrak{H}^{(1)} : C \to \mathfrak{C}_1$ and $\mathfrak{H}^{(2)} : C \to \mathfrak{C}_2$ be homotopies from $\mathfrak{f}$ to $\mathfrak{g}$ and from $\mathfrak{g}$ to $\mathfrak{h}$, respectively. We put

$$\mathfrak{C}' = \{(y_1, y_2) \in \mathfrak{C}_1 \oplus \mathfrak{C}_2 \mid \mathrm{Eval}_1 y_1 = \mathrm{Eval}_0 y_2\}.$$

It is easy to see that $\mathfrak{C}'$ is a model of $[0,1] \times C'$. We define $\mathfrak{H} : C \to \mathfrak{C}'$ by

$$\mathfrak{H}_k(x_1, \ldots, x_k) = (\mathfrak{H}_k^{(1)}(x_1, \ldots, x_k), \mathfrak{H}_k^{(2)}(x_1, \ldots, x_k)).$$

It is easy to see that $\mathfrak{H}$ is a homotopy from $\mathfrak{f}$ to $\mathfrak{h}$. We have thus proved that $\sim$ is transitive. The other part of the proof of Proposition 3 is similar and is omitted. (See [33] §4.3.) □

Using Proposition 3 we can define the notion of two A_∞ algebras being homotopy equivalent to each other and also the notion of A_∞ homomorphism being a homotopy equivalence, in an obvious way.

We can then prove the following two basic results of homotopy theory of A_∞ algebras.

Theorem 8. *If* $\mathfrak{f} : (C, \mathfrak{m}) \to (C', \mathfrak{m})$ *is an* A_∞ *homomorphism such that* $\mathfrak{f}_1 : (C, \mathfrak{m}_1) \to (C', \mathfrak{m})$ *is a chain homotopy equivalence, then* $\mathfrak{f}$ *is a homotopy equivalence. Namely there exists* $\mathfrak{g} : (C', \mathfrak{m}) \to (C, \mathfrak{m})$ *such that* $\mathfrak{g} \circ \mathfrak{f}$ *and* $\mathfrak{f} \circ \mathfrak{g}$ *are homotopic to the identity.*

This theorem seems to be known to experts. The proof based on our definition of homotopy is in [33] §4.5.

Theorem 9. *Let* $(C, \mathfrak{m})$ *be an* A_∞ *algebra. Let* $C' \subset C$ *be a subchain complex of* $(C, \mathfrak{m}_1)$ *such that the inclusion* $(C', \mathfrak{m}_1) \to (C, \mathfrak{m}_1)$ *is a chain homotopy equivalence.*

Then there exists a sequence of operators $\mathfrak{m}'_k$ *for* $k \geq 2$ *such that* $\mathfrak{m}'_k$ *and* $\mathfrak{m}'_1 = \mathfrak{m}_1$ *define a structure of* A_∞ *algebra on* C'.

Moreover there exists $\mathfrak{f}_k : B_k C'[1] \to C[1]$ *for* $k \geq 2$ *such that* $\mathfrak{f}_k$ *together with* $\mathfrak{f}_1 = \text{inclusion}$ *define a homotopy equivalence* $C' \to C$.

We put $H(C) = \operatorname{Ker} \mathfrak{m}_1 / \operatorname{Im} \mathfrak{m}_1$.

Corollary 2. *If* R *is a field (or* $H(C)$ *is a free* R*-module) then there exists a structure of* A_∞ *algebra on* $H(C)$ *for which* $\mathfrak{m}_1 = 0$ *and which is homotopy equivalent to* $(C, \mathfrak{m})$.

Theorem 9 and Corollary 2 have a long history which starts with [42]. Theorem 9 and Corollary 2 are proved in [33] §5.4.4. (The proof in [33] is similar to one in [45].)

We also refer to [61] and references therein for more results on homological algebra of A_∞ structures.

We now discuss relations of the algebraic machinery described above to geometry. We first remark that we can define the notion of homotopy for two A_∞ maps between A_∞ spaces. Moreover we can prove that, if two A_∞ maps are homotopic to each other, then the induced A_∞ homomorphisms are also homotopic. We omit the proof of this since we do not use it.

Let us consider the case of A_∞ correspondences. We consider two A_∞ correspondences (55), (56) and two morphisms

$$\begin{array}{ccccc} & & \mathcal{F}_{k+1} & & \\ & & \uparrow{\scriptstyle \pi_0} & & \\ M^k & \xleftarrow[\pi_2]{} & \mathfrak{F}_{k+1} & \xrightarrow[\pi_1]{} & M' \end{array} \tag{68}$$

$$\begin{array}{ccccc} & & \mathcal{F}_{k+1} & & \\ & & \uparrow{\scriptstyle \pi_0} & & \\ M^k & \xleftarrow[\pi_2]{} & \mathfrak{F}'_{k+1} & \xrightarrow[\pi_1]{} & M' \end{array} \tag{69}$$

between them.

Definition 11. *A homotopy from* (68) *to* (69) *is a sequence of diagrams*

$$\begin{array}{ccccc} & & \mathcal{F}_{k+1} & & \\ & & \uparrow{\scriptstyle \pi_0} & & \\ M^k & \xleftarrow[\pi_2]{} & \mathfrak{H}_{k+1} & \xrightarrow[\pi_1]{} & M' \times [0,1] \end{array} \tag{70}$$

together with the smooth maps

$$\begin{aligned} \circ_{\mathfrak{mh}} : \mathfrak{M}_{l+1}\ {}_{\pi_2}\times_{\pi_1,\cdots,\pi_1} (\mathfrak{H}_{k_1+1} \times_{[0,1]} \cdots \\ \cdots \times_{[0,1]} \mathfrak{H}_{k_l+1}) \to \mathfrak{H}_{k_1+\cdots+k_l+1} \end{aligned} \tag{71}$$

$$\circ_{\mathfrak{hm},i} : \mathfrak{H}_{(k-j+i)+1}\ {}_{ev_i}\times_{ev_0} \mathfrak{M}_{(j-i+1)+1} \to \mathfrak{H}_{k+1} \tag{72}$$

with the following properties. (We remark that in (71) *we take fiber product over* $[0,1]$ *using the* $[0,1]$ *factor of* π_1*.)*

(1) $\pi_1^{-1}(M' \times \{0\}) = \mathfrak{F}_{k+1}$, $\pi_1^{-1}(M' \times \{1\}) = \mathfrak{F}'_{k+1}$. *The restriction of the maps* (71), (72) *are the maps* (58), (59), *respectively.*

(2) *The following diagrams commute and are cartesian.*

$$\begin{array}{ccc} \mathcal{M}_{l+1} \times \mathcal{F}_{k_1+1} \times \cdots \mathcal{F}_{k_l+1} & \xrightarrow[\circ_{\mathrm{mf}}]{} & \mathcal{F}_{k_1+\cdots+k_l+1} \\ {\scriptstyle \pi_0\times\cdots\times\pi_0}\uparrow & & {\scriptstyle \pi_0}\uparrow \\ \mathfrak{M}'_{l+1}\ {}_{\pi_2}\times (\mathfrak{H}_{k_1+1} \times_{[0,1]} \cdots \times_{[0,1]} \mathfrak{H}_{k_l+1}) & \xrightarrow[\circ_{\mathfrak{mh}}]{} & \mathfrak{H}_{k_1+\cdots+k_l+1} \end{array} \tag{73}$$

$$\begin{array}{ccc} \mathcal{F}_{(k-j+i)+1} \times \mathcal{M}_{(j-i+1)+1} & \xrightarrow[\circ_{\mathfrak{fm},i}]{} & \mathcal{F}_{k+1} \\ {\scriptstyle \pi_0\times\pi_0}\uparrow & & {\scriptstyle \pi_0}\uparrow \\ \mathfrak{H}_{(k-j+i)+1}\ {}_{ev_i}\times_{ev_0} \mathfrak{M}_{(j-i+1)+1} & \xrightarrow[\circ_{\mathfrak{hm},i}]{} & \mathfrak{H}_{k+1} \end{array} \tag{74}$$

(3) *Formulae* (49), (50), (51), (52) *holds after replacing* $\circ_{\mathrm{fm},i}$, $\circ_{\mathrm{mf}}$ *by* $\circ_{\mathfrak{hm},i}$, $\circ_{\mathfrak{mh}}$.

(4) *The following diagrams commute. We put* $k = k_1 + \cdots + k_l$

$$\begin{array}{ccccc} M^k & \xleftarrow[\pi_2]{} & \mathfrak{H}_{k+1} & \xrightarrow[\pi_1]{} & M' \times [0,1] \\ \| & & \uparrow{\scriptstyle \circ_{\mathrm{mf}}} & & \| \\ M^k & \longleftarrow & \mathfrak{M}'_{l+1}\ {}_{\pi_2}\times (\mathfrak{H}_{k_1+1} \times_{[0,1]} \cdots \times_{[0,1]} \mathfrak{H}_{k_l+1}) & \longrightarrow & M' \times [0,1] \end{array} \tag{75}$$

$$\begin{array}{ccccc} M^k & \xleftarrow[\pi_2]{} & \mathfrak{H}_{k+1} & \xrightarrow[\pi_1]{} & M' \times [0,1] \\ \| & & \uparrow{\scriptstyle \circ_{\mathfrak{hm},i}} & & \| \\ M^k & \longleftarrow & \mathfrak{H}_{(k-j+i)+1} \ {}_{ev_i}\times_{ev_0} \mathfrak{M}_{(j-i+1)+1} & \xrightarrow[\pi_1 \circ \mathrm{pr}_1]{} & M' \times [0,1] \end{array} \tag{76}$$

(5) *The union of the images of* $\circ_{\mathfrak{mh}}$ *and of* $\circ_{\mathfrak{hm},i}$ *is the boundary of* $\mathfrak{H}_{k+1}$. *Those images intersect only at their boundaries.*
(6) *The identification* (5) *preserves orientations.*

Now we have

Theorem 10. *A homotopy* $\mathfrak{H}$ *between two* A_∞ *correspondences* $\mathfrak{F}$ *and* $\mathfrak{F}'$ *induces a homotopy between the two* A_∞ *homomorphisms induced by Theorem 6.*

The proof can be extracted from [33] §7.2.10–13.

9 Filtered A_∞ algebra and Filtered A_∞ correspondence.

So far we have developed a machinery which works at least to construct A_∞ algebras defined by the intersection theory on a manifold. (See Example 2.) To apply the machinery to the case when we use moduli spaces (of pseudo-holomorphic curves for example), we need some generalizations. One generalization we need is related to the fact that the structure constants of the algebraic system we will construct are not numbers but a kind of formal power series. The reason why we need to consider formal power series lies in the following fact. In the case of Gromov-Witten theory, for example, the moduli space $\mathfrak{M}_{k+1}$ which appears in the definition of correspondence is not compact and does not have a good compactification either. We need to impose some energy bounds to prove the compactness of the moduli space of pseudo-holomorphic curves. Filtered A_∞ algebras (and its cousins) will be used in order to take care of this problem.

Definition 12. *A proper submonoid* G *is a submonoid of* $\mathbb{R}_{\geq 0} \times 2\mathbb{Z}$ *with the following properties.*

(1) *If* $(0,\mu) \in G$ *then* $\mu = 0$.
(2) *For each* $E_0 \in \mathbb{R}_{\geq 0}$ *the set* $\{(E,\mu) \in G \mid E \leq E_0\}$ *is finite.*

This definition is closely related to Gromov compactness in the theory of pseudo-holomorphic curve.

Let $E : G \to \mathbb{R}_{\geq 0}$, $\mu : G \to \mathbb{Z}$ be the projections to the first and second factors. We define the *Novikov ring* Λ_G^R ([52]) associated with G as the set of all such (formal) series

$$\sum_{\beta \in G} a_i T^{E(\beta)} e^{\mu(\beta)/2} \tag{77}$$

where T and e are formal generators of degree 0 and 2 respectively, and $a_i \in R$ (a commutative ring with unit, which we fixed at the beginning). It is easy to see that, by Definition 12 (2), we can define sum and product between two elements of the form (77), and Λ_G^R becomes a ring.

The ring Λ_G^R is contained in the *universal Novikov ring* $\Lambda_{0,nov}^R$ which is the set of all the (formal) series

$$\sum_{\beta \in G} a_i T^{E_i} e^{\mu_i} \tag{78}$$

where $E_i \in \mathbb{R}_{\geq 0}$ and $\mu_i \in \mathbb{Z}$ are sequences such that $\lim_{i\to\infty} E_i = \infty$. (The fact that Novikov ring is a natural coefficient ring of Floer homology was first observed by Floer. It was used by [38] and [58].)

Remark 5. We consider a monoid G together with a partial order $\leq$ such that the following holds.

(1) $g \leq g'$, $h \leq h'$ implies $g \cdot h \leq g' \cdot h'$.
(2) For any g_0 there exists only a finite number of $g \in G$ with $g \leq g_0$.
(3) We have $\beta_0 \leq g$ for any $g \in G$. Here $\beta_0 = (0,0)$.

We then take the completion of its group ring

$$\hat{R}(G) = \left\{ \sum_{g \in G} a_g[g] \;\middle|\; a_g \in R, \quad \text{infinite sum} \right\}.$$

By (2) we can define a products of two elements of $\hat{R}(G)$ in an obvious way. $\hat{R}(G)$ then is a ring.

In case of our $G \subset \mathbb{R} \times 2\mathbb{Z}$, we have $\hat{R}(G) = \Lambda_G^R$.

In the case when noncommutative G appears (such as the case we consider Lagrangian submanifolds with noncommutative fundamental group in Floer theory) to use appropriate $\hat{R}(G)$ with noncommutative G may give more information.

The reason why we use universal Novikov ring $\Lambda_{0,nov}^R$ here is that, in our application, the monoid G depends on various choices (such as almost complex structure in the case Gromov-Witten or Floer theory). So to state the independence of the structure of the choices, it is more convenient to use $\Lambda_{0,nov}^R$ which contains all of Λ_G^R.

Let $\overline{C}$ be a free graded R module. We put $C = \overline{C} \widehat{\otimes}_R \Lambda_{0,nov}^R$. Here $\widehat{\otimes}_R$ is the completion of the algebraic tensor product $\otimes$ with respect to the non-Archimedean norm defined by the ideal generated by T.

Definition 13. *A structure of G-gapped filtered A_∞ algebra on C is defined by a family of the operations*

$$\mathfrak{m}_{k,\beta} : B_k\overline{C}[1] \to \overline{C}[1]$$

of degree $1 - \mu(\beta)$, *for* $\beta \in G$ *and* $k = 0, 1, \cdots$, *satisfying the following conditions.*

(1) $\mathfrak{m}_{k,\beta_0} = 0$ *if* $\beta_0 = (0,0)$ *and* $k = 0$.
(2) *We define*

$$\mathfrak{m}_k = \sum_{\beta \in G} T^{E(\beta)} e^{\mu(\beta)/2} \mathfrak{m}_{k,\beta} : B_k C[1] \to C[1].$$

Then it satisfies (33).

Definition 14. *Let* $(C, \mathfrak{m})$, $(C', \mathfrak{m}')$ *be G-gapped filtered* A_∞ *algebras. A G-gapped filtered* A_∞ *homomorphism* $\mathfrak{f} : C \to C'$ *is defined by a family of R module homomorphisms*

$$\mathfrak{f}_{k,\beta} : B_k \overline{C}[1] \to \overline{C}'[1]$$

of degree $-\mu(\beta)$, *for* $\beta \in G$ *and* $k = 0, 1, \cdots$, *satisfying the following conditions.*

(1) $\mathfrak{m}_{k,\beta_0} = 0$ *if* $\beta_0 = (0,0)$ *and* $k = 0$.
(2) *We define*

$$\mathfrak{f}_k = \sum_{\beta} T^{E(\beta)} e^{\mu(\beta)/2} \mathfrak{f}_{k,\beta} : B_k C[1] \to C'[1].$$

Then it satisfies (46).

Remark 6. We remark that in the case of *filtered* A_∞ algebra, the maps $\mathfrak{m}_0$ or $\mathfrak{f}_0$ may be nonzero. Filtered A_∞ algebra (resp. homomorphism) is said to be strict if $\mathfrak{m}_0 = 0$ (resp. $\mathfrak{f}_0 = 0$).

We can develop homotopy theory of filtered A_∞ algebra in the same way as that of A_∞ algebra. Namely Propositions 2, 3 and Theorems 7, 8, 9 hold without change. (See [33] for their proofs.) (We can do it for each fixed G.)

We next explain how to obtain a filtered A_∞ algebra and filtered A_∞ homomorphisms by smooth correspondence. We first define $\mathcal{M}_{k+1,\beta}$ for each (k, β) such that $k \geq 2$ or $\beta \neq \beta_0$. We consider $\Sigma = \bigcup_{a \in A} D_a^2$ and $z_i \in \partial \Sigma$ satisfying the same condition as the definition of $\mathcal{M}_{k+1}$ *except the definition of stability.* Let $\beta(\cdot) : A \to G$ be a map with $\beta = \sum \beta(a)$. We define the following stability condition for $(\Sigma; z_1, \cdots, z_k; \beta(\cdot))$.

Definition 15. *For each component* D_a *either one of the following holds.*
(1) D_a *contains at least three marked or singular points.* (2) $\beta(a) \neq (0,0)$.

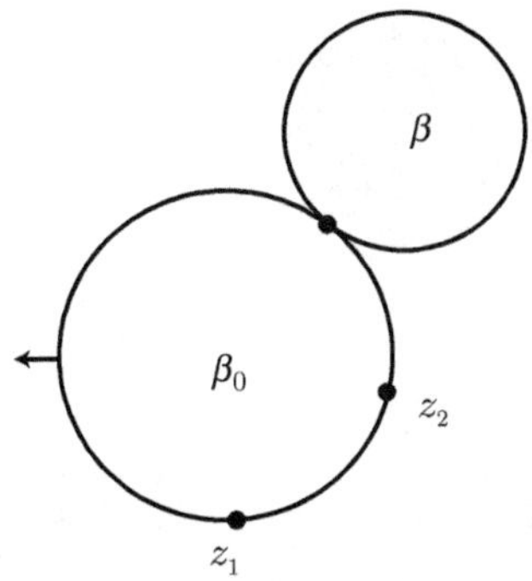

Fig. 13.

$\mathcal{M}_{k+1,\beta}$ is the set of all the isomorphism classes of such $(\Sigma; z_0, \cdots, z_k; \beta(\cdot))$. This definition is closely related to the notion of stable map [44].

We can define a topology on $\mathcal{M}_{k+1,\beta}$ in an obvious way. Namely, in a neighborhood of $(\Sigma; z_0, \cdots, z_k; \beta(\cdot))$ there are $(\Sigma'; z'_0, \cdots, z'_k; \beta'(\cdot))$ where Σ' is obtained by resolving a singularity $p \in \Sigma$. If a component D_a of Σ' is obtained by gluing two components D_{a_1} and D_{a_2} of Σ at p, then we put $\beta'(a) = \beta(a_1) + \beta(a_2)$.

By a similar gluing as §4, we obtain a continuous map :

$$\circ_i : \mathcal{M}_{k+1,\beta_1} \times \mathcal{M}_{l+1,\beta_2} \to \mathcal{M}_{k+l,\beta_1+\beta_2}.$$

We remark here that the topology on $\mathcal{M}_{k+1,\beta}$ is rather pathological. Namely it is *not* Hausdorff. Let us exhibit it by an example. We consider $\overset{\circ}{\mathcal{M}}_{2+1,\beta}$ (which is the set of elements of $\mathcal{M}_{2+1,\beta}$ with no singularity), for $\beta \neq (0,0)$. Actually $\overset{\circ}{\mathcal{M}}_{2+1,\beta} \cong \overset{\circ}{\mathcal{M}}_{2+1}$ is a point. On the other hand

$$\mathcal{M}_{0+1,\beta} \circ_3 \mathcal{M}_{3+1,\beta_0} \subset \mathcal{M}_{2+1,\beta}.$$

(Here $\beta_0 = (0,0)$.) The left hand side is diffeomorphic to $\mathcal{M}_{3+1}$ and is an interval $[0,1]$.

Thus *any* neighborhood of *any* point of $[0,1] \cong \mathcal{M}_{0+1,\beta} \circ_3 \mathcal{M}_{4+1,\beta_0}$ contains the point $\overset{\circ}{\mathcal{M}}_{2+1,\beta}$. Thus $\mathcal{M}_{2+1,\beta}$ is not Hausdorff.

Remark 7. An appropriate language to describe this situation is (Artin) stack. We do not use the notion of stack later in this article. So the reader can skip this remark safely if he wants.

Let us consider the example we discussed above. We consider an element $(\Sigma; z_0, z_1, z_2, z_3; \beta(\cdot)) \in \mathcal{M}_{0,\beta} \circ_3 \mathcal{M}_{3+1,\beta_0}$. The group of its automorphisms is the group $\mathrm{Aut}(D^2, \{1\})$ which consists of the biholomorphic maps $u : D^2 \to D^2$ with $u(1) = 1 \in \partial D^2$. This group is isomorphic to

$$\left\{ \begin{pmatrix} a & b \\ 0 & a^{-1} \end{pmatrix} \,\middle|\, a \in \mathbb{R}_+,\ b \in \mathbb{R} \right\} \subset PSL(2;\mathbb{R}) \cong \mathrm{Aut}(D^2).$$

The (infinitesimal) neighborhood of the element $(\Sigma; z_0, z_1, z_2, z_3; \beta(\cdot))$ in $\mathcal{M}_{2+1,\beta}$ is, by definition, a quotient of

$$\mathcal{M}_{0+1,\beta} \circ_3 \mathcal{M}_{3+1,\beta_0} \ \times\ [0,\epsilon)$$

by the action of the group $\mathrm{Aut}(D^2, \{1\})$.

Note the parameter a above acts on $[0,\epsilon)$ factor by $t \mapsto at$. Also the element $\partial/\partial b$ in the Lie algebra of $\mathrm{Aut}(D^2, \{1\})$ moves the position of third marked point z_3 of the $\mathcal{M}_{3+1,\beta_0}$ factor, when $[0,\epsilon)$ factor is positive. Thus we have

$$\frac{\mathcal{M}_{0+1,\beta} \circ_3 \mathcal{M}_{3+1,\beta_0} \ \times\ [0,\epsilon)}{\mathrm{Aut}(D^2,\{1\})} = (\mathcal{M}_{0+1,\beta} \circ_3 \mathcal{M}_{3+1,\beta_0} \times \{0\}) \cup \{\text{one point}\}.$$

Here $\{\text{one point}\}$ in the right hand side is the quotient of $\mathcal{M}_{0+1,\beta} \circ_3 \mathcal{M}_{3+1,\beta_0} \times (0,\epsilon)$ by $\mathrm{Aut}(D^2, \{1\})$ action and is identified with $\overset{\circ}{\mathcal{M}}_{2+1,\beta}$.

Thus the neighborhood of $(\Sigma; z_0, z_1, z_2, z_3; \beta(\cdot))$ in $\mathcal{M}_{3+1,\beta_0}$ is as we mentioned above.

The fact that $\mathcal{M}_{3+1,\beta_0}$ is not Hausdorff is a consequence of the noncompactness of the group $\mathrm{Aut}(D^2, \{1\})$.

Now we define the notion of G-gapped filtered A_∞ correspondence. Actually the definition is almost the same as Definition 4. Suppose we have a family of the following commutative diagrams.

$$\begin{array}{ccccc} & & \mathcal{M}_{k+1,\beta} & & \\ & & \uparrow{\scriptstyle \pi_0} & & \\ M^k & \xleftarrow[\pi_2]{} & \mathfrak{M}_{k+1}(\beta) & \xrightarrow[\pi_1]{} & M \end{array} \tag{79}$$

such that $\mathfrak{M}_{k+1}(\beta)$ is a smooth (Hausdorff) manifold with boundary or corners and that

$$\dim \mathfrak{M}_{k+1}(\beta) = \dim M + \mu(\beta) + k - 2. \tag{80}$$

We assume also that we have a family of smooth maps

$$\circ_{\mathfrak{m},i} : \mathfrak{M}_{k+1}(\beta_1) \ {}_{ev_i}\times_{ev_0} \mathfrak{M}_{l+1}(\beta_2) \to \mathfrak{M}_{k+l}(\beta_1 + \beta_2). \tag{81}$$

To state one of the conditions (cartesian axiom (2) below) we need some notations. Let us consider the space $\mathcal{M}_{k+l,\beta}$. It is not Hausdorff as mentioned before. It is decomposed into union of *smooth manifolds* according to its combinatorial structure. Namely if we collect all of the elements $(\Sigma; z_0, \cdots, z_k; \beta(\cdot))$ which is homeomorphic to a given element of $\mathcal{M}_{k+l,\beta}$ then it is a smooth manifold. (This manifold is actually a ball.) We write this stratum $\mathcal{M}_{k+l,\beta}(\mathbf{S})$, where $\mathbf{S}$ stands for a homeomorphism type of $(\Sigma; z_0, \cdots, z_k; \beta(\cdot))$. (We remark that $\mathcal{M}_{k+l,\beta}(\mathbf{S})$ is Hausdorff.)

Definition 16. *A system of objects as in* (79), (80), (81) *is said to be a G-gapped filtered A_∞ correspondence if the following holds.*

(1) *The following diagram commutes.*

$$\begin{array}{ccc} \mathcal{M}_{k+1,\beta_1} \times \mathcal{M}_{l+1,\beta_2} & \xrightarrow[\circ_i]{} & \mathcal{M}_{k+l,\beta_1+\beta_2} \\ {\scriptstyle \pi_0\times\pi_0}\uparrow & & {\scriptstyle \pi_0}\uparrow \\ \mathfrak{M}_{k+1}(\beta_1)\ {}_{ev_0}\times_{ev_i} \mathfrak{M}_{l+1}(\beta_2) & \xrightarrow[\circ_{\mathfrak{m},i}]{} & \mathfrak{M}_{k+l}(\beta_1+\beta_2) \end{array} \tag{82}$$

(2) *The inverse image*

$$\pi_0^{-1}(\mathcal{M}_{k+l,\beta}(\mathbf{S})) \subset \mathfrak{M}_{k+l}(\beta) \tag{83}$$

of each such stratum $\mathcal{M}_{k+l,\beta}(\mathbf{S})$ of $\mathcal{M}_{k+l,\beta}$ is a smooth submanifold of $\mathfrak{M}_{k+l}(\beta)$. Its codimension is the number of singular points of $\mathbf{S}$. *We denote* (83) *by $\mathfrak{M}_{k+l}(\beta;\mathbf{S})$. Restriction of ev_0 to each such stratum $\mathfrak{M}_{k+l}(\beta;\mathbf{S})$ is a smooth map $\mathfrak{M}_{k+l}(\beta;\mathbf{S}) \to \mathcal{M}_{k+l,\beta}(\mathbf{S})$. Diagram* (82) *is a cartesian diagram as a diagram of sets.*
(3) *Formulae* (39) *and* (40) *hold.*
(4) *The following diagram commutes.*

$$\begin{array}{ccccc} M^{k+l} & \xleftarrow[\pi_2]{} & \mathfrak{M}_{k+l}(\beta_1+\beta_2) & \xrightarrow[\pi_1]{} & M \\ \| & & \uparrow{\scriptstyle \circ_{\mathfrak{m},i}} & & \| \\ M^{k+l} & \longleftarrow & \mathfrak{M}_{k+1}(\beta_1)\ {}_{ev_i}\times_{ev_0} \mathfrak{M}_{l+1}(\beta_2) & \xrightarrow[\pi_1\circ \mathrm{pr}_2]{} & M \end{array} \tag{84}$$

(5) *For each n the boundary of $\mathfrak{M}_{n+1}(\beta)$ is a union of*

$$\circ_{\mathfrak{m},i}(\mathfrak{M}_{k+1}(\beta_1)\ {}_{ev_i}\times_{ev_0} \mathfrak{M}_{l+1}(\beta_2))$$

for various k,l,i,β_1,β_2 with $k+l=n$, $i=1,\cdots,l$, $\beta_1+\beta_2=\beta$. They intersect each other only at their boundaries.
(6) *The identification in* (5) *preserves orientation.*

Note the axiom (2) above is more complicated than the corresponding axiom in Definition 4. This is because $\mathcal{M}_{k+1,\beta}$ is not Hausdorff and hence we can not say the Diagram (82) being cartesian in the category of topological spaces. One might say that the diagram (82) is cartesian in the sense of stacks. (The author wants to avoid using the notion of stack here since he is not familiar with it.)

We can define the notion of (G-gapped filtered) morphism between two G-gapped filtered A_∞ correspondences in the same way as Definition 8. The homotopy between two morphisms are defined in the same way as Definition 11. We then have :

Theorem 11. *G-gapped filtered A_∞ correspondence on M induces a structure of G-gapped filtered A_∞ algebra on a cochain complex representing the cohomology group of M.*

A morphism between G-gapped filtered A_∞ correspondences induces a G-gapped filtered A_∞ homomorphism. A homotopy between two morphisms induces a homotopy between G-gapped filtered A_∞ homomorphisms.

The proof of this theorem can be extracted from [33] §7.2.

10 Kuranishi correspondence.

To study correspondences, manifolds form an overly restrictive category of spaces, since we cannot take fiber products in general, for example. As a consequence, composition of correspondences is defined only under some transversality assumptions. We can use the notion of Kuranishi structure to resolve this problem. Moreover, Kuranishi structure is a general framework to handle various transversality problems and to study moduli spaces arising in differential geometrys in a uniform way. In this section we define the notion of A_∞ Kuranishi correspondence and use it furthermore to generalize Theorem 11.

We first review briefly the notion of Kuranishi structure. (See [30] Chapter 1 or [33] Appendix 1 for more details.) The notion of Kuranishi structure is simple and elementary. The author believes that the main obstacle to understanding it is somewhat psychological. Actually the definition of Kuranishi structure is very similar to the definition of manifold.

We consider a space Z which is Hausdorff and compact.

Definition 17. *A Kuranishi chart is a quintuple (V, E, Γ, s, ψ) such that*

(1) *$V \subset \mathbb{R}^n$ is an open set and $E = V \times \mathbb{R}^m$. Here n, m are nonnegative integers which may depend on the chart.*
(2) *Γ is a finite group acting effectively on V and has a linear action on the fiber $\mathbb{R}^m$ of E.*
(3) *$s : V \to \mathbb{R}^m$ is a Γ-equivariant map.*
(4) *ψ is a homeomorphism from $s^{-1}(0)/\Gamma$ to an open subset of Z.*

If $p \in \psi(s^{-1}(0))$ we call (V, E, Γ, s, ψ) a *Kuranishi neighborhood* of p. Sometimes we call V a Kuranishi neighborhood, by abuse of notation. We call E the *obstruction bundle.* s is called the *Kuranishi map.* We remark that in case Z is a moduli space s is actually a Kuranishi map in the usual sense.

A Kuranishi chart is said to be oriented if there is an orientation of $\Lambda^{\mathrm{top}} TV \otimes \Lambda^{\mathrm{top}} E$ which is preserved by the Γ-action.

Remark 8. Roughly speaking, a Kuranishi neighborhood of p gives a way to describe a neighborhood of p in Z as a solution of an equation $s(x) = (s_1(x), \dots, s_m(x)) = 0$. In case s_i is a polynomial, it defines a structure of scheme as follows. (More precisely, since the finite group action is involved it gives a structure of Deligne-Mumford stack.) Let us consider the quotient ring R_s of the polynomials ring $R[X_1, \cdots, X_n]$ by the ideal which is generated by the polynomials $s_i(X_1, \cdots, X_n)$ $(i = 1, \cdots, m)$. Then we obtain a ringed space $\mathrm{Spec}(R_s)$, that is the affine scheme defined by the ring R_s. By gluing them we obtain a scheme.

If we try to apply this construction of algebraic geometry to differential geometry (that is our situation), then we will be in the trouble. We can indeed construct a sheaf of rings (of smooth functions modulo the components of s) on Z and then Z becomes a ringed space (that is a space together with a sheaf of local ring). However the structure of ringed space does not seem to hold enough information we need. For example, since the Krull dimension of the ring of the germs of smooth functions is infinite, it follows that the dimension (that is $n - m$ in case of (1) of Definition 17) does not seem to be determined from the structure of ringed space. So it seems difficult to obtain the notion of (virtual) fundamental chains using the structure of ringed space.

Therefore, in place of using the structure of ringed space, we 'remember' the equation $s = 0$ itself as a part of the structure and glue the Kuranishi chart in that sense as follows.

Let $(V_i, E_i, \Gamma_i, s_i, \psi_i)$ and $(V_j, E_j, \Gamma_j, s_j, \psi_j)$ be two Kuranishi charts.

Definition 18. *A coordinate change from $(V_i, E_i, \Gamma_p, s_i, \psi_i)$ to $(V_j, E_j, \Gamma_j, s_j, \psi_j)$ consists of a Γ_i-invariant open subset $V_{ji} \subset V_i$, maps $\phi_{ji} : V_{ji} \to V_j$, $\hat{\phi}_{ji} : E_i|_{V_{ji}} \to E_j$ and a homomorphism $h_{ji} : \Gamma_i \to \Gamma_j$ with the following properties.*

(1) *ϕ_{ji} is an h_{ji}-equivariant smooth embedding.*
(2) *$\hat{\phi}_{ji}$ is an h_{ji}-equivariant embedding of the vector bundles over ϕ_{ji}.*
(3) *h_{ji} is defined and is injective if $V_{ji} \neq \emptyset$.*
(4) *$s_j \circ \phi_{ji} = \hat{\phi}_{ji} \circ s_i$.*
(5) *$\psi_j \circ \phi_{ji} = \psi_i$ on $(s_i^{-1}(0) \cap V_{ji})/\Gamma_j$.*
(6) *$\psi_i((s_i^{-1}(0) \cap V_{ji})/\Gamma_i)$ contains a neighborhood of $\psi_i(s_i^{-1}(0)/\Gamma_i) \cap \psi_j(s_j^{-1}(0)/\Gamma_j)$ in Z.*
(7) *If $\gamma(\phi_{ji}(V_{ji})) \cap \phi_{ji}(V_{ji}) \neq \emptyset$ and $\gamma \in \Gamma_j$, then $\gamma \in h_{ji}(\Gamma_i)$.*

Remark 9. We can state (7) also as $h_{ij}((\Gamma_j)_p) = (\Gamma_i)_{\phi_{ij}(p)}$. (Here $(\Gamma_j)_p$ etc. is the isotropy group.) This condition is assumed in [30] as a part of the assumption that $V_{ij}/\Gamma_j \to V_i/\Gamma_i$ is an embedding of orbifold. Therefore Definition 18 (including (7)) is equivalent to [30] Definition 5.3.

Definition 19. *A Kuranishi structure on a compact metrizable space Z is $((V_i, E_i, \Gamma_i, s_i, \psi_i); i \in I)$ with the following properties. Here $(I, \preceq)$ is a partially ordered set.*

(1) *If $i \preceq j$, then we have a coordinate change $(\phi_{ji}, \hat{\phi}_{ji}, h_{ji})$.*
(2) *If $\psi_i(s_i^{-1}(0)/\Gamma_i) \cap \psi_j(s_j^{-1}(0)/\Gamma_j) \neq \emptyset$ then either $i \preceq j$ or $j \preceq i$ holds.*

(3) *If* $i \preceq j \preceq k$ *and* $V_{ki} \cap \phi_{ji}^{-1}(V_{kj}) \neq \emptyset$ *then there exists* $\gamma_{kji} \in \Gamma_k$ *such that*

$$\phi_{kj} \circ \phi_{ji} = \gamma_{kji}\phi_{ki}, \quad \hat{\phi}_{kj} \circ \hat{\phi}_{ji} = \gamma_{kji}\hat{\phi}_{ki}, \quad (h_{kj} \circ h_{ji})(\mu) = \gamma_{kji} h_{ki}(\mu)\gamma_{kji}^{-1}$$

on $V_{ki} \cap \phi_{ji}^{-1}(V_{kj})$ *and for* $\mu \in \Gamma_i$.
(4) $\psi_i(s_i^{-1}(0)/\Gamma_i)$ *(*$i = 1, \cdots, I$*) is an open covering of* Z.

We call $((V_i, E_i, \Gamma_i, s_i, \psi_i); i \in I)$ *a Kuranishi atlas.*

Note Kuranishi atlas Definition 19 is called a good coordinate system in [30] Definition 6.1. Hence by [30] Lemma 6.3 the above definition of Kuranishi structure is equivalent to one in [30].

We consider Kuranishi map s_j in an neighborhood of $\psi_{ji}(V_{ji})$ and its derivation to the normal direction. It induces a linear map

$$Ds_j : \frac{\phi_{ji}^* TV_j}{TV_{ji}} \longrightarrow \frac{\phi_{ji}^* E_j}{E_i|_{ji}}. \tag{85}$$

Kuranishi structure is said to have *tangent bundle* if (85) is an isomorphism.

Kuranishi structure is said to be *oriented* if (85) is compatible with orientations of $\Lambda^{\text{top}} TV_i \otimes \Lambda^{\text{top}} E_i$ and of $\Lambda^{\text{top}} TV_j \otimes \Lambda^{\text{top}} E_j$.

We remark that the following diagram commutates :

$$\begin{array}{ccccc} \dfrac{\phi_{ji}^* TV_j}{TV_{ji}} & \longrightarrow & \dfrac{\phi_{ki}^* TV_k}{TV_i} & \longrightarrow & \dfrac{\phi_{kj}^* TV_k}{TV_j} \\ \cong \big\downarrow & & \cong \big\downarrow & & \cong \big\downarrow \\ \dfrac{E_j}{\hat{\phi}_{ji}(E_i)} & \longrightarrow & \dfrac{E_k}{\hat{\phi}_{ki}(E_i)} & \longrightarrow & \dfrac{E_k}{\hat{\phi}_{kj}(E_j)}. \end{array} \tag{86}$$

We thus defined a 'space with Kuranishi structure which has a tangent bundle and is oriented'. Since the notation in the quote is rather lengthy we call it oriented *K-space* from now on.

We remark that since (85) is an isomorphism

$$\dim V_i - \operatorname{rank} E_i \tag{87}$$

is independent of $i \prec j \prec k$ (if Z is connected). We call it the *dimension* of the Kuranishi structure or K-space.

We next define a map from K-space to a manifold.

Definition 20. *Let* M *be a manifold and* Z *be a K-space. A system* $f = (f_i)$ *of maps* $f_i : V_i \to N$ *is said to be a strongly smooth map if each of* f_i *is smooth and* $f_j \circ \phi_{ji} = f_i$.

$f = (f_i)$ *is said to be weakly submersive if each of* $f_i : V_i \to N$ *is a submersion.*

We remark that strongly smooth map induces a continuous map $Z \to N$ in an obvious way.

We also define K-space with boundary and corners as follows. If we replace the condition 'V is an open subsets in $\mathbb{R}^n$' in (1) of Definition 17, by 'V is an n dimensional submanifold with corners in $\mathbb{R}^n$' it will be the definition of Kuranishi neighborhood with corners. We then proceed in the same way as Definitions 18, 19 we define (oriented) K-space with corners.

A point x of K-space is said to be in the codimension k corner if $x = \psi_i(y)$ with y in the codimension k corner of V_i/Γ_i. We can easily show that the set of all codimension k corner of a given K-space Z has a structure of K-space with corners.

For our purpose to study correspondence, the notion of fiber product of K-space is important. Let Z and Z' be K-spaces with their Kuranishi atlas $((V_i, E_i, \Gamma_i', s_i, \psi_i); i \in I)$, $((V_{i'}', E_{i'}', \Gamma_{i'}, s_{i'}', \psi_{i'}'); i' \in I')$, respectively.

Let (f_i) and $(f_{i'}')$ be weakly submersive strongly smooth maps from Z to N and Z' to N, respectively. Here N is a smooth manifold. We first take a fiber product $Z \times_N Z'$ in the category of topological space. The next lemma is in [33] §A1.2.

Lemma 6. *$Z \times_N Z'$ has a structure of K-space.*

Proof : We consider the fiber products

$$V_{(i,i')} = V_i \; {}_{f_i}\!\times_{f_{i'}'} V_{i'}'.$$

By the assumption that (f_i) and $(f_{i'}')$ are weakly submersive, the above fiber product is well-defined in the category of smooth manifold. We define $E_{(i,i')}$ as the pull back of the exterior product of E_i and $E_{i'}'$. The group $\Gamma_{(i,i')} = \Gamma_i \times \Gamma_{i'}'$ acts on $V_{(i,i')}$ and $E_{(i,i')}$ as the restriction of the direct product action. Using weakly submersivity of f_i, $f_{i'}'$ we can prove that this action is effective. We put $s_{(i,i')}(x,y) = (s_i(x), s_{i'}'(y))$. It is easy to see that

$$s_{(i,i')}^{-1}(0) = s_i^{-1}(0) \; {}_{f_i}\!\times_{f_{i'}'} s_{i'}'^{-1}(0).$$

Hence we obtain $\phi_{(i,i')}$. Thus we have a Kuranishi chart

$$(V_{(i,i')}, E_{(i,i')}, \Gamma_{(i,i')}, s_{(i,i')}, \phi_{(i,i')}).$$

It is easy to see that we can glue coordinate transformation and construct a K-space. □

Remark 10. In the above construction, it may happen that $i_1 \prec i_2$, $i_1' \prec i_2'$, $\psi_{(i_1,i_2')}(s_{(i_1,i_2')}^{-1}(0)) \cap \psi_{(i_1',i_2)}(s_{(i_1',i_2)}^{-1}(0)) \neq \emptyset$, but neither $(i_1, i_2') \prec (i_1', i_2)$ nor $(i_1', i_2) \prec (i_1, i_2')$. In this case Definition 19 (2) is not satisfied. However we can shrink $V_{(i_1',i_2)}$ and $V_{(i_1,i_2')}$ in the way as in Figure 14 below so that Definition 19 (2) is satisfied.

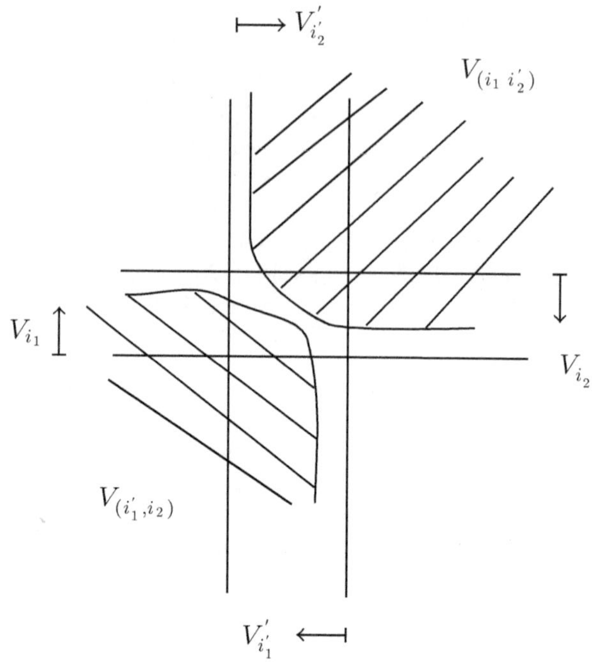

Fig. 14.

In the situation of Lemma 6, we assume that we also have a strongly smooth map $g : Z' \to N'$ such that $f' \times g : Z' \to N \times N'$ are weakly submersive. Then, it is easy to see that g induces a strongly smooth map $g : Z \times_N Z' \to N'$ which is weakly submersive.

A few more notations are in order.

Let Z be a K-space with corner and $p, q \in \partial Z$. We say that they are in the same *component* of ∂Z and write $p \sim q$ if there exists a sequence of Kuranishi charts $(V_i, E_i, \Gamma_i, s_i, \psi_i)$ $(i = 0, \cdots, l)$ such that

(1) $p = \psi_1(x_1) \in \psi_1(s_1^{-1}(0)/\Gamma_1)$ and $q = \psi_l(x'_l) \in \psi_l(s_l^{-1}(0)/\Gamma_l)$.
(2) Either $x_i = \phi_{i(i+1)}(x'_i)$ or $x'_i = \phi_{(i+1)i}(x'_i)$. Here in the first case $i \prec i+1$ and $x_i \in \partial V_{i(i+1)}$. In the second case $i+1 \prec i$ and $x'_i \in \partial V_{(i+1)i}$.
(3) x'_i and x_{i+1} can be joined by a path which is contained in ∂V_{i+1}.

A component of ∂Z is a closure of $\sim$ equivalence class. It has a structure of K-space.

We say that 'p and q are connected by a path contained in the set of boundary points in the Kuranishi neighborhood' if above condition is satisfied.

Let $p, q \in \partial Z \setminus$ corner. We say that they are in the same *stratum* of ∂Z and write $p \sim' q$, if p and q are connected by a path which is contained in the set of boundary points in the Kuranishi neighborhood and does not intersect with corner points. A stratum is a closure of $\sim'$ equivalence class. It has a structure of K-space.

We can define stratum of codmension d corner of Z in the same way. It has a structure of K-space also.

Let M be a closed and oriented manifold. We assume that we have a diagram

$$\begin{array}{ccccc} & & \mathcal{M}_{k+1,\beta} & & \\ & & \uparrow{\scriptstyle \pi_0} & & \\ M^k & \underset{\pi_2}{\longleftarrow} & \mathfrak{M}_{k+1}(\beta) & \underset{\pi_1}{\longrightarrow} & M. \end{array} \tag{88}$$

Here $\mathfrak{M}_{k+1}(\beta)$ is a K-space and

$$(\pi_0, \pi_1, \pi_2) : \mathfrak{M}_{k+1}(\beta) \to \mathcal{M}_{k+1,\beta} \times M^{k+1}$$

is assumed to be strongly smooth and is *weakly submersive.* We assume

$$\dim \mathfrak{M}_{k+1}(\beta) = \dim M + \mu(\beta) + k - 2. \tag{89}$$

Definition 21. (88) *is said to be a G-gapped filtered Kuranish* A_∞ *correspondence, if there exists a map*

$$\circ_{\mathfrak{m},i} : \mathfrak{M}_{k+1}(\beta_1) \ {}_{ev_i}\times_{ev_0} \mathfrak{M}_{l+1}(\beta_2) \to \mathfrak{M}_{k+l}(\beta_1 + \beta_2). \tag{90}$$

which identifies the left hand side with a union of finitely many stratum of the boundary of the right hand side as K-spaces, such that the following holds.

(1) *The following diagram commutes.*

$$\begin{array}{ccc} \mathcal{M}_{k+1,\beta_1} \times \mathcal{M}_{l+1,\beta_2} & \underset{\circ_l}{\longrightarrow} & \mathcal{M}_{k+l,\beta_1+\beta_2} \\ {\scriptstyle \pi_0\times\pi_0}\uparrow & & {\scriptstyle \pi_0}\uparrow \\ \mathfrak{M}_{k+1}(\beta_1) \ {}_{ev_i}\times_{ev_0} \mathfrak{M}_{l+1}(\beta_2) & \underset{\circ_{\mathfrak{m},i}}{\longrightarrow} & \mathfrak{M}_{k+l}(\beta_1+\beta_2) \end{array} \tag{91}$$

(2) *The inverse image*

$$\pi_0^{-1}(\mathcal{M}_{k+l,\beta}(\mathbf{S})) \subset \mathfrak{M}_{k+l}(\beta) \tag{92}$$

of each such stratum $\mathcal{M}_{k+l,\beta}(\mathbf{S})$ of $\mathcal{M}_{k+l,\beta}$ is a union of strata of codimension d corner of $\mathfrak{M}_{k+l}(\beta)$. Here d is the number of singular points of $\mathbf{S}$. *We denote* (92) *by* $\mathfrak{M}_{k+l}(\beta;\mathbf{S})$. *Restriction of* ev_0 *to each such stratum* $\mathfrak{M}_{k+l}(\beta;\mathbf{S})$ *is a strongly smooth weakly submersive map* $\mathfrak{M}_{k+l}(\beta;\mathbf{S}) \to \mathcal{M}_{k+l,\beta}(\mathbf{S})$. *Diagram* (91) *is a cartesian diagram as a diagram of sets.*
(3) *The Formulae* (24), (25) *holds after replacing* $\circ_i$ *by* $\circ_{\mathfrak{m},i}$
(4) *The following diagram commutes.*

$$\begin{array}{ccccc} M^{k+l} & \xleftarrow[\pi_2]{} & \mathfrak{M}_{k+l}(\beta_1+\beta_2) & \xrightarrow[\pi_1]{} & M \\ \| & & \uparrow{\circ_{\mathfrak{m},i}} & & \| \\ M^{k+l} & \longleftarrow & \mathfrak{M}_{k+1}(\beta_1)\ {}_{ev_i}\times_{ev_0} \mathfrak{M}_{l+1}(\beta_2) & \xrightarrow[\pi_1\circ \mathrm{pr}_1]{} & M \end{array} \tag{93}$$

(5) *For each n the boundary of $\mathfrak{M}_{n+1}(\beta)$ is a union of*

$$\circ_{\mathfrak{m},i}(\mathfrak{M}_{k+1}(\beta_1)\ {}_{ev_i}\times_{ev_0} \mathfrak{M}_{l+1}(\beta_2))$$

for various $k, l, i, \beta_1, \beta_2$ with $k+l = n$, $i = 1, \cdots, l$, $\beta_1+\beta_2 = \beta$. They intersect each other only at their boundaries.
(6) *The identification* (6) *preserves orientations, with signs which will be described by Definition* 27.

In the rest of this article we say Kuranishi correspondence sometimes in place of G-gapped filtered Kuranish A_∞ correspondence for simplicity.

Now, in a similar way as the Definitions 8 and 11, we can define morphism between Kuranishi correspondences, and homotopy between morphisms. To rewrite Definitions 8 and 11 to our situation is straightforward and hence we omit them here.

As we mentioned before we can define composition of morphisms between Kuranishi correspondences, as follows. Let

$$\begin{array}{ccccc} & & \mathcal{M}_{k+1,\beta} & & \\ & & \uparrow{\pi_0} & & \\ M_i^k & \xleftarrow[\pi_2]{} & \mathfrak{M}^{(i)}_{k+1}(\beta) & \xrightarrow[\pi_1]{} & M_i. \end{array} \tag{94}$$

be Kuranishi correspondences for $i = 1, 2, 3$. Let

$$\begin{array}{ccccc} & & \mathcal{F}_{k+1,\beta} & & \\ & & \uparrow{\pi_0} & & \\ M_i^k & \xleftarrow[\pi_2]{} & \mathfrak{F}^{(ij)}_{k+1}(\beta) & \xrightarrow[\pi_1]{} & M_j \end{array} \tag{95}$$

be morphisms between Kuranishi correspondences for $(ij) = (12), (23)$. We will define a morphism of Kuranishi correspondence $(\mathfrak{F}^{(13)}_{k+1}(\beta))$ which is a composition of $(\mathfrak{F}^{(12)}_{k+1}(\beta))$ and $(\mathfrak{F}^{(23)}_{k+1}(\beta))$ as follows.

In the same way as (64) we can define

$$\begin{aligned} \mathrm{Comp}^{\beta_0,\beta_1,\cdots,\beta_l}_{k_1,\cdots,k_l;k'_0,\cdots;k'_l} : \mathcal{F}_{l+k'+1,\beta_0} & \\ \times (\mathcal{F}_{k_1+1,\beta_1} \times \cdots \times \mathcal{F}_{k_l+1,\beta_l}) & \to \mathcal{F}_{k+k'+1,\beta_0+\cdots+\beta_l} \end{aligned} \tag{96}$$

where $k' = k'_0 + \cdots + k'_l$ and $k_1 + \cdots + k_l = k$. By a filtered analogue of Lemma 5, the images of (96) (for various l, k_i, k'_i, β_i with $k + k' = n$, $\sum \beta_i = \beta$) decompose $\mathcal{F}_{n+1,\beta}$.

Now we consider the fiber product

$$\mathfrak{F}^{\beta_0,\beta_1,\cdots,\beta_l}_{k_1,\cdots,k_l;k'_0,\cdots;k'_l} = \mathfrak{F}^{(12)}_{l+k'+1}(\beta_0) \times_{M_2^l} (\mathfrak{F}^{(23)}_{k_1+1}(\beta_1) \times \cdots \times \mathfrak{F}^{(23)}_{k_l+1}(\beta_l)). \tag{97}$$

Here the map

$$\mathfrak{F}_{k_1+1}(\beta_1) \times \cdots \times \mathfrak{F}_{k_l+1}(\beta_l) \to M_2^l$$

is $(ev_0, \cdots, ev_0)$ and

$$\mathfrak{F}_{l+k'+1}(\beta_0) \to M_2^l$$

is $(ev_{k'_0+1}, ev_{k'_1+2}, \cdots, ev_{k'_{l-1}+l})$. (See (65).) Now by a filtered analogue of Lemma 5, we can glue spaces $\mathfrak{F}^{\beta_0,\beta_1,\cdots,\beta_l}_{k_1,\cdots,k_l;k'_0,\cdots;k'_l}$ for various l, k_i, k'_i, β_i with $k + k' = n$, $\sum \beta_i = \beta$ along their boundaries to obtain a K-space $\mathfrak{F}^{(13)}_{n+1}(\beta)$. Moreover we can define

$$\pi_0 : \mathfrak{F}^{(13)}_{n+1}(\beta) \to \mathcal{F}^{(13)}_{n+1,\beta}$$

such that (96), (97) commute with π_0. Furthermore we can define π_1 and π_2 such that

$$\begin{array}{ccccc} & & \mathcal{F}_{k+1,\beta} & & \\ & & \uparrow{\scriptstyle \pi_0} & & \\ M_1^k & \xleftarrow[\pi_2]{} & \mathfrak{F}^{(13)}_{k+1}(\beta) & \xrightarrow[\pi_1]{} & M_3 \end{array} \tag{98}$$

is a morphism between Kuranishi correspondences.

Definition 22. (98) *is a composition of* $(\mathfrak{F}^{(12)}_{k+1}(\beta))$ *and* $(\mathfrak{F}^{(23)}_{k+1}(\beta))$. *We write it as :*

$$(\mathfrak{F}^{(13)}_{k+1}(\beta)) = (\mathfrak{F}^{(23)}_{k+1}(\beta)) \circ (\mathfrak{F}^{(12)}_{k+1}(\beta)).$$

It is easy to see that the notion of composition between morphisms are compatible with the notion of homotopy of morphisms.

Lemma 7. *Composition of morphisms are homotopy associative. Namely, the morphisms of Kuranishi correspondence*

$$(\mathfrak{F}^{(34)}_{k+1}(\beta)) \circ \Big((\mathfrak{F}^{(23)}_{k+1}(\beta)) \circ (\mathfrak{F}^{(12)}_{k+1}(\beta))\Big)$$

and

$$\Big(\mathfrak{F}^{(34)}_{k+1}(\beta) \circ (\mathfrak{F}^{(23)}_{k+1}(\beta))\Big) \circ (\mathfrak{F}^{(12)}_{k+1}(\beta))$$

are homotopic to each other.

Proof : We remark that in the definition of (66), the time allocation of the component of the stable curve Σ' which comes from the the first factor Σ lies in $[0, 1/2]$. For the component which comes from the second factors Σ_i, its time allocation is in $[1/2, 1]$. Hence, for the same reason as the nonassociativity of the product in loop space (see §4), our composition is not strictly associative. However, for the same reason that the product in loop space is homotopy associative, we can easily construct the required homotopy by using the homotopy between parametrizations. □

Definition 23. *We define a category $\mathfrak{HAKCorr}_G$ as follows. An object is a manifold M together with a G-filtered Kruranishi correspondence on it. A morphisms between objects is a homotopy class of morphisms of Kuranishi correspondences. The composition is defined by Definition 22. It is well-defined by Lemma 7. We call it the homotopy category of G-filtered Kuranishi correspondence.*

We define a category $\mathfrak{HAlg}_G^{\mathbb{Q}}$ as follows. Its objects are a G-filtered A_∞ algebras with $\mathbb{Q}$-coefficients. A morphism is a homotopy class of G-filtered A_∞ homomorphisms. We call it the homotopy category of G-filtered A_∞ algebras.

Now the main theorem of this paper is as follows.

Theorem 12. *There exists a functor $\mathfrak{HAKCorr}_G \to \mathfrak{HAlg}_G^{\mathbb{Q}}$.*

The proof can be extracted from [33]. It is proved also in §12, §13 over $\mathbb{R}$ coefficient.

Remark 11. In Definition 23, we take the *homotopy class* of the morphism of Kuranishi correspondence as a morphism of our categories. One of the reasons we do so is the fact that the associativity holds only up to homotopy. On the other hand, as we explained in the proof of Lemma 7, the way the associativity breaks down is the same as the way the associativity of the product in loop space breaks down. Therefore, it is very likely that we can define an appropriate notion of 'A_∞ category' (or ∞ category) in place of taking the quotient by the homotopy. Note that A_∞ category as defined in [20] is a category where the set of morphisms has a structure of chain complex. The 'A_∞ category' above is different from that. Namely the set of morphisms does not have a structure of chain complex. Its relation to the A_∞ category in [20] is similar to the relation of A_∞ space to A_∞ algebra.

On the other hand, there is a notion of 2-category of A_∞ category. (See [48] §7.) In particular, there is a 2-category of A_∞ algebras. Since, in the world of A_∞ structures, we can define 'homotopies of homotopies of ... of homotopies of ...' in a natural way (see [33] §7.2.12), it is also very likely that we can define ∞ category (or 'A_∞ category') an object of which is an A_∞ category or an A_∞ algebra and whose morphisms are A_∞ functors or A_∞ homomorphisms.

Then it seems very likely that we can generalize Theorem 12 to the existence of an 'A_∞ functor' (or ∞ functor) in an appropriate sense.

11 Floer theory of Lagrangian submanifolds.

In this section we explain briefly how the general construction of the earlier sections were used in [33] to study Floer homology of Lagrangian submanifolds.

Let (M,ω) be a compact symplectic manifold and L be a Lagrangian submanifold. We assume that L is oriented and is relatively spin. Here L is said to be relatively spin if its second Stiefel-Whitney class lifts to a cohomology class in $H^2(M;\mathbb{Z}_2)$. Moreover we fix a relative spin structure. (See [33] §8.1 for its definition.) For example, if L is spin, the choice of spin structure determines a choice of its relative spin structure.

We denote by $\mu : \pi_2(M;L) \to \mathbb{Z}$ the Maslov index. (See [3] or [33] §2.1.1 for its definition.) Since L is oriented its image is contained in $2\mathbb{Z}$. We next define $E : \pi_2(M,L) \to \mathbb{R}$ by

$$E(\beta) = \int_\beta \omega.$$

This is well-defined since L is a Lagrangian submanifold. We put

$$G_+(L) = \mathrm{Im}(E,\mu) \subset \mathbb{R} \times 2\mathbb{Z}. \tag{99}$$

Note this does not satisfy the conditions of Definition 12.

We next take and fix a compatible almost complex structure J on M. Let Σ be a Riemann surface (which may have boundary). We say that $u : \Sigma \to M$ is J-holomorphic if

$$J \circ du = du \circ j_\Sigma$$

where j_Σ is the complex structure of Σ. Now we define

$$G_0(J) = \left\{ (E(\beta),\mu(\beta)) \,\middle|\, \begin{array}{l} \text{There exists a } J\text{-holomorphic map} \\ u : (D^2,\partial D^2) \to (M,L) \text{ in the homotopy class } \beta \end{array} \right\}.$$

By Gromov compactness [36], the monoid $G(J)$ generated by $G_0(J)$ satisfies the conditions of Definition 12.

Now for $\beta \in G(J)$ we define a moduli space $\mathfrak{M}_{k+1}(\beta)$ as follows. Let us consider an element $(\Sigma; z_0,\dots,z_k;\beta(\cdot)) \in \mathcal{M}_{k+1,\beta}$. Let $\Sigma = \bigcup_{a\in A} D^2_a$ be its decomposition. (See §9.) We consider a continuous map $u : (\Sigma,\partial\Sigma) \to (M,L)$ such that

(1) $u : D^2_a \to M$ is J-holomorphic.
(2) $\left(E\left([u|_{D^2_a}]\right), \mu\left([u|_{D^2_a}]\right)\right) = \beta(a)$.

Let $\overset{\circ}{\mathfrak{M}}_{k+1}(\beta)$ be the set of isomorphism classes of all such $(\Sigma; z_0,\dots,z_k;$ $\beta(\cdot); u)$. We can compactify it by including the stable maps with sphere bubbles. Let $\mathfrak{M}_{k+1}(\beta)$ be the compactification. (See [33] §2.1.2.)

We define

$$(ev_0,\dots,ev_k) : \mathfrak{M}_{k+1}(\beta;J) \to L^{k+1}$$

by putting

$$ev_i(\Sigma;z_0,\dots,z_k;\beta(\cdot);u) = u(z_i)$$

and extending it to the compactification. We also define

$$\pi_0 : \mathfrak{M}_{k+1}(\beta;J) \to \mathcal{M}_{k+1,\beta}$$

by putting

$$\pi_0(\Sigma;z_0,\dots,z_k;\beta(\cdot);u) = (\Sigma;z_0,\dots,z_k;\beta(\cdot))$$

and extending it to the compactification. We next define

$$\circ_{\mathfrak{m},i} : \mathfrak{M}_{k+1}(\beta_1;J) \ {}_{ev_i}\times_{ev_0} \mathfrak{M}_{l+1}(\beta_2;J) \to \mathfrak{M}_{k+l}(\beta_1+\beta_2;J) \tag{100}$$

as follows. Let

$$\mathcal{S} = (\Sigma;z_0,\cdots,z_k;\beta(\cdot);u) \in \mathfrak{M}_{k+1}(\beta_1;J)$$
$$\mathcal{S}' = (\Sigma';z'_0,\cdots,z'_l;\beta'(\cdot);u') \in \mathfrak{M}_{l+1}(\beta_2;J)$$

with

$$ev_0(\mathcal{S}') = u'(z_0) = u(z_i) = ev_i(\mathcal{S}). \tag{101}$$

We identify $z_i \in \Sigma$ and $z'_0 \in \Sigma'$ to obtain Σ''. By (101) we obtain a J-holomorphic map $u'' : (\Sigma'',\partial\Sigma'') \to (M,L)$ by putting $u'' = u$ on Σ and $u'' = u'$ on Σ'. We set.

$$\left(z''_0,\dots,z''_{k+l-1}\right) = (z_0,\dots,z_{i-1},z'_1,\dots,z'_l,z_{i+1},\dots,z_k).$$

We now define :

$$\left(\Sigma'';z''_0,\dots,z''_{k+l-1};\beta''(\cdot);u''\right) = \mathcal{S}\circ_{\mathfrak{m};i}\mathcal{S}' \in \mathfrak{M}_{k+l}(\beta_1+\beta_2;J).$$

We have thus defined (100).

Proposition 4. *$\mathfrak{M}_{k+1}(\beta;J)$ is a $G(J)$-gapped filtered A_∞ Kuranishi correspondence.*

This is [33] Propositions 7.1.1 and 7.1.2. (We remark that the moduli space $\mathfrak{M}_{k+1}(\beta;J)$ here is denoted by $\mathcal{M}^{\rm main}_{k+1}(\beta;J)$ in [33].)

Theorem 12 and Proposition 4 (together with the filtered analogue of Corollary 2) imply that $H\left(L;\Lambda^{\mathbb{Q}}_{0,\rm nov}\right)$ has a structure of $G(J)$-gapped filtered A_∞ algebra, which we write $\mathfrak{m}^J$.

We next explain its independence of the almost complex structure J. (We remark that $\mathfrak{m}^J$ may also depend on the various choices (other than J) which we make during the constructions. However Theorem 12 implies that it is independent of such choices up to homotopy equivalence.) Let J_0, J_1 be two compatible almost complex structures. Since the set of all compatible almost complex structures is contractible, it follows that there exists a path $t \mapsto J_t$ of almost complex structures joining J_0 to J_1. We denote this path by $\mathcal{J}$. We are going to associate a morphism of Kuranishi correspondences to $\mathcal{J}$. We define a set

$$G_0(\mathcal{J}) = \bigcup_{t \in [0,1]} G(J_t).$$

Let $G(\mathcal{J})$ be the monoid generated by it. Again by Gromov compactness $G(\mathcal{J})$ satisfies the conditions of Definition 12.

Now we consider $(\Sigma; z_0, \dots, z_k; \beta(\cdot); \rho(\cdot)) \in \mathcal{F}_{k+1,\beta}$. Here we remark that $(\Sigma; z_0, \dots, z_k; \beta(\cdot)) \in \mathcal{M}_{k+1,\beta}$ and $\rho : A \to [0,1]$ is a time allocation. We decompose Σ as $\Sigma = \bigcup_{a \in A} D_a^2$. We consider a continuous map $u : (\Sigma, \partial\Sigma) \to (M, L)$ such that :

(1) $u : D_a^2 \to M$ is $J_{\rho(a)}$-holomorphic.
(2) $\left(E\left(\left[u|_{D_a^2}\right]\right), \mu\left(\left[u|_{D_a^2}\right]\right)\right) = \beta(a)$.

Let $\mathring{\mathfrak{F}}_{k+1}(\beta; \mathcal{J})$ be the set of all isomorphism classes of such objects $(\Sigma; z_0, \dots, z_k; \beta(\cdot); \rho(\cdot); u)$. By adding the stable maps with sphere bubbles we can compactify it and obtain $\mathfrak{F}_{k+1}(\beta; \mathcal{J})$. (See [33] §4.6.) (We remark that $\mathfrak{F}_{k+1}(\beta; \mathcal{J})$ is denoted by $\mathcal{N}_{k+1}(\beta; \mathcal{J})$ in [33].)

Proposition 5. *$\mathfrak{F}_{k+1}(\beta; \mathcal{J})$ is a morphism between the two $G(\mathcal{J})$-gapped filtered A_∞ Kuranishi correspondences $\mathfrak{M}_{k+1}(\beta; J_0)$ and $\mathfrak{M}_{k+1}(\beta; J_1)$.*

We remark that $G(J_i) \subset G(\mathcal{J})$. Hence every $G(J_i)$-gapped filtered A_∞ Kuranishi correspondence may be regarded as a $G(\mathcal{J})$-gapped filtered A_∞ Kuranishi correspondence.

Thus, using Theorem 12, we obtain a filtered A_∞ homomorphism

$$\mathfrak{f}^{\mathcal{J}} : \left(H\left(L; \Lambda_{0,nov}^{\mathbb{Q}}\right), \mathfrak{m}^{J_0}\right) \to \left(H\left(L; \Lambda_{0,nov}^{\mathbb{Q}}\right), \mathfrak{m}^{J_1}\right). \tag{102}$$

We remark that if $\beta = \beta_0 = (0,0)$ then

$$\mathfrak{F}_{k+1}(\beta_0; \mathcal{J}) = \mathcal{F}_{k+1} \times L, \tag{103}$$

since every J-holomorphic map u with $\int u^*\omega = 0$ is necessary constant. Using this fact and the filtered version of Theorem 8 (that is [33] Theorem 4.2.45) we can prove that $\mathfrak{f}^{\mathcal{J}}$ is a homotopy equivalence.

We next assume that there are two paths $\mathcal{J}$ and $\mathcal{J}'$ of almost complex structures joining J_0 to J_1. Again, since the set of compatible almost complex structures is contractible, it follows that there is a two-parameter family $\widehat{\mathcal{J}}$ of almost complex structures interpolating $\mathcal{J}$ and $\mathcal{J}'$. Using it we can prove the following:

Proposition 6. *There exists $G(\widehat{\mathcal{J}}) \supseteq G(\mathcal{J}) \cup G(\mathcal{J}')$ and a homotopy $\mathfrak{H}_{k+1}(\beta; \widehat{\mathcal{J}})$ of morphisms of filtered $G(\widehat{\mathcal{J}})$-gapped Kuranishi correspondences between $\mathfrak{F}_{k+1}(\beta; \mathcal{J})$ and $\mathfrak{F}_{k+1}(\beta; \mathcal{J}')$.*

See [33] §4.6.2.

Thus, by Theorem 12 and the filtered version of Corollary 2, we have the following :

Theorem 13. *To each relatively spin Lagrangian submanifold L of a compact symplectic manifold M we can associate a structure of filtered A_∞ algebra on $H\left(L;\Lambda_{0,nov}^{\mathbb{Q}}\right)$.*

It is independent of the choices up to homotopy equivalence. The homotopy class of the homotopy equivalences is also independent of the choices.

This is Theorem A of [33]. We remark that since $\mathfrak{m}_{1,\beta_0}=0$ on $H\left(L;\Lambda_{0,\mathrm{nov}}^{\mathbb{Q}}\right)$, it follows that any homotopy equivalence between two filtered A_∞ structures on it is an isomorphism, that is, a filtered A_∞ homomorphism which has an inverse. In [33] it is also proved that the filtered A_∞ algebra we obtain is unital.

The structure obtained in Theorem 13 is highly nontrivial. We gave various calculations in [34] §37 and §55. See [14, 15] for some other calculations. We also gave various applications to symplectic topology in [33] Chapter 6, [34] Chapter 8 etc.. Since, in this article, we concentrate on foundations, we do not explain calculations or applications here.

12 Transversality.

In §12 and §13, we will prove Theorem 12. Once stated appropriately, this theorem can be proved by the argument we wrote in §6 as a proof of Theorem 4 *except for transversality and orientation.* So in §12 and §13 we focus on these two points. In this section we discuss transversality.

We first remark that, after 1996, the transversality problem (in the theory of pseudo-holomorphic curves for example) became a problem of finite dimensional topology rather than one of (linear or nonlinear) analysis. In the early days of gauge theory or pseudo-holomorphic curve theory, various kinds of perturbations were introduced and used by various authors for various purposes. In those days, the heart of the study of the transversality problem was to find an appropriate geometric parameter, via which we have enough room to perturb the partial differential equations so that the relevant transversality is achieved. Therefore the transversality problem was closely tied to the analysis of the particular nonlinear differential equation being studied. This situation changed once the virtual fundamental chain technique was introduced. We now can reduce the problem to one of finite dimensional topology in quite general situations, including *all* the cases in pseudo-holomorphic curve theory. So the main point to work out is a finite dimensional problem. One of the main outcomes of the discussion of the preceding sections is a formulation of this finite dimensional problem in a precise and rigorous way. (Of course finding explicit geometric parameters for perturbation can be interesting since it may give additional information on the algebraic system we obtain and may have geometric applications.)

When our situation is 'Morse type' and not 'Bott-Morse' type, the transversality can be achieved in general by taking abstract multivalued perturbation, that is by applying [30] Theorem 3.11 and Lemma 3.14, *directly*. Here 'Morse type' in our situation means that the correspondence we study is a correspondence between 0 dimensional spaces (that are discrete sets). Thus in this case the transversality problem had been solved by the method of [30].

In the 'Bott-Morse' case, the problem is more involved. Namely in case we study correspondence between manifolds of positive dimension, we need to perform the construction of virtual fundamental chains more carefully. This is the point we focus in this section. We refer [33] §7.2.2 (especially right after Situation 7.2.7) for the explanation of the reason why Bott-Morse case is harder to study.

As far as the author knows at the time of writing this article, there are two methods to deal with Bott-Morse case, both of which works in all the situations that are important for the applications to pseudo-holomorphic curve theory. One uses a kind of singular (co)homology and the other uses de Rham cohomology.

The first method was worked out in detail in [33] §7.2. As far as the author knows, this is the only way which works over $\mathbb{Q}$ (or $\mathbb{Z}$ sometimes) coefficient, in the general situation. The other advantage of this method is that singular homology is more flexible and so is useful for explicit calculations. (See [34] §56, §57 for some examples of calculations using singular homology.) The disadvantage of this method is that it destroys various symmetry of the problem.

The first method is summarized as follows. We first choose a countable set of smooth singular chains on our manifold M (in the case of Theorem 4 for example) and perturb the moduli space $\mathfrak{M}_{k+1}$ etc. so that the fiber product (43) is transversal for each $(P_1, \ldots, P_k)$ with P_i in the set we choosed above. We then define the operations by Formula (43). The trouble is that the chain which is an output of the operation, may not be in the chain complex generated by the chains we start with. So we increase our chain complex by adding those outputs. We next perturb again the moduli space $\mathfrak{M}_{k+1}$ to achieve transversality with those newly added chains. One important point is that we need to perturb $\mathfrak{M}_{k+1}$ in a way depending the chains P_i on M. We continue this process countably many times and obtain a structure we want. One needs to work out rather delicate argument to organize the induction so that we can take such perturbations in a way so that they are all compatible to each other. We omit the detail and refer [33].

The second approach, using de Rham theory, works only over $\mathbb{R}$ coefficient. It is however somewhat simpler than the first one. In fact, for example, to prove Corollary 1 over $\mathbb{R}$ coefficient using de Rham theory, there is nothing to do in geometric side. Namely de Rham complex has a ring structure which is associative in the chain level. Therefore, by applying Corollary 2, we immediately obtain Corollary 1 over $\mathbb{R}$ coefficient. The case of Kuranishi correspondence is not such easy but is somewhat simpler than working with singular homology. The method using de Rham cohomology is somewhat similar to the discussion by Ruan in [59] and also to the argument of [27] §16. It was used systematically in [33] §7.5 and in [29]. Another advantage of this method is that it is easier to keep symmetry of the problem. For example we can prove the cyclic symmetry of the A_∞ algebra in Theorem 13 in this way. We will explain this method more later in this section.

We remark that there is a third method which works under some restrictions. It is the method to use Morse homology [19] (see for example [60]) or Morse homotopy ([20, 6, 23]). Let us discuss this method briefly.

We first consider the case of Theorem 4. We take *several* functions f_i on M (in the case of Theorem 4 for example) so that $f_i - f_j$ for $i \neq j$ are Morse functions and the gradient flow of $f_i - f_j$ are Morse-Smale. We also assume that the stable and unstable manifolds of $f_i - f_j$ for various $i \neq j$ are transversal to each other. (Of course the stable manifold of $f_i - f_j$ is not transversal to itself. So we exclude this case.) We then regard the stable manifolds of $f_i - f_{i+1}$ as a chain P_i and consider (44). Then we obtain an operator

$$\mathfrak{m}_k : C^*(M; f_0 - f_1) \otimes \cdots \otimes C^*(M; f_{k-1} - f_k) \to C^*(M; f_0 - f_k). \qquad (104)$$

Here $C^*(M; f_1 - f_2)$ is the Morse-Witten complex of $f_1 - f_2$. (It is the complex (19) for *Morse* function $f = f_1 - f_2$. See for example [60].) More precisely, since we need to squeeze our structure to the Morse-Witten complex which is much smaller than singular chain complex, we need to combine the construction above with the proof of Theorem 2 as is done in [45] §6.4. Then the structure constant of the operation (104) turns out to be obtained by counting the order of appropriate sets of maps from a metric rooted ribbon tree to M such that each edge will be mapped to a gradient line of $f_i - f_j$. (See [20] §3,4 and [31] §12, §13.) It satisfies the relation (33) and hence defines a (topological) A_∞ *category*. Since $f_i = f_j$ is excluded. It is difficult to define A_∞ *algebra* in this way, directly.

We can generalize this construction to the case of Theorem 13 under some restrictions. In the situation of Theorem 13, operators

$$\mathfrak{m}_{k,\beta} : C^*(L; f_0 - f_1) \otimes \cdots \otimes C^*(L; f_{k-1} - f_k) \to C^*(L; f_0 - f_k) \qquad (105)$$

are defined by counting a map from the configuration as in Figure 15 below to L.

Here we put functions $f_0, \dots, f_k$ on $D^2 \setminus$ tree according to the counter clockwise order. The small circles in the figure are mapped to the boundary value of a pseudo-holomorphic discs which bounds L. We assume that the sum of the homology classes of those pseudo-holomorphic disc is β. If e is an edge of the tree, then we assume that e is mapped to a gradient line of $f_i - f_j$. Here e is between two domains on which f_i and f_j are put. (Figure 15.1 is copied from page 429 of [23]. In [23] the case when the Lagrangian submanifold is a diagonal of the direct product $M \times M$ of symplectic manifold M was

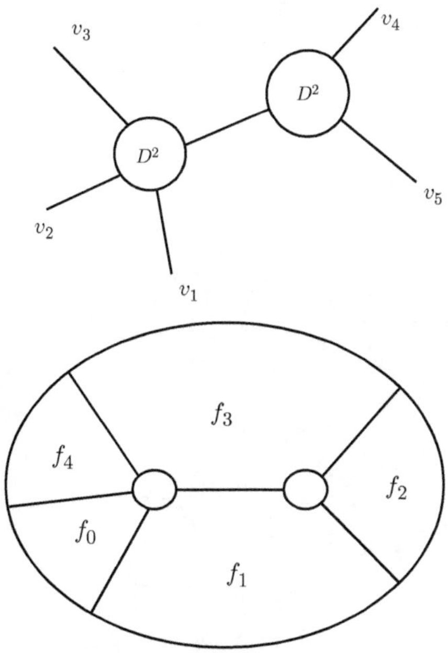

Fig. 15.

discussed. Y.-G. Oh [56] page 260 generalized it to more general Lagrangian submanifold L in the case of $\mathfrak{m}_1$, and also pointed out in [56] page 264 that it can be generalized to higher $\mathfrak{m}_k$ in some case.)

An important point of this construction is a cancellation of the two potential boundary of the moduli space of such maps. One of them corresponds to the shrinking of the edge, and the other is a splitting of the pseudo-holomorphic disc into a union of two discs. (See Figure 16.) This point was used for example in [32] page 290 for this purpose. (It was written there as the cancellation of (A4.70.2) and (A4.70.4).)

We need however to put a restriction on our Lagrangian submanifold L to make this argument rigorous. Namely for general Lagrangian submanifold L for which $\mathfrak{m}_0 \neq 0$, we need to include tad pole (such as in the Figure 17 below). This causes some problem to rigorously define (104).

In case when our Lagrangian submanifold is monotone with minimal maslov index ≥ 2, we can exclude such phenomenon. This fact was proved by Y.-G. Oh in [55] who established Floer homology of Lagrangian submanifold under this assumption. Under the same condition, Buhovsky [9] recently studied multiplicative structure of Floer homology using Morse homotopy.

Now we will discuss transversality problem in more detail using de Rham cohomology. We consider the situation of Kuranishi correspondence over M, that is the situation of Definition 21. Let $\Lambda^d(M)$ be the set of all smooth d forms on M. Using the correspondence (88) we want to construct

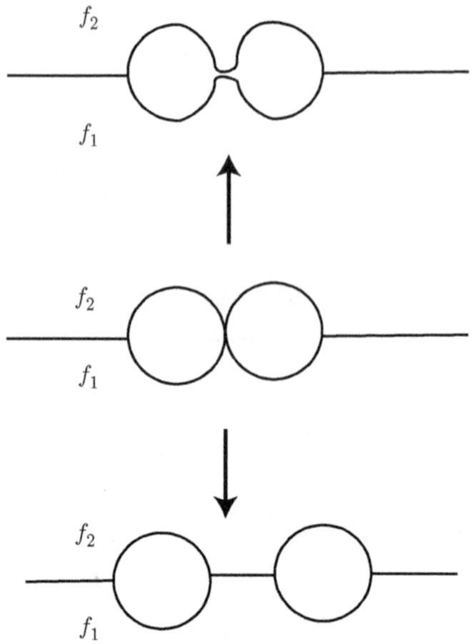

Fig. 16.

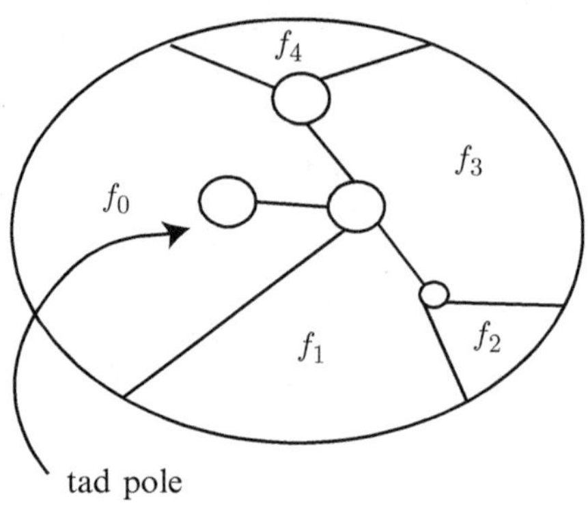

Fig. 17.

a homomorphism $\mathfrak{m}_{k,\beta} : \Lambda^d(M^k) \to \Lambda^{d+1-\mu(\beta)}(M)$. Intuitively we might take

$$\mathfrak{m}_{k,\beta} \text{ ``}=\text{'' } \pi_1! \circ \pi_2^*, \tag{106}$$

where π_2 is pull back of the differential form and $\pi_1!$ is integration along fiber. We remark however that integration along fiber is not well-defined as a smooth form unless π_1 is a submersion. Therefore we need to take appropriate smoothing of the virtual fundamental chain to make sense of (106). We need to perform smoothing in a way compatible with the operadic structure of our K-spaces, which describes how they are glued. We explain this construction in three steps. At the first step, we work in one Kuranishi chart. At the second step, we work on each K-space $\mathfrak{M}_{k+1}(\beta)$. Finally we explain the way to make it compatible for various k, β.

Step 1 : Let $\mathcal{V} = (V, E, \Gamma, s, \psi)$ be a Kuranishi chart of $\mathfrak{M}_{k+1}(\beta)$. We first review the notion of multisection which was introduced [30] Definition 3.1, 3.2.

Let $E = V \times \mathbb{R}^{n\nu}$. We denote by meas$(\mathbb{R}^{n\nu})$ the space of all compactly supported Borel measures on $\mathbb{R}^{n\nu}$. Let $\mathfrak{k} \in \mathbb{Z}_{>0}$. We denote by $\text{meas}_{\mathfrak{k}}(\mathbb{R}^{n\nu})$ the set of measures of the form $\frac{1}{\mathfrak{k}}\sum_{c=1}^{\mathfrak{k}} \delta_{v_c}$ where δ_v is a delta measure on $\mathbb{R}^{n\nu}$ supported at v. The Γ action on $\mathbb{R}^{n\nu}$ induces a Γ action on $\text{meas}_{\mathfrak{k}}(\mathbb{R}^{n\nu})$.

Definition 24. *A $\mathfrak{k}$-multisection of E is by definition a Γ equivariant map $\mathfrak{s} : V \to \text{meas}_{\mathfrak{k}}(\mathbb{R}^{n\nu})$.*

It is said to be smooth if, for any sufficiently small $U \subset V$, we have smooth maps $s_c : U \to \mathbb{R}^{n\nu}$ $(c = 1, \cdots, k)$ such that

$$\mathfrak{s}(x) = \frac{1}{\mathfrak{k}}\sum_{c=1}^{\mathfrak{k}} \delta_{\mathfrak{s}_c(x)} \tag{107}$$

We say $\mathfrak{s}_c$ a branch of $\mathfrak{s}$.

We remark that we do not require $\mathfrak{s}_c$ to be Γ equivariant. Namely the Γ action may exchange them.

Remark 12. The multisection is introduced in [30] in a slightly different but equivalent way as above. The smoothness of multisection is a bit tricky thing to define. Here we assume that the branch $\mathfrak{s}_c$ exists locally. This is related to the notion liftability discussed in [30]. The liftable and smooth $\mathfrak{k}$-multisection in the sense of [30] is a smooth $\mathfrak{k}$-multisection in the sense above.

In case each branch is transversal to 0 the inverse image of 0 of multisection looks like the following Figure 18. (Figure 18 is a copy of [30] Figure 4.8.)

Since we assumed that (π_0, π_1, π_2) is weakly submersive, it follows that

$$(\pi_1, \pi_2) : V \to M^{k+1}$$

is a submersion.

Let $W_{\mathcal{V}}$ be a manifold which is oriented and without boundary. We do not assume $W_{\mathcal{V}}$ is compact. (We choose that the dimension of $W_{\mathcal{V}}$ is huge.) We consider smooth $\mathfrak{k}$-multisection

$$\mathfrak{s}_{\mathcal{V}} : V \times W_{\mathcal{V}} \to \text{meas}_{\mathfrak{k}}(\mathbb{R}^{n\nu})$$

of the pull back of E to $V \times W_{\mathcal{V}}$. The action of Γ on $W_{\mathcal{V}}$ is the trivial action.

Definition 25. *We say that $\mathfrak{s}_{\mathcal{V}}$ is strongly submersive if the following holds.*
(1) For each $(x, w) \in V \times W_{\mathcal{V}}$ we may choose its branches $\mathfrak{s}_{\mathcal{V},c}$ $(c = 1, \cdots, \mathfrak{k})$ on a neighborhood U of (x, w), such that 0 is a regular vales of it.

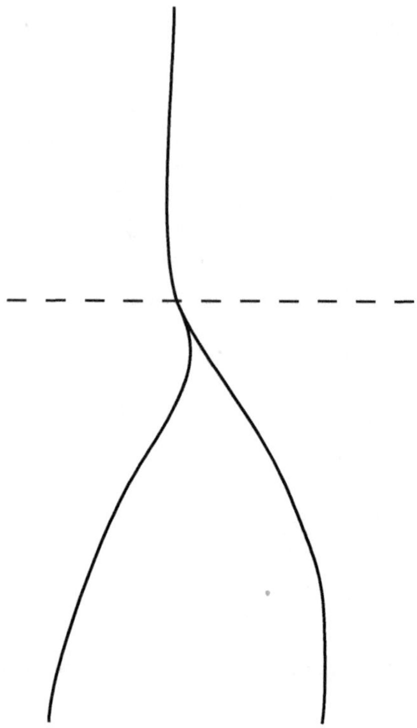

Fig. 18.

(2) *We put*

$$\mathfrak{s}_{\mathcal{V},c}^{-1}(0) \cap U = \{(x,w) \in U \mid \mathfrak{s}_{\mathcal{V},c}(x,w) = 0\},$$

which is a smooth manifold by (1). *Then*

$$(\pi_1, \pi_2) : \mathfrak{s}_{\mathcal{V},c}^{-1}(0) \cap U \to M^{k+1}.$$

is a submersion.

Hereafter we say multisection in place of $\mathfrak{k}$-multisection in case we do not need to specify $\mathfrak{k}$.

Lemma 8. *We may choose $W_{\mathcal{V}}$ so that for any ϵ there exists a smooth multisection such that each branch of it is in the ϵ neighborhood of s point-wise.*

Proof : We first choose $W_{\mathcal{V}}$ huge and find a single valued section $\mathfrak{s}_\epsilon : V \times W_{\mathcal{V}} \to \mathbb{R}^{n_{\mathcal{V}}}$ which approximate s and that 0 is its regular value. We then put

$$\mathfrak{s}(x,w) = \frac{1}{\#\Gamma} \sum \delta_{\gamma \mathfrak{s}_\epsilon(\gamma^{-1}x,w)}. \tag{108}$$

It is straightforward to see that (108) has the required properties. □

We take a smooth multisection $\mathfrak{s}_{\mathcal{V}}$ which is strongly submersive. We next take a smooth provability measure $\omega_{\mathcal{V}}$ on $W_{\mathcal{V}}$ which is Γ invariant and has compact support. We put

$$\mathcal{W}_{\mathcal{V}} = (W_{\mathcal{V}}, \mathfrak{s}_{\mathcal{V}}, \omega_{\mathcal{V}}).$$

We next choose an open covering U_α of $V \times W_{\mathcal{V}}$ such that $\mathfrak{s}_{\mathcal{V}}$ has an expression

$$\mathfrak{s}_{\mathcal{V}} = \frac{1}{\mathfrak{k}_\alpha} \sum_{c=1}^{\mathfrak{k}_\alpha} \delta_{\mathfrak{s}_{\mathcal{V},\alpha,c}} \tag{109}$$

on U_α, and choose a partition of unity χ_α subordinate to U_α.

We regard $\omega_{\mathcal{V}}$ as a differential form of degree $\dim W_{\mathcal{V}}$ and pull it back to $\mathfrak{s}_{\mathcal{V},\alpha,c}^{-1}(0)$. We denote it by the same symbol. Now we define perturbed correspondence $\mathrm{Corr}_{\mathcal{V}}^{\mathcal{W}_{\mathcal{V}}} : \Lambda^d(M^k) \to \Lambda^{d+1-\mu(\beta)}(M)$ by

$$\mathrm{Corr}_{\mathcal{V}}^{\mathcal{W}_{\mathcal{V}}}(u) = \pm \sum_\alpha \frac{1}{\#\Gamma \cdot \mathfrak{k}_\alpha} \sum_i (\pi_{1+}!) \left((\chi_\alpha \pi_{2+}^*(u) \wedge \omega_{\mathcal{V}})|_{\mathfrak{s}_{\mathcal{V},\alpha,c}^{-1}(0)} \right). \tag{110}$$

Here $\pi_{1+} : \mathfrak{s}_{\mathcal{V}}^{-1}(0) \to M$ is a composition of projection with π_1, and the map π_{2+} is defined in the same way. The integration along fiber in (110) is well-defined since π_{1+} is a submersion and $\omega_{\mathcal{V}}$ is of compact support. We do not discuss sign in this section. See §13.

We remark that the right hand side of (110) is independent of the choices of the covering U_α, decomposition (109), and the partition of unity, but depend only on the data encoded in $\mathcal{W}_{\mathcal{V}}$. So hereafter we write the right hand side of (110) as

$$\pm \frac{1}{\#\Gamma} (\pi_{1+}!) \left((\pi_{2+}^*(u) \wedge \omega_{\mathcal{V}})|_{\mathfrak{s}_{\mathcal{V}}^{-1}(0)} \right), \tag{111}$$

for simplicity.

We remark that $\mathfrak{s}_{\mathcal{V}}$ may be regarded as a family of multisections $\mathfrak{s}_{\mathcal{V},w}(\cdot) = \mathfrak{s}_{\mathcal{V}}(\cdot, w)$ parametrized by $w \in W_{\mathcal{V}}$. The correspondence (111) is an average of the correspondences by $\mathfrak{s}_{\mathcal{V},w}^{-1}(0)$ by the smooth probability measure $\omega_{\mathcal{V}}$. The technique of multisection uses finitely many perturbations and its average. Here we take a family of uncountably many perturbations and use its average.

Step 2 : We next combine and glue the correspondence for the Kuranishi charts in a given Kuranishi atlas of $\mathfrak{M}_{k+1}(\beta)$. Let $\mathcal{V}_i = (V_i, E_i, \Gamma_i, s_i, \psi_i)$, $i = 1, \cdots, I$ be a Kuranishi atlas. We may enumerate them so that $i \prec j$ implies $i < j$.

We will construct $\mathcal{W}_{\mathcal{V}_i}$ by induction on i as follows. Suppose we constructed them for each j with $j < i$. We define $W_{0,i}$ and $W_{\mathcal{V}_i}$ inductively. We put

$$W_{\mathcal{V}_i} = \left(\prod_{j<i} W_{0,j} \right) \times W_{0,i}$$

By induction hypothesis, $W_{0,j}$ $(j < i)$ are defined. We will define $W_{0,i}$ later.

We can extend $\mathfrak{s}_{\mathcal{V}_j}$ uniquely to a section $\mathfrak{s}_{i,\mathcal{V}_j}$ of E_j on $\Gamma_i(\varphi_{ij}(V_{ij})) \times W_{\mathcal{V}_j}$ so that it is Γ_i invariant. Note that we use Condition (7) in Definition 18 here. We may regard it as a section on

$$\Gamma_i(\varphi_{ij}(V_{ij})) \times W_{\mathcal{V}_i} \tag{112}$$

by composing it with an obvious projection. We next extend it to a tubular neighborhood of (112) as follows. By (85) we can identify

$$\varphi_{ij}^* E_i \cong E_j \oplus N_{\varphi_{ij}(V_{ij})} V_i,$$

where $N_{\varphi_{ij}(V_{ij})} V_i$ is he normal bundle. A point in a tubular neighborhood of $\varphi_{ij}(V_{ij})$ can be written as $(\varphi_{ij}(x), v)$ here v is in the fiber of $N_{\varphi_{ij}(V_{ij})} V_j$. We now put

$$\mathfrak{s}_{i,\mathcal{V}_j}((\varphi_{ji}(x), v), (w_k)_{k \le i}) = \mathfrak{s}_{i,\mathcal{V}_j}(\varphi_{ji}(x), (w_k)_{k \le j}) \oplus v.$$

(More precisely we take branch of our multivalued section $\mathfrak{s}_{i,\mathcal{V}_j}$ and apply the above formula to each branch.)

We extend it by using Γ_i invariance. We denote it by the same symbol $\mathfrak{s}_{i,\mathcal{V}_j}$. By using induction hypothesis, the sections we constructed above for various j, coincide at the part where the tubular neighborhoods of $\varphi_{ij}(V_{ij})$ for different j intersect to each other. Thus we obtain a desired section on a neighborhood of the union of the images φ_{ij} for various j. Now we choose $W_{0,i}$ and extend this section so that it satisfies the conditions (1) (2) (3) of Step 1. Moreover we can choose

$$\omega_{\mathcal{V}_i} = \prod \omega_{0,j} \times \omega_{0,i},$$

where $\omega_{0,j}$ is a provability measure on $W_{0,j}$ chosen in earlier stage of induction. We thus have

$$\mathcal{W}_{\mathcal{V}_i} = (W_{\mathcal{V}_i}, s_{\mathcal{V}_i}, \omega_{\mathcal{V}_i}).$$

Now we define

$$\mathrm{Corr}^{\mathcal{W}_{\mathcal{V}_i}}_{\mathfrak{M}_{k+1}(\beta)}(u) = \sum_i \pm \frac{1}{\#\Gamma_i} (\pi_{1+}!) \left((\pi_{2+}^*(u) \wedge \omega_{\mathcal{V}_i})|_{\mathcal{V}_i^o \cap \mathfrak{s}_{\mathcal{V}_i}^{-1}(0)} \right). \tag{113}$$

Here

$$\mathcal{V}_i^o = \mathcal{V}_i \setminus \bigcup_{i \prec l} (V_{li} \times W_i).$$

We remark that precisely speaking the right hand side of (113) should be written in a way similar to (110) using partition of unity and branches.

By construction, we have the following equality.

$$\begin{aligned}&\frac{1}{\#\Gamma_j}(\pi_{1+}!)\left((\pi_{2+}^*(u)\wedge\omega_{\mathcal{V}_i})|_{(V_{ij}\times W_j)\cap\mathfrak{s}_{\mathcal{V}_i}^{-1}(0)}\right)\\&=\frac{1}{\#\Gamma_i}(\pi_{1+}!)\left((\pi_{2+}^*(u)\wedge\omega_{\mathcal{V}_i})|_{(\mathrm{Tube}(\Gamma_i\cdot\varphi(V_{ij}))\times W_i)\cap\mathfrak{s}_{\mathcal{V}_i}^{-1}(0)}\right).\end{aligned}\tag{114}$$

Here $\mathrm{Tube}(\Gamma_i\cdot\varphi(V_{ij}))$ is the tubular neighborhood of $\Gamma_i\cdot\varphi(V_{ij})$. We can use this fact to prove (113) is smooth.

(114) also imply that when we apply Stokes' formula, the boundary term is the integral on the boundary of $\mathcal{M}_{k+1}(\beta)$.

Step 3 : Now we will explain the way how we perform the above construction for various K-spaces, $\mathfrak{M}_{k+1}(\beta)$ in a way so that they are compatible for various k,β.

We first explain the reason why, by the method of continuous family of perturbations, we can construct the compatible system of virtual fundamental cochains inductively, in the situation where fiber product appears.

Let $\mathcal{V}=(V,E,\Gamma,s,\psi)$, $\mathcal{V}'=(V',E',\Gamma',s',\psi')$ be Kuranishi neighborhoods of $p\in Z$ and $p'\in Z'$ respectively. We consider the diagram

$$M_1\xleftarrow{\pi_1}V\xrightarrow{\pi_2}M\xleftarrow{\pi_1'}V'\xrightarrow{\pi_2'}M_2\tag{115}$$

of smooth manifolds such that $(\pi_1,\pi_2):V\to M_1\times M$, $(\pi_1',\pi_2'):V'\to M\times M_2$ are submersions.

We take $\mathcal{W}_{\mathcal{V}}=(W_{\mathcal{V}},\mathfrak{s}_{\mathcal{V}},\omega_{\mathcal{V}})$, $\mathcal{W}_{\mathcal{V}'}=(W_{\mathcal{V}'},\mathfrak{s}_{\mathcal{V}'},\omega_{\mathcal{V}'})$ as in step one. Namely we assume that $(\pi_1,\pi_2):\mathfrak{s}_{\mathcal{V}}^{-1}(0)\to M_1\times M$ is a submersion and we put a similar assumption for (π_1',π_2'). Then, in the same way as (110), we obtain homomorphisms

$$\mathrm{Corr}_{\mathcal{V}}^{\mathcal{W}_{\mathcal{V}}}:\Lambda(M_1)\to\Lambda(M),\qquad\mathrm{Corr}_{\mathcal{V}'}^{\mathcal{W}_{\mathcal{V}'}}:\Lambda(M)\to\Lambda(M_2)\tag{116}$$

by correspondences. The composition of them is obtained by a correspondence which is a fiber product of $\mathcal{V}$ and $\mathcal{V}'$, as follows. We consider

$$\mathcal{V}\times_M\mathcal{V}'=(V\,{}_{\pi_1}\!\times_{\pi_2}V',E\times E',\Gamma\times\Gamma',s\times s')$$

and

$$(W_{\mathcal{V}}\times W_{\mathcal{V}'},\mathfrak{s}_{\mathcal{V}}\times\mathfrak{s}_{\mathcal{V}'},\omega_{\mathcal{V}}\times\omega_{\mathcal{V}'}).\tag{117}$$

We write (117) as $\mathcal{W}_{\mathcal{V}}\times\mathcal{W}_{\mathcal{V}'}$. We use them in the same way as Step 1 and obtain $\mathrm{Corr}_{\mathcal{V}\times_M\mathcal{V}'}^{\mathcal{W}_{\mathcal{V}}\times\mathcal{W}_{\mathcal{V}'}}$. It is easy to see that :

$$\mathrm{Corr}_{\mathcal{V}\times_M\mathcal{V}'}^{\mathcal{W}_{\mathcal{V}}\times\mathcal{W}_{\mathcal{V}'}}=\pm\mathrm{Corr}_{\mathcal{V}'}^{\mathcal{W}_{\mathcal{V}}}\circ\mathrm{Corr}_{\mathcal{V}}^{\mathcal{W}_{\mathcal{V}'}}.\tag{118}$$

(We do not discuss sign in this section. See the next section.) The formula (118) plays a crucial role to prove the A_∞ relation for the operations which we will define by the smooth correspondence as in (110) satisfies the A_∞ relation. In order to use (118) for this purpose, we need to choose the continuous family of perturbations so that at the boundary of each of the spaces $\mathfrak{M}_{k+1}(\beta)$, the perturbation is obtained as the fiber product such as (117). Here we remark that by (5) of Definition 21, the boundary of $\mathfrak{M}_{k+1}(\beta)$ is decomposed to a union of fiber products of various $\mathfrak{M}_{k'+1}(\beta')$. We thus proceed inductively and construct continuous family of perturbations.

To carry out the idea described above, we start with defining the order by which we organize the induction. The following definition is taken from [33] Definitions 7.2.61 and 7.2.63. Let G be as in Definition 12.

Definition 26. *For $\beta \in G$, we put :*

$$\|\beta\| = \sup\{n \,|\exists \beta_i \in G \setminus \{(0,0)\}\ \beta_1 + \cdots + \beta_n = \beta\} + [E(\beta)] - 1.$$

Here $[x]$ is the largest integer $\le x$. We put $\|(0,0)\| = -1$.

We define a partial order $<$ on $G \times \mathbb{Z}_{\ge 0}$ as follows. Let $\beta_1, \beta_2 \in G$, $k_1, k_2 \in \mathbb{Z}_{\ge 0}$. We define $>$ so that $(\beta_1, k_1) > (\beta_2, k_2)$ if and only if one of the following holds.

(1) $\|\beta_1\| + k_1 > \|\beta_2\| + k_2$. (2) $\|\beta_1\| + k_1 = \|\beta_2\| + k_2$ *and* $\|\beta_1\| > \|\beta_2\|$.

We will define the continuous family of perturbations on $\mathfrak{M}_{k+1}(\beta)$ according to the (partial) order $<$ of (β, k).

We assume that we have a continuous family of perturbations for all $\mathfrak{M}_{k'+1}(\beta')$ with $(\beta', k') < (\beta, k)$ and construct a perturbation on $\partial\mathfrak{M}_{k+1}(\beta)$. Let $\mathbf{x} \in \partial\mathfrak{M}_{k+1}(\beta)$. We assume that $\mathbf{x}$ is contained in the codimension d corner of $\mathfrak{M}_{k+1}(\beta)$ (but is not in the codimension $d+1$ corner of it). (Here $d \ge 1$.) We put

$$\mathcal{S} = (\Sigma; z_0, \cdots, z_k; \beta(\cdot)) = \pi_0(\mathbf{x}) \in \mathcal{M}_{k+1,\beta}.$$

Σ has exactly d singular points. Let $\Sigma = \cup_{a \in A} D^2_{\mathrm{a}}$ be the decomposition of Σ. (Here $\#A = d + 1$.)

For each $a \in A$, we define

$$\mathcal{S}_a = (D^2_{\mathrm{a}}; z_{a;0}, \cdots, z_{a;k_a}; \beta_a(\cdot)) \in \mathcal{M}_{k_a+1,\beta_a}$$

as follows. D^2_{a} is the disc. The marked points of D^2_{a} are singular or marked points of Σ which are on D^2_{a}. We put $\beta_a = \beta(a)$; $\beta(\cdot)$ is the map which assigns β_a to the unique component of D^2_{a}. The 0-th marked point $z_{a;0}$ is defined as follows. If $z_0 \in D^a_2$ then $z_{a;0} = z_0$. If not there is a unique D^2_b such that $a < b$ and $D^2_b \cap D^2_{\mathrm{a}} \ne \emptyset$. Here $<$ is the order on A which is defined during the proof of Theorem 5. (See Figure 19.) Then $z_{a;0}$ is the unique point in $D^2_{\mathrm{a}} \cap D^2_b$.

We can use (1), (2) and (3) of Definition 21 repeatedly to find a unique element $\mathbf{x}_a \in \mathfrak{M}_{k_a+1}(\beta_a)$ such that $\pi_0(\mathbf{x}_a) = \mathcal{S}_a$ and that $\mathbf{x}_a$ is sent to $\mathbf{x}$ after applying $\circ_{\mathrm{m},*}$ repeatedly. In fact $\mathcal{S}$ is obtained from $\mathcal{S}_a$ by applying $\circ_i$ several times. We apply $\circ_{\mathrm{m},*}$ to $\mathbf{x}_a$ in the same way as $\circ_i$ is applied to $\mathcal{S}_a$. Then Definition 21 (3) implies that this composition is independent of the order used in applying it.

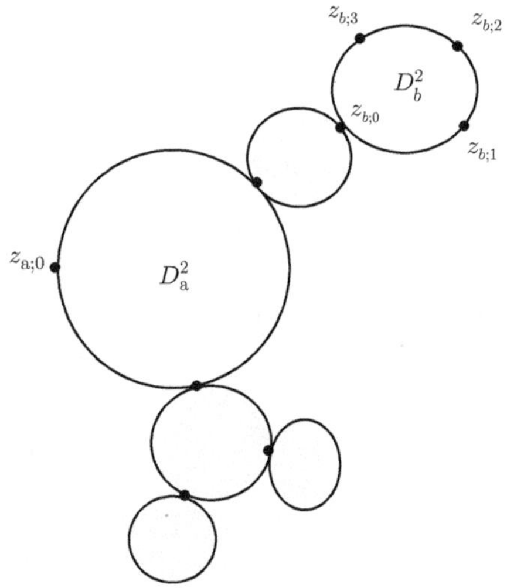

Fig. 19.

We have

Lemma 9. $(k_a, \beta_a) < (k, \beta)$ *for each* a.

The proof is elementary and is omitted. (See [34] Lemma 7.2.47.)

Now we fix a Kuranishi neighborhood $\mathcal{V}_a = (V_a, E_a, \Gamma_a, s_a, \psi_a)$ of $\mathbf{x}_a$ for each $a \in A$. Then by applying (3) and (4) of Definition 21 repeatedly, we find that a Kuranishi neighborhood of $\mathbf{x}$ is obtained as follows:

$$\mathcal{V} = \prod_{a, M^{d-1}} \mathcal{V}_a = \left(\prod_{a, M^{d-1}} V_a \times [0, \epsilon)^{d-1}, \prod_a E_a, \prod s_a, \prod_a \psi_a \right) \tag{119}$$

Here $\prod_{a,M^{d-1}} V_a$ is an appropriate fiber product of V_a over M^{d-1}. (See [33] §7.1.1 for the detailed description of this fiber product.) The factor $[0, \epsilon)^d$ appears since the point $\mathbf{x}$ lies on the codimension d corner.

By the inductive hypothesis, we already defined a continuous family of perturbations $\mathcal{W}_{\mathcal{V}_a} = (W_{\mathcal{V}_a}, \mathfrak{s}_{\mathcal{V}_a}, \omega_{\mathcal{V}_a})$ for each a. We then define $W_{\mathcal{V}} = \prod_a W_{\mathcal{V}_a}$. (Note that some of the factors $W_{\mathcal{V}_a}$ coincide with each other. In that case we repeat the same factor as many times as $W_{\mathcal{V}_a}$ appears in $W_{\mathcal{V}}$.)

We also define $\mathfrak{s}_{\mathcal{V}}$ etc. by restricting the direct product of $\mathfrak{s}_{\mathcal{V}_a}$ etc. to $\prod_{a,M^{d-1}} V_a \times \{0\}^{d-1}$. We thus obtain the required continuous family of perturbations $\mathcal{W}_{\mathcal{V}}$ on the set of codimension d corners of the Kuranishi neighborhood of $\mathbf{x}$. We can extend it to its neighborhood by composing with the obvious projection of the $[0, \epsilon)^{d-1}$ factor to $\{0\}$. We thus obtain a continuous family of perturbations in a Kuranishi neighborhood of $\mathbf{x}$.

By construction it is obvious that the system of family of perturbations above is compatible with the way we glued them at the earlier stage of induction using Step 2. So we obtain a continuous family of perturbations in a neighborhood of the boundary. Then we use Steps 1 and 2 to extend it to the whole $\mathfrak{M}_{k+1}(\beta)$.

Now since the continuous family of perturbations we constructed is consistent with the decomposition of the boundary given in Definition 21 (6), it follows from (118) and Stokes' formula that the operations $\mathfrak{m}_{k,\beta}$ obtained in this way satisfy the A_∞ formula (33). We thus obtain a filtered A_∞ structure on the de Rham complex.

This is the transversality part of the proof of Theorem 12. (We discussed only the construction of operations. The construction of morphisms and their homotopies is similar.)

Remark 13. Actually there is another serious problem to be taken care of in order to rigorously establish a filtered A_∞ structure. This problem is pointed out in [33] §7.2.3, discussed in detail in [33] §7.2 and is summarized as follows. We took continuous families of perturbations inductively. The zero set $\mathfrak{s}_{\mathcal{V}}^{-1}(0)$ should be in a small neighborhood of the original moduli space $s^{-1}(0)$, since if it runs out of the Kuranishi neighborhood we will be unable to use Stokes' theorem to prove the A_∞ formula. By taking the perturbation small, we can do it without difficulty as long as we have only finitely many steps to work out. However in an actual geometric situation, we will define an infinite number of operations $\mathfrak{m}_{k,\beta}$. This causes trouble in the following way. Let us consider the argument of Step 3 above. We first choose ϵ small enough so that $\mathfrak{s}_{\mathcal{V}}^{-1}(0)$ which we construct at the first stage of the induction lies in an ϵ-neighborhood of the original moduli space. Then in the k-th step, the perturbation we find on the boundary or corner is a fiber product of k perturbations of the earlier steps. So the perturbation which is already defined is away from the original moduli space by a distance something like $k\epsilon$. We remark that, in the fiber product decomposition like (119), the same factor (which was already determined at the first stage of the induction, for example) may appear many times. And once we fixed the perturbation at some stage of the induction, we are not supposed to change it later. Thus the zero set of the perturbed section runs out of the Kuranishi neighborhood at some finite stage.

The idea to overcome this difficulty is as follows. For each fixed (n, K) we can choose our perturbations so that we can continue the construction for each (β, k) with $(\beta, k) \le (n, K)$. (Here $<$ is the order defined in Definition 26.) We next define an appropriate notion of $A_{n,K}$ structure. Then, in this way, we can construct an $A_{n,K}$ structure for any but fixed (n, K). We can also prove that the $A_{n,K}$ structure which is constructed above is homotopy equivalent to the $A_{n',K'}$ structure as an $A_{n,K}$ structure, for arbitrary n', K'. We finally use homological algebra to show that this implies that the $A_{n,K}$ structure can be extended to an A_∞ structure.

The argument outlined above is carried out in detail in [33] §7.2. The same trouble seems to occur frequently for the rigorous constructions of various topological field theories by the Kuranishi correspondence. The method we explained above seems to work in all cases. (At the time of writing this chapter, the author does not know any other way to resolve this problem.) An earlier example where a similar problem appeared is the study of Floer homology of periodic Hamiltonian systems in a monotone symplectic manifold. K. Ono [58] resolved this difficulty in that case in a similar but slightly different way, using the projective limit. (This is the reason why the result of [58] is more general than that of [38].)

In order to prove Theorem 12, we need to work out the same argument for morphisms and homotopies. In order to carry it out in the case of homotopies, we need to show compatibility of $A_{n,K}$ homotopies between various different n, K. This requires us to study the notion of homotopy of homotopies. Two (rather heavy) subsections §7.2.12 and §7.2.13 of [33] are devoted to this point.

13 Orientation.

In this section we discuss signs or orientations. The problem of orientation and sign appears in two related but different ways in the construction of topological field theory.

(1) To prove that the K-space of the appropriate moduli problem is orien*table*. To find and describe the geometric data which determines the orientation of the K-space.
(2) To organize the orientations of several fiber products appearing in the construction in a consistent way. To fix the sign convention of the algebraic system involved. To prove that the system of orientations organized above is consistent with the sign convention of the algebraic systems.

The point (1) is a problem of family index theory. (This observation goes back to [17].) For example, in the case of the moduli space of pseudo-holomorphic discs which bounds a given Lagrangian submanifold, it is proved that the moduli space is orientable in case L is relatively spin, in [32] and [63]. We also remark that even in case we can prove that the moduli space involved is orientable (in a way consistent with the fiber products as in Definition 21 (5)), it is a different problem to specify the geometric data which determines the orientation. In other words, proving the *existence* of a coherent orientation is not enough to complete this step. Actually there can be several different choices of coherent orientations, in general. (See [13] for explicit example of this phenomenon in Lagrangian Floer theory.)

Such a phenomenon already appears in classical Morse theory as follows. Let M be a smooth manifold with $H^1(M;\mathbb{Z}_2) \neq 0$ and $f : M \to \mathbb{R}$ be a Morse function. To define Morse homology we need to specify the orientation of the moduli space $\mathcal{M}(p,q)$ of the gradient lines joining two given critical points p, q. The system of orientations of them for various p, q is said to be coherent if it is consistent with the fiber product structure in Formula (17). We can find such a coherent orientation for each representation : $\pi_1(M) \to \mathbb{Z}_2$ and different choices induce different homology groups. So we need to find some way to distinguish the trivial representation from other ones, to find a system of orientations which gives the ordinary homology.

The point (2) is of a different nature. At first sight, it might look like rather a technical problem that could be resolved by 'patience' and 'carefulness'. (For this reason the importance of this point is frequently overlooked.) Indeed, in early days, when the structure involved was rather simple, one could fix the sign convention by hand once the point (1) was understood. However as time elapsed and as the structure to deal with became more advanced, it got harder to find a correct sign convention. Especially the work to check whether it coincides with the sign or orientation of geometric origin becomes more and more cumbersome. (It seems that the amount of the work to study the sign grows exponentially as the complexity of the structure we deal with grows.) Then one arrives at the point where fixing sign and orientation only by patience and carefulness becomes impossible. We thus need some 'principle' to fix the sign convention and to show that it coincides with one of geometric origin. In other words, studying the sign is related to the procedure of ensuring that the construction is sufficiently canonical.

In this section, we do not discuss point (1) since it is related to the geometric origin of the moduli space (or K-space) and so is not a part of the *general* theory we are building. Our focus in this section is point (2). The major part of [33] Chapter 8 is actually devoted to this point. There we still gave an explicit choice of signs and of the orientations of the moduli spaces and its fiber products. Though there is some 'principle' behind each of our choices, it is hard to state it in a mathematical and rigorous way, so it was rarely mentioned explicitly. And the proof in [33] Chapter 8 of the consistency of the orientation and sign was based on calculations.

The purpose of this section is to explain the way in which we translate the discussion of [33] Chapter 8 to the more abstract situation of this chapter. (In [33] Chapter 8 the situation of Lagrangian Floer theory is discussed.) Along the way, we state precisely the compatibility condition of orientations among various spaces $\mathfrak{M}_{k+1}(\beta)$. This point was postponed in Definition 21.

We first introduce some notations. Let $\mathfrak{S}_k$ be the symmetric group of order $k!$. We put $\mathfrak{M}^+_{k+1}(\beta) = \mathfrak{S}_k \times \mathfrak{M}_{k+1}(\beta)$ on which $\mathfrak{S}_k$ acts by the left multiplication of the first factor. There is a one to one correspondences between the set of orientations on $\mathfrak{M}_{k+1}(\beta)$ and the set of orientations on $\mathfrak{M}^+_{k+1}(\beta)$ such that the action of σ is orientation preserving if and only if $\sigma \in \mathfrak{S}_k$ is an even permutation. We hereafter identify them.

Remark 14. In the case when $\mathfrak{M}_{k+1}(\beta)$ is the moduli space $\mathfrak{M}_{k+1}(\beta; J)$ of pseudo-holomorphic discs (which we introduced in §11), the space $\mathfrak{M}^+_{k+1}(\beta)$ is regarded as a compactification of the set of $(D^2; z_0, \cdots, z_k; u)$ such that $z_i \in \partial D^2$ and that u is J-holomorphic map with $(\int u^*(\omega), \eta([u])) = \beta$. Note we do *not* require the points $z_0, \cdots, z_k$ to respect the cyclic order. The $\mathfrak{S}_k$ action is defined by $(z_0, z_1, \cdots, z_k) \mapsto (z_0, z_{\sigma(1)}, \cdots, z_{\sigma(k)})$. The geometric meaning of the discussion below becomes clearer if the reader keeps this example in his mind.

The map

$$(\pi_1, \pi_2) : (ev_0, e_1, \cdots, ev_k) : \mathfrak{M}_{k+1}(\beta) \to M^{k+1}$$

is extended to $\mathfrak{M}^+_{k+1}(\beta)$ by

$$ev_i(\sigma, \mathbf{x}) = ev_{\sigma(i)}(\mathbf{x}) \quad (i \neq 0), \qquad ev_0(\sigma, \mathbf{x}) = ev_0(\mathbf{x}).$$

We extend $\circ_{\mathfrak{m},i}$ to

$$\circ_{\mathfrak{m},i} : \mathfrak{M}^+_{k+1}(\beta_i)_{ev_i} \times_{ev_0} \mathfrak{M}^+_{l+1}(\beta_2) \to \mathfrak{M}^+_{k+l}(\beta_1 + \beta_2)$$

by

$$(\sigma_1, \mathbf{x}_1) \circ_{\mathfrak{m},i} (\sigma_2, \mathbf{x}_2) = (\sigma, \mathbf{x}_1 \circ_{\mathfrak{m},\sigma_1(i)} \mathbf{x}_2)$$

where σ is defined by

$$\sigma(j) = \begin{cases} \sigma_1(j) & j < i, \quad \sigma_1(j) < \sigma_1(i) \\ \sigma_1(j) + l - 1 & j < i, \quad \sigma_1(j) > \sigma_1(i) \\ \sigma_2(j - i + 1) + \sigma_1(i) - 1 & i \leq j < i + l - 1 \\ \sigma_1(j - l + 1) & j \geq i + l, \quad \sigma_1(j - l + 1) < \sigma_1(i) \\ \sigma_1(j - l + 1) + l - 1 & j \geq i + l, \quad \sigma_1(j - l + 1) > \sigma_1(i) \end{cases} \tag{120}$$

$\mathfrak{M}^+_{k+1}(\beta)$ is a K-space whose boundary stratum is a union of the images of $\circ_{\mathfrak{m},i}$. Namely

$$\circ_{\mathfrak{m},i} \left(\mathfrak{M}^+_{k+1}(\beta_i) \,{}_{ev_i}\times_{ev_0} \mathfrak{M}^+_{l+1}(\beta_2)\right) \subset \partial\mathfrak{M}^+_{k+l}(\beta_1 + \beta_2)$$

The compatibility condition of the orientations is defined as follows.

Definition 27. *We say the orientations of $\mathfrak{M}_{k+1}(\beta)$ for various k, β are compatible if the embedding*

$$\circ_{\mathfrak{m},i} : \mathfrak{M}^+_{k+1}(\beta_1)_{ev_1} \times_{ev_0} \mathfrak{M}^+_{l+1}(\beta_2) \subset (-1)^{(k-1)(l-1)+(\dim M + k)} \partial\mathfrak{M}^+_{k+l}(\beta_1 + \beta_2)$$

is orientation preserving for every k, l, β_1, β_2.

Actually this is a copy of the conclusion of [33] Proposition 8.3.3.

Remark 15. (1) We remark that, in Definition 27, the condition is put only on the fiber product by ev_1. Using the action of symmetric group, the compatibility of the orientations of other cases (namely the case when the fiber product is taken by ev_i) are induced automatically. This is the reason why we introduced $\mathfrak{M}^+_{k+1}(\beta_1)$.
(2) Actually we need to fix several conventions to discuss signs. In particular, we need to specify the orientation of the fiber product. Here we omit them and refer to [33] Chapter 8.

Now we consider chains (P_i, f_i) on M. Here P_i is a smooth manifold and f_i is a smooth map $P_i \to M$. Under the appropriate transversality condition we consider the fiber product of them with $\mathfrak{M}_{k+1}(\beta)$ and obtain a compact oriented and smooth manifold. We define an orientation of it by

$$(-1)^* \mathfrak{M}_{k+1}(\beta) \times_{M^k} (P_1 \times \cdots \times P_k) \tag{121}$$

with

$$* = (\dim M + 1) \sum_{j=1}^{k-1} \sum_{i=1}^{j} (\dim M - \dim P_i).$$

This definition is taken from [33] §8.5. Now we can translate the proof of [33] Proposition 8.5.1 *word* for *word* to our more abstract situation and show the A_∞ relation for the operations $\mathfrak{m}_{k,\beta}$ defined by (121), as far as the transversality condition is satisfied. We remark that the notation of this section and that in [33] Chapter 8 correspond as follows :

$$\mathfrak{M}^+_{k+1}(\beta) \longleftrightarrow \mathcal{M}_{k+1}(\beta), \qquad \mathfrak{M}_{k+1}(\beta) \longleftrightarrow \mathcal{M}^{\mathrm{main}}_{k+1}(\beta).$$

In §12, to discuss the transversality problem, we use the de Rham complex. So the orientation problem which is required to work out the proof of §12 is fixing the sign of the operations defined by (111), etc, on differential forms on M. We can reduce this problem to the problem of signs on the operations among the chains (P_i, f_i) in M as follows. (This is explained in more detail in [33] §8.10.3.)

In §12 we constructed the operations by using a continuous family of perturbations as follows. We took W a huge parameter space and considered the correspondence

$$M^k \xleftarrow{\ \pi_2\ } V \times W \xrightarrow{\ \pi_1\ } M.$$

Here we have an obstruction bundle E over V and have a multisection $\mathfrak{s}$: $V \times W \to E \times W$. We fixed a branch $\mathfrak{s}_c$ and took $\mathfrak{s}_c^{-1}(0) \subset V \times W$. We also use that ω is a top form on W. We pull it back to $\mathfrak{s}_c^{-1}(0)$ and the operation is defined by

$$u \mapsto \pm \pi_1! \left(\pi_2^*(u) \wedge \omega|_{\mathfrak{s}_c^{-1}(0)} \right), \tag{122}$$

where u is a differential form on M^k. The other part of the construction such as taking partitions of unity, etc., does not affect the problem of signs. Therefore, we only need to find a way to define the sign $\pm$ in (122) so that the resulting operation satisfies the A_∞ relation.

To reduce this problem to the orientation of the fiber product as in (121), we proceed as follows. We can approximate our smooth form u by a current realized by the product of chains (P_i, f_i). So while discussing the orientation problem we only need to consider the case $u = (u_1, \ldots, u_k)$ where u_i is realized by (P_i, f_i). We next take generic $w \in W$. Then the fiber product

$$(\mathfrak{s}_c^{-1}(0) \cap (V \times \{w\})) \times_{M^k} (P_1 \times \cdots \times P_k) \tag{123}$$

is well-defined. (Namely, transversality holds.) For such w we can define the sign by the same formula (121). In fact $\mathfrak{s}_c^{-1}(0) \cap (V \times \{w\})$ is (an open subset of) a perturbation of our moduli space $\mathfrak{M}_{k+1}(\beta)$.

We next regard ω as a smooth *measure*. We remark that we need to fix an orientation of W for this purpose. We did this already when we identify the smooth measure on W with a differential form of degree $\dim W$ in §12.

Using this smooth probability measure we average the current which is obtained by pushing down (123) to M by ev_0. It is easy to see that the average coincides with (122). Thus using (121) we can fix the sign in (122). The A_∞ relation (with sign) of the operation given by (121) implies the A_∞ relation *with sign* of the operation defined by using (122).

This is the argument to reduce the problem of signs in Theorem 12 to the result of [33] Chapter 8. We discussed only the case of construction of filtered A_∞ algebras. The orientation problem in the construction of the filtered A_∞ homomorphisms and homotopies between them can be reduced to [33] Chapter 8 in the same way. The proof of Theorem 12 is now complete.

We finally go back to a point mentioned before, that is, the data used to determine the sign of our A_∞ algebra. Using the consistency condition as in Definition 27, the orientation of the K-spaces $\mathfrak{M}_{k+1}(\beta)$ is determined by $\mathfrak{M}_{k'+1}(\beta')$ for other k', β' with $(\beta', k') < (\beta, k)$. Here the order $<$ is introduced in Definition 26. So the choice of orientation of $\mathfrak{M}_{k+1}(\beta)$ for which (β, k) is minimal determines the orientation of the other $\mathfrak{M}_{k+1}(\beta)$. (More precisely we can slightly modify $<$ to $<'$ so that $(\beta', k') <' (\beta, k)$ if and only if $\mathfrak{M}_{k'+1}(\beta')$ appears in the boundary of $\mathfrak{M}_{k+1}(\beta)$. $<'$ above implies $<$ in Definition 26. But the converse may not be true.) The minimal (β, k) is $(\beta_0, 2)$ and $(\beta, 0)$ where $\beta_0 = (0, 0)$ and β is a primitive element of G.

In the situation of §11, $\mathfrak{M}_{2+1}(\beta_0)$ is L itself. We can fix the orientation of it so that $\mathfrak{m}_{2,\beta_0}$ is induced by usual cup product as in Example 1.

The orientation of $\mathfrak{M}_1(\beta)$ is more involved. It depends on the geometric data such as relative spin structure in the case of Lagrangian Floer theory. We remark that in general we can not choose orientations of various $\mathfrak{M}_1(\beta)$ with β primitive independently, because then the compatibility condition may not be satisfied. In fact if $\beta_1 + \beta_2 = \beta_1' + \beta_2' = \beta$ are decompositions of β to different sum of primitive elements, then by looking the consistency at $\mathfrak{M}_1(\beta)$, the choice of the orientations of three of $\mathfrak{M}_1(\beta_1)$, $\mathfrak{M}_1(\beta_2)$, $\mathfrak{M}_1(\beta_1')$, $\mathfrak{M}_1(\beta_2')$ determine the orientation of the fourth one automatically. This kind of phenomenon occurs since our monoid G may not be free.

If G is free (namely is isomorphic to $\mathbb{Z}_{\geq 0}^m$ for some m), then the choice of the orientation of $\mathfrak{M}_1(\beta)$ for the generators β of our monoid G corresponds one to one to the choice of system of orientations of all $\mathfrak{M}_{k+1}(\beta)$ satisfying the compatibility condition. In such a situation there is a simpler proof of the existence of consistent system of orientations and signs. See [27] §7 for such an argument.

We remark that we choose G satisfying Definition 12, because of Gromov compactness and is related to nonlinear analysis of pseudo-holomorphic curve theory. The problem of orientation is related to index theory and to linear analysis. Therefore during the discussion of the sign, we can replace G by a bigger monoid. Actually we can take $G = G_+(L)$ in (99). This might be a way to reduce the problem to the case when G is free.

14 Variations and generalizations.

There are many directions we can generalize the construction of this article. We mention some of them briefly below. Many of them are subjects of the future research and the argument of this section is rather brief. Proof of none of them are regarded to be completed except those which are proved in the reference rigorously.

14.1 Unitality

There is a unital version of the notion of A_∞ space and A_∞ algebra. Usually the unital version is called A_∞ space in the literature. So the version in §4 should be called non-unital A_∞ space. Let M be a space with a base point $*$. Then we require

$$\mathsf{m}_k(a; x_1, \cdots, x_{i-1}, *, x_{i+1}, \cdots, x_k) = \mathsf{m}_{k-1}(a; x_1, \cdots, x_{i-1}, x_{i+1}, \cdots, x_k)$$

in addition to define the notion of a unital A_∞ space. For A_∞ algebra, its unit $\mathbf{e}$ is an element of degree 0 (before shifted) such that :

$$\mathfrak{m}_k(x_1, \cdots, x_{i-1}, \mathbf{e}, x_{i+1}, \cdots, x_k) = 0$$

for $k \neq 2$ and

$$\mathfrak{m}_2(x, \mathbf{e}) = (-1)^{\deg x}\mathfrak{m}_2(\mathbf{e}, x) = x.$$

There is also the notion of homotopy unit. (See [33] §3.3. See also [61] §2 and references therein for various versions of unit or homotopy unit and the relationships among them.)

The singular homology of a unital A_∞ space is a unital A_∞ algebra with the 0-chain $*$ as its unit. This can be proved in the same way as Theorem 3.

The situation is different for Kuranishi correspondences. The candidate for the unit is the fundamental chain which is regarded as a degree 0 cochain by Poincaré duality. In the case of de Rham theory which we worked out in §12, it is a 0-form and is the constant function $\equiv 1$. This is actually the case when there is a map

$$\mathfrak{forget}_i : \mathfrak{M}_{k+1}(\beta) \to \mathfrak{M}_k(\beta)$$

which is compatible with the map $\mathfrak{forget} : \mathcal{M}_{k+1} \to \mathcal{M}_k$ forgetting the ith marked point and which is compatible with $ev_j : \mathfrak{M}_{k+1}(\beta) \to M$. We can prove this in the same way as [33] Lemma 7.3.8. In this way we can prove that our A_∞ algebra of Theorem 13 on the de Rham complex of a Lagrangian submanifold has a strict unit.

14.2 Module and Category

There is a notion of A_∞ (bi)module and also A_∞ category. See [33] §12 and Chapter 15 for the bimodule and [26] etc. for the A_∞ category. We can modify Theorem 12 to include these cases, in a straightforward way.

14.3 Local coefficient

In the case of Bott–Morse theory, we constructed a structure (that is, a higher boundary operator $\partial_{\mathcal{M}}$) on the direct sum $\bigoplus_a C(R_a; \Theta_a^-)$ of chain complexes of singular chains on R_a *with local coefficients.* In order to include such a situation in our machinery, we consider a correspondence such as

$$(M, \Theta_M)^k \xleftarrow{ev_1} \mathfrak{M}_{k+1} \xrightarrow{ev_2} (M, \Theta_M). \tag{124}$$

In this case, in place of assuming the K-space $\mathfrak{M}_{k+1}$ to be oriented, we assume that it has a relative orientation. Namely we assume that

$$ev_1^*(\Theta_M \otimes \cdots \otimes \Theta_M) \otimes ev_2^*\Theta_M \otimes \Lambda^{\rm top}TV \otimes \Lambda^{\rm top}E \tag{125}$$

has a trivialization. Here V is a Kuranishi neighborhood and E is an obstruction bundle. We also assume that the trivialization of (125) is compatible with coordinate change. Namely we assume it is compatible with Diagram (86).

This situation appears when we study the Bott–Morse version of Lagrangian Floer homology for a pair of Lagrangian submanifolds with clean intersection. See [33] §3.7.5 and §8.8, for details. The argument there can be directly generalized to our abstract situation.

14.4 Family version

For a family of Lagrangian submanifolds L in M we can study family Floer homologies. (See [25, 39].) An abstract version of this construction can be formulated as follows.

We can generalize homotopy of Kuranishi correspondence (that is $[0,1]$ parametrized family of Kuranishi correspondences) to a family parametrized by an arbitrary manifold. Namely we can consider the following situation. Let $M \to X \xrightarrow{\pi} B$ be a family of manifolds M parametrized by a manifold B. We consider

$$\begin{array}{ccccc} & & \mathcal{M}_{k+1,\beta} & & \\ & & \uparrow \pi_0 & & \\ X^k & \xleftarrow[\pi_2]{} & \mathfrak{M}_{k+1}(\beta) & \xrightarrow[\pi_1]{} & X. \end{array} \tag{126}$$

such that

$$(\pi_1, \pi_2)(\mathfrak{M}_{k+1}(\beta)) \subseteq \{(x_0, \cdots, x_k) \in X^{k+1} \mid \pi(x_0) = \cdots = \pi(x_k)\} \subset X^{k+1}.$$

Then by replacing the fiber product $\mathfrak{M}_{k+1}(\beta_1)\ {}_{ev_i}\times_{ev_0} \mathfrak{M}_{l+1}(\beta_2)$ over M (in (1) Definition 21) by the fiber product over X we can generalize the Definition 21 to a B parametrized version. We call it the B parametrized family of Kuranishi correspondence.

Let us briefly describe the corresponding algebraic object. Let B be a simplicial complex. We define the notion of B parametrized family of A_∞ algebra as follows. For each simplex σ there is a (filtered) A_∞ algebra $C(\sigma)$ whose homology group is one of M. If σ_i is the i-th face of σ there is a linear A_∞ homomorphism

$$\mathrm{Eval}_{\partial_i} : C(\sigma) \to C(\sigma_i)$$

which is a homotopy equvalence. Let σ_{ij} be the set of all codimension 2 simplex of σ. Then we require the existence of the following exact sequence

$$C(\sigma) \to \bigoplus_i C(\sigma_i) \to \bigoplus_{ij} C(\sigma_{ij}).$$

See [33] Definition 7.2.188 for a similar exact sequence for rectangle.

We can associate B parametrized family of A_∞ algebra to B parametrized family of Kuranishi correspondence in a way similar to the proof of Theorem 12.

14.5 Group action and localization to fixed point set

In the study of Gromov-Witten invariant, localization to the fixed point set plays an important role. Gromov-Witten invariant of a manifold M is a family of numbers parametrized by homology classes of M and by homology classes of the Deligne-Mumford compactification of moduli space of Riemann surfaces. The localization formula gives a way to reduce its calculation to the study of neighborhood of the fixed point locus of the moduli space of pseudo-holomorphic curves, when a group acts on it.

A problem to extend it to our story, for example to the study of Lagrangian Floer theory, lies in the fact that it is rather hard to find a correct statement of the (expected) result. This is because the structure constant of the algebraic system (which is the number obtained by counting the order of the moduli space in an appropriate sense) itself is not well-defined.

The result of this paper gives a way to formulate such a statement.

Let us exhibit a way to do so by considering the special case where the group is S^1 and the action of it on M is trivial. Let $\mathfrak{M}_{k+1}(\beta)$ be a Kuranishi correspondence on M. We assume S^1 acts on $\mathfrak{M}_{k+1}(\beta)$ such that the structure map and evaluation map is S^1 equivalent. (We put trivial action on M and on $\mathcal{M}_{k+1,\beta}$.) We put

$$(\mathfrak{M}_{k+1}(\beta))^{S^1} = \{x \in \mathfrak{M}_{k+1}(\beta) \mid \forall g \in S^1 \;\; gx = x\}. \tag{127}$$

We will define a K-space $(\mathfrak{M}_{k+1}(\beta))^{S^1+}$ as follows. Let $(V_i, E_i, \Gamma_i, \psi_i)$ be a Kuranishi chart. Let $V_i^{S^1}$ be the set of S^1 fixed point of V_i. We fix sufficiently small ϵ and take an ϵ neighborhood $B_\epsilon(V_i^{S^1})$ of $V_i^{S^1}$. Here we use S^1 invariant Riemannian metric on V_i. We identify $B_\epsilon(V_i^{S^1})$ with an open subset of the normal bundle. We consider the sphere bundle which is a boundary of $B_\epsilon(V_i^{S^1})$ and denote it by $S_\epsilon(V_i^{S^1})$. Our group S^1 acts on it so that the isotropy group is finite. We take a quotient of $S_\epsilon(V_i^{S^1})$ by the S^1 action and glue the quotient with $B_\epsilon(V_i^{S^1}) \setminus S_\epsilon(V_i^{S^1})$. We denote the resulting space by $\mathbb{P}V_i^{S^1}$. It is an orbifold. Our obstruction bundle E_i induces an orbibundle E_i' on it. The Kuranishi map s_i induces a section s_i' of E_i'. Let $Z_i = s_i'^{-1}(0)/\Gamma_i$. We can glue them using coordinate transformation to obtain a space Z. By covering Z_i with open subset of V_i and using the restriction of E_i', s_i' there, we obtain a Kuranishi chart for each points on Z_i. We can glue them in an obvious way to obtain a Kuranishi chart on Z. We thus obtain a Kuranishi structure on Z. We denote the space Z together with the above Kuranishi structure by $(\mathfrak{M}_{k+1}(\beta))^{S^1+}$.

Using the fact that the structure map and evaluation map of $\mathfrak{M}_{k+1}(\beta)$ S^1 equivalent, we can show that $(\mathfrak{M}_{k+1}(\beta))^{S^1+}$ is regarded as a Kuranishi correspondence on M.

In the next theorem we use $\mathbb{R}$ coefficient.

Theorem 14. *The filtered A_∞ structure associated to $(\mathfrak{M}_{k+1}(\beta))^{S^1+}$ by Theorem 12 is homotopy equivalent to one associated to $\mathfrak{M}_{k+1}(\beta)$.*

Sketch of the proof : Let δ be a positive number sufficiently smaller than ϵ. For each Kuranishi neighborhood V_i we consider $V_i \setminus B_\delta(V_i^{S^1})$. Since S^1 action is locally free there the quotient space $\frac{V_i \setminus B_\delta(V_i^{S^1})}{S^1}$ is an orbifold. The bundle E_i induces an orbibundle $\overline{E}_i$ on it. In the same way as §12, we can take a continuous family of multisection $\overline{E}_i$ of this bundle and lift it to $V_i \setminus B_\delta(V_i^{S^1})$. We thus have a continuous family of S^1 equivariant multisection on $V_i \setminus B_\delta(V_i^{S^1})$ which are transversal to zero. Obviously we can do it in a way compatible with the coordinate change. We can extend this multisection to $B_\delta(V_i^{S^1})$ so that it is transversal to zero but is not necessary S^1 invariant. We use this continuous family of multisections to define the operators $\mathfrak{m}_{k,\beta}$ as in §12.

Since evaluation map ev is S^1 equivariant, we can show that the contribution of the part to $V_i \setminus B_\delta(V_i^{S^1})$ to $\mathfrak{m}_{k,\beta}$ is 0. Hence the theorem. □

14.6 Other operad or prop

As we mentioned several times, the construction works for other operads or props than A_∞ operads. The argument of §12 can be generalized with littile change. For the part of the proof we gave in §13, we need certain modification.

The construction of appropriate *differentiable* operad or prop is also a nontrivial problem. It seems to the author that claims in the talks [5], [70] etc. can be reinterpreted as an existence of a differential prop associated to the moduli space of higher genus Riemann surface. Namely master-equation claimed in those talks are Maurer-Cartan equation (30), which is an important part of the axiom of differential PROP. (See Definition 2.)

14.7 Gravitational descendant

In the usual theory of operads, spaces $\mathcal{P}(n)$ (or $\mathcal{M}_{k+1}$) are assumed to be contractible. However in the situation of several 'operads' or 'props' appearing in topological field theory there is a situation where they have a nontrivial homotopy type. In the case of A_∞ operad, $\mathcal{M}_{k+1}$ is contractible. The important case where nontrivial homotopy type appears is the case of higher genus Riemann surface and/or the case where interior marked point is included.

We can modify our construction of the structure to include the nontrivial homotopy type of operad or PROP. Various related ideas are discussed by various people (See for example [11].) mainly from the algebraic side.

An example of such a construction is as follow. We consider the direct sum

$$\Lambda(\mathcal{M}) = \bigoplus_k \Lambda(\mathcal{M}_{k+1})$$

of the de Rham complexes of our differential operads. We use Maurer-Cartan axiom (30) to obtain a homomorphism

$$\Delta : \Lambda(\mathcal{M}) \to \Lambda(\partial\mathcal{M}) \to \Lambda(\mathcal{M}) \mathbin{\hat{\otimes}} \Lambda(\mathcal{M}), \tag{128}$$

by restriction. Here $\hat{\otimes}$ is the tensor product in the sense of Fréchet-Schwartz. space.

We call a sequence $c_m \in \Lambda(\mathcal{M})$ a *multiplicative sequence* ([37] §1) if $c_0 = 1$ and

$$\Delta c_m = \sum_{i=0}^{m} c_i \otimes c_{m-i}, \qquad dc_i = 0. \tag{129}$$

When a multiplicative sequence c_m is given, we use it to replace (111) by

$$\sum_{m=0}^{\infty} \pm \frac{s^m}{\#\Gamma}(\pi_{1+}!)\left((\pi_{2+}^*(u) \wedge \omega_{\mathcal{V}} \wedge \pi_0^* c_m)|_{\mathfrak{s}_{\mathcal{V}}^{-1}(0)}\right), \tag{130}$$

where s is another formal parameter. We then obtain a formal deformation of our structure parametrized by s.

Unfortunately in case of A_∞ operad, the space $\mathcal{M}_{k+1}$ is contractible. Therefore there is no multiplicative sequence other than trivial one. However in case we include higher genus Riemann surface and interior marked points, non trivial example is obtained by using Mumford-Morita class. ([10].)

14.8 Infinite dimensional M

In the situation of String topology ([12]) and the loop space formulation of Lagrangian Floer theory ([29, 16]), the correspondence we use is slightly different from those discussed in this paper and can be described by a diagram:

$$(\Omega M)^k \xleftarrow{(ev_1,\cdots,ev_k)} \mathfrak{M}_{k+1}(\beta) \xrightarrow{ev_0} \Omega M. \tag{131}$$

Here ΩL is the free loop space and is of infinite dimension. The structure map is

$$\mathfrak{M}_{k+1}(\beta_1)\ {}_{ev_*\circ ev_i}\times_{ev_*\circ ev_0} \mathfrak{M}_{l+1}(\beta_2) \to \mathfrak{M}_{k+l+1}(\beta_1+\beta_2) \tag{132}$$

Here $ev_* : \Omega(M) \to M$ is the map $\ell \mapsto \ell(*)$ and $* \in S^1$ is the base point. The interesting new point (due to Chas and Sullivan) appearing here is that we take fiber product over M and not over $\Omega(M)$.

We need several modifications of the argument of this paper to include this case. We however remark that the method in [29] to realize transversality in the case of loop space is very similar to one in §12 of this paper.

We may also consider the case of gauge theory (of 4 manifolds, for example) where our M is an infinite dimensional space consisting of gauge equivalence classes of connections. There seems to be much more works to be done to extend the frame work of this paper to include gauge theory.

References

1. J. Adams *Infinite Loop Space* Annals of Mathematics Studies **90** Princeton Univ. Press Princeton (1978).
2. M.F. Atiyah *Topological quantum field theories* Publ. IHES **68** (1989) 175–186.
3. V. Arnold and A. Givental *Symplectic Geometry* Encyclopedia of Math. Sci., Dynamical System V, ed. V. Arnold and S. Novikov Springer Berlin (1980) 4–136.
4. D. Austin and P. Braam *Morse-Bott theory and equivariant cohomology* The Floer Memorial Volume, Progr. Math. **133** ed. H. Hofer, C. Taubes, A. Weinstein and E. Zehnder, Birkhäuser Basel (1995) 123–183.
5. S. Barannikov *Batalin-Vilkovisky geometry of Gromov-Witten invariants* Talk at ESI Workshop on Homological Mirror Symmetry (2006).

6. M. Betz and R. Cohen *Graph moduli spaces and cohomology operations* Turkish J. Math. **18** (1994) 23–41.
7. J. M. Boardman and R. M. Vogt *Homotopy Invariant Algebraic Structures on Toplogical Spaces* Lecture Notes in Math. **347** (1973) Springer-Verlag.
8. R. Bott *Nondegenerate critical manifolds* Ann. of Math. **60** (1954) 248–261.
9. L. Buhovsky *Multiplicative structures in lagrangian Floer homology* Preprint. (2006).
10. T. Coates and A. Givental *Quantum Riemann - Roch, Lefschetz and Serre* math.AG/0110142.
11. K. Costello *Topological conformal field theories and Calabi-Yau categories* math.QA/0412149.
12. M. Chas and D. Sullivan *String topology* math.GT/9911159, (1999).
13. C.-H. Cho *Holomorphic discs, spin structures and Floer cohomology of the Clifford tori* Internat. Math. Res. Notices (2004) **35** 1803–1843.
14. C.-H. Cho *Products of Floer cohomology of torus fibers in toric Fano manifolds* Commun. Math. Phys. **260** (2005) 613–640, math.SG/0412414
15. C.-H. Cho and Y.-G.Oh *Floer cohomology and disc instantons of Lagrangian torus fibers in Fano toric manifolds* Asian J. Math. **10** (2006) 773 - 814.
16. K. Cieliebak and J. Latschev *Symplectic field theory and string topology.* working draft (2006).
17. S. Donaldson *An application of gauge theory to the topoology of 4-manifolds* J. Differential Geom. **18** (1983) 269 - 316.
18. Y. Eliashberg, A. Givental and H. Hofer *Introduction to symplectic field theory* Geom. and Func. Analysis **special volume** (2000) 560–673.
19. A. Floer *Morse theory for Lagrangian intersections* J. Differential Geom. **28** (1988) 513–547.
20. K. Fukaya *Morse homotopy, A^∞-categories, and Floer homologies* Proc. of the 1993 Garc Workshop on Geometry and Topology ed. H. J. Kim, Lecture Notes series **18**, Seoul Nat. Univ. Seoul (1993) 1–102 (http://www.math.kyoto-u.ac.jp/~fukaya/ fukaya.html).
21. K. Fukaya *Topological field theory and Morse theory* [translation of Sūgaku **46** (1994), 289–307] Sugaku Expositions **10** (1997) 19–39.
22. K. Fukaya *Floer homology of connected sum of homology 3-spheres* Topology **35** (1996) 89–136.
23. K. Fukaya *Morse homotopy and its quantization* Geometry and Topology ed. W. Kazez, International Press (1997) 409–441.
24. K. Fukaya *Floer homology for 3-manifolds with boundary I.* **Never to appear**: (http://www.math.kyoto-u.ac.jp/~fukaya/fukaya.html) (1997).
25. K. Fukaya *Floer homology for families—a progress report* Integrable Systems, Topology, and Physics (Tokyo, 2000) Comtemp. Math. **309** (2002) 33–68.
26. K. Fukaya *Floer homology and mirror symmetry. II* Minimal Surfaces, Geometric Analysis and Symplectic Geometry (Baltimore, MD, 1999) Adv. Stud. Pure Math. **34** (2002) 31–127.
27. K. Fukaya *Mirror symmetry of abelian varieties and multi-theta functions* J. Algebraic Geom. **11** (2002) 393–512.
28. K. Fukaya *Deformation theory, homological algebra and mirror symmetry* Geometry and Physics of Branes (Como, 2001) Ser. High Energy Phys. Cosmol. Gravit. IOP (2003) Bristol 121–209.

29. K. Fukaya *Application of Floer homology of Lagrangian submanifolds to symplectic topology* Morse Theoretic Methods in Nonlinear Analysis and in Symplectic Topology, NATO Science Series II: Mathematics, Physics and Chemistry, **217** (2005) 231–276.
30. K. Fukaya, and K. Ono *Arnold conjecture and Gromov-Witten invariant,* Topology **38** (1999) 933–1048.
31. K. Fukaya and Y.-G. Oh *Zero-loop open strings in the cotangent bundle and Morse homotopy* Asian J. Math **1** (1997) 99–180.
32. K. Fukaya, Y.-G. Oh, H. Ohta and K. Ono *Lagrangian intersection Floer theory : anomaly and obstructions*, Preprint (2000), http://www.kusm.kyoto-u.ac.jp/
33. K. Fukaya, Y.-G. Oh, H. Ohta and K. Ono *Lagrangian intersection Floer theory : anomaly and obstructions*, Preprint (2006) AMS/IP studies in Advanced Mathematics 46, Amer. Math. Soc./International Press, 2009.
34. K. Fukaya, Y.-G. Oh, H. Ohta and K. Ono *Supplement to the 2006 preprint version of [33]*, Preprint (2007).
35. P. Griffiths and J. W. Morgan *Rational Homotopy Theory and Differential Forms* Progress in Math. **16** Birkhäuser Basel (1981).
36. M. Gromov *Pseudoholomorphic curves in symplectic geometry* Invent. Math. **82** (1985) 307–347.
37. F. Hirzebruch *Topological methods in Algebraic geometry* Springer (1956)
38. H. Hofer and D. Salamon *Floer homology and Novikov rings* The Floer Memorial Volume Progr. Math., **133** H. Hofer, C. Taubes, A. Weinstein and E. Zehnder, Birkhäuser Basel (1995) 483–524
39. M. Hutchings *Floer homology of families I* math.SG/0308115. Algebra of Topol. 8(2008), 435–492.
40. T. Kimura, A. Voronov, and G. Zuckerman *Homotopy Gerstenhaber algebra and topological field theory* Contemporary Math. **202** (1997) 305–333.
41. N. Iwase *On the ring structure of $K^*(XP^n)$*. Master thesis, Kyushu Univ. (1983) (in Japanese).
42. T.V. Kadeišvili *The algebraic structure on the homology of an A_∞-algebra* Soobshch. Akad. Nauk Gruzin. SSR **108** (1982) 249–252 (in Russian)
43. M. Kontsevich *Feynman diagrams and low-dimensional topology* Proc. of the first European Congr. Math. **I** (1994) 97–122.
44. M. Kontsevich *Enumeration of rational curves via torus actions* Moduli Space of Surface ed. by H. Dijkgraaf, C. Faber and G.v.d. Geer Birkhäuser Basel (1995) 335–368.
45. M. Kontsevich and Y. Soibelman *Homological mirror symmetry and torus fibrations* Symplectic Geometry and Mirror Symmetry ed. K. Fukaya, Y.-G. Oh, K. Ono and G. Tian, World Sci. Publishing River Edge (2001) 203–263.
46. K. Lefèvre-Hasegawa *Sur les A_∞ catégories* Thése Doctorat Universite Paris 7 (2003).
47. J. Li and G. Tian *Virtual moduli cycles and Gromov-Witten invariants of general symplectic manifolds* Topics in Symplectic 4-Manifolds (Irvine, CA, 1996). First Int. Press Lect. Ser. **1** Internat. Press Cambridge, MA, USA (1998) 47–83.
48. V. Lyubashenko *Category of A_∞-categories* math.CT/0210047.
49. M. Markl, S. Shnider and J. Stasheff, *Operads in Algebra, Topology and Physics* Math. Surveys and Monographs **96** Amer. Math. Soc. (2002).
50. J.P. May *The geometry of iterated loop spaces* Lecture note in Math. **271** Springer-Verlag, New-York (1972).

51. J.E. McClure *On the chain-level intersection pairing for PL manifolds* Preprint (2004) math.QA/0410450.
52. S. Novikov *Multivalued functions and functional - an analogue of the Morse theory* Sov. Math. Dokl. **24** (1981) 222–225
53. H. Nakajima *Lectures on Hilbert schemes of points on surfaces* University Lecture Series, **18**, American Mathematical Society, Providence, RI, (1999).
54. H. Nakajima *22nd century geometry suggested by string theory: generating space* Suugaku seminar (1997) (in Japanese).
55. Y.-G. Oh *Floer cohomology of Lagrangian intersections and pseudo- holomorphic disks I, & II* Comm. Pure and Appl. Math. **46** (1993) 949–994 & 995–1012.
56. Y.-G. Oh *Relative Floer and quantum cohomology and the symplectic topology of Lagrangian submanifolds* Contact and Symplectic Geometry ed. C. B. Thomas, Cambridge Univ. Press, Cambridge UK (1996) 201–267.
57. H. Ohta *Obstruction to and deformation of Lagrangian intersection Floer cohomology* Symplectic Geometry and Mirror Symmetry (Seoul, 2000) 281–310.
58. K. Ono *On the Arnold conjecture for weakly monotone symplectic manifolds* Invent. Math. **119** (1995) 519–537.
59. Y. Ruan *Virtual neighborhood and pseudoholomorphic curve* Turkish J. Math. (1999) 161–231.
60. M. Schwarz *Morse homology* Progress in Mathematics, **111** Birkhäuser Verlag, Basel (1993).
61. P. Seidel *Fukaya categories and Picard-Lefschetz theory* Preprint, to appear in the ETH Lecture Notes series of the European Math. Soc. (2006).
62. B. Siebert *Gromov-Witten invariants for general symplectic manifolds* dg-ga /9608005.
63. V. de Silva *Products on symplectic Floer homology* Thesis, Oxford Uinv. (1997).
64. V.A. Smirnov *Simplicial and Operad Methods in Algebraic Topology* (Translated from the Russian manuscript by G. L. Rybnikov) Translations of Mathematical Monographs, **198** American Mathematical Society, Providence (2000).
65. J. Stasheff *Homotopy associativity of H-spaces I, II* Trans. Amer. Math. Soc. **108** (1963) 275–312.
66. J. Stasheff *Higher homotopy algebra: string field theory and Drinfeld's quasi Hopf algebra*, Proc. of XXth International Conference on Diff. Geometric method in Theoretical Physics (New York 1991) **1** (1992) 408 – 425.
67. K. Strebel *Quadratic Differentials* Springer (1984).
68. M. Sugawara *A condition that a space is group-like* J. Okayama Univ. **7** (1957) 123–149.
69. D. Sullivan *Infinitesimal computations in topology* Publ. Math. IHES **74** (1978) 269–331.
70. A. Voronov *The master equation of open-closed sigma-model* Talk at Stringy Topology in Morelia (2006).
71. S. Wilson *On the algebra and geometry of manifolds chains and cochains* Thesis, SUNY Stony Brook (2005).
72. E. Witten *Noncommutative geometry and string field theory* Nuclear Phys. B **268** (1986) 253–294.
73. E. Witten *Topological quantum field theory* Comm. Math. Phys., **118** (1988) 411–449.
74. B. Zwiebach *Quantum open string theory with manifest closed string factorization* Phys. Lett. **B256** (1991) 22–29.

Notes on the Self-Reducibility of the Weil Representation and Higher-Dimensional Quantum Chaos

Shamgar Gurevich* and Ronny Hadani†

* Department of Mathematics, University of California, Berkeley, CA 94720, USA
shamgar@math.berkeley.edu
† Department of Mathematics, University of Chicago, IL, 60637, USA
hadani@math.uchicago.edu

Summary. In these notes we discuss the *self-reducibility property* of the Weil representation. We explain how to use this property to obtain sharp estimates of certain higher-dimensional exponential sums which originate from the theory of quantum chaos. As a result, we obtain the Hecke quantum unique ergodicity theorem for a generic linear symplectomorphism A of the torus $\mathbb{T} = \mathbb{R}^{2N}/\mathbb{Z}^{2N}$.

Key words: Hannay–Berry Model, Quantum Unique Ergodicity, Bounds on Exponential Sums, Weil Representation, Self-Reducibility.

AMS codes: 11F27, 11L07, 81Q50.

1 Introduction

1.1 The Weil representation

In his celebrated 1964 Acta paper [34] Weil constructed a certain (projective) unitary representation of a symplectic group over a local field k (for example k could be $\mathbb{R}$, $\mathbb{C}$, or a p-adic field). This representation has many fascinating properties which have gradually been brought to light over the last few decades. It now appears that this representation is a central object, bridging various topics in mathematics and physics, including number theory, the theory of theta functions and automorphic forms, invariant theory, harmonic analysis, and quantum mechanics. Although it holds such a fundamental status, it is satisfying to observe that the Weil representation already appears in the study of functions on linear spaces. Given a k-linear space L, there exists an associated (polarized) symplectic vector space $V = L \times L^{*}$. The Weil representation of the group $Sp = Sp(V, \omega)$ can be realized on the Hilbert space

Ö. Ceyhan et al. (eds.), *Arithmetic and Geometry Around Quantization*,
Progress in Mathematics 279, DOI: 10.1007/978-0-8176-4831-2_7,

$\mathcal{H} = L^2(L, \mathbb{C})$. Interestingly, some elements of the group Sp act by certain kinds of generalized Fourier transforms. In particular, there exists a specific element $\mathrm{w} \in Sp$ (called the Weyl element) whose action is given, up to a normalization, by the standard Fourier transform. From this perspective, the classical theory of harmonic analysis seems to be devoted to the study of a particular operator in the Weil representation.

In these notes we will be concerned only with the case of the Weil representations of symplectic groups over finite fields. The main technical part is devoted to the study of a specific property of the Weil representation—the *self-reducibility property*. Briefly, this is a property concerning a relationship between the Weil representations of symplectic groups of different dimensions. In parts of these notes we devoted some effort to developing a general theory. In particular, the results concerning the self-reducibility property apply also to the Weil representation over local fields.

We use the self-reducibility property to bound certain higher-dimensional exponential sums which originate from the theory of quantum chaos, thereby obtaining a proof of one of the main statements in the field—the Hecke quantum unique ergodicity theorem for a generic linear symplectomorphism of the $2N$-dimensional torus.

1.2 Quantum chaos problem

One of the main motivational problems in quantum chaos is [2, 3, 26, 30] describing eigenstates

$$\widetilde{H}\Psi = \lambda\Psi, \quad \Psi \in \mathcal{H},$$

of a chaotic Hamiltonian

$$\widetilde{H} = Op(H) : \mathcal{H} \to \mathcal{H},$$

where $\mathcal{H}$ is a Hilbert space. We deliberately use the notation $Op(H)$ to emphasize the fact that the quantum Hamiltonian $\widetilde{H}$ is a quantization of a classical Hamiltonian $H : M \to \mathbb{C}$, where M is a classical symplectic phase space (usually the cotangent bundle of a configuration space $M = T^*X$, in which case $\mathcal{H} = L^2(X)$). In general, describing Ψ is considered to be an extremely complicated problem. Nevertheless, for a few mathematical models of quantum mechanics rigorous results have been obtained. We shall proceed to describe one of these models.

Hannay–Berry model

In [18] Hannay and Berry explored a model for quantum mechanics on the two-dimensional symplectic torus $(\mathbb{T}, \omega)$. Hannay and Berry suggested to quantize simultaneously the functions on the torus and the linear symplectic group $\Gamma \simeq SL_2(\mathbb{Z})$. One of their main motivations was to study the phenomenon of quantum chaos in this model [26, 28]. More precisely, they considered an ergodic discrete dynamical system on the torus which is generated by a hyperbolic automorphism $A \in \Gamma$. Quantizing the system, the classical phase space $(\mathbb{T}, \omega)$ is replaced by a *finite dimensional* Hilbert space $\mathcal{H}$, classical observables, i.e., functions $f \in C^\infty(\mathbb{T})$, by operators $\pi(f) \in \mathrm{End}(\mathcal{H})$, and classical symmetries by a unitary representation $\rho : \Gamma \to U(\mathcal{H})$.

Shnirelman's theorem

Analogous with the case of the Schrödinger equation, consider the following eigenstates problem

$$\rho(A)\Psi = \lambda\Psi.$$

A fundamental result, valid for a wide class of quantum systems which are associated to ergodic classical dynamics, is Shnirelman's theorem [31], asserting that in the semi-classical limit almost all (in a suitable sense) eigenstates become equidistributed in an appropriate sense.

A variant of Shnirelman's theorem also holds in our situation [4]. More precisely, we have that in the semi-classical limit $\hbar \to 0$ for almost all (in a suitable sense) eigenstates Ψ of the operator $\rho(A)$ the corresponding *Wigner distribution* $\langle \Psi | \pi(\cdot)\Psi \rangle : C^\infty(\mathbb{T}) \to \mathbb{C}$ approaches the phase space average $\int_{\mathbb{T}} \cdot |\omega|$. In this respect, it seems natural to ask whether there exist exceptional sequences of eigenstates? Namely, eigenstates that do not obey the Shnirelman's rule (*scarred* eigenstates). It was predicted by Berry [2, 3] that *scarring* phenomenon is not expected to be seen for quantum systems associated with *generic* chaotic classical dynamics. However, in our situation the operator $\rho(A)$ is not generic, and exceptional eigenstates were constructed. Indeed, it was confirmed mathematically in [8] that certain $\rho(A)$-eigenstates might localize. For example, in that paper a sequence of eigenstates Ψ was constructed, for which the corresponding Wigner distribution approaches the measure $\frac{1}{2}\delta_0 + \frac{1}{2}|\omega|$ on $\mathbb{T}$.

Hecke quantum unique ergodicity

A quantum system that obeys Shnirelman's rule is also called quantum ergodic. Can one impose some natural conditions on the eigenstates so that no exceptional eigenstates will appear? Namely, *quantum unique ergodicity* will hold. This question was addressed in a paper by Kurlberg and Rudnick [25]. In that paper, they formulated a rigorous notion of Hecke quantum unique ergodicity for the case $\hbar = 1/p$. The following is a brief description of that work. The basic observation is that the degeneracies of the operator $\rho(A)$ are coupled with the existence of symmetries. There exists a commutative group of operators that commutes with $\rho(A)$, which can in fact be computed. In more detail, the representation ρ factors through the quotient group $Sp = SL_2(\mathbb{F}_p)$. We denote by $T_A \subset Sp$ the centralizer of the element A, now considered as an element of the quotient group. The group T_A is called (cf. [25]) the *Hecke torus* corresponding to the element A. The Hecke torus acts semisimply on $\mathcal{H}$. Therefore, we have a decomposition

$$\mathcal{H} = \bigoplus_{\chi : T_A \to \mathbb{C}^\times} \mathcal{H}_\chi,$$

where $\mathcal{H}_\chi$ is the Hecke eigenspace corresponding to the character χ. Consider a unit eigenstate $\Psi \in \mathcal{H}_\chi$ and the corresponding Wigner distribution $\mathcal{W}_\chi : C^\infty(\mathbb{T}) \to \mathbb{C}$, defined by the formula $\mathcal{W}_\chi(f) = \langle \Psi | \pi(f) \Psi \rangle$. The main statement in [25] proves an explicit bound on the semi-classical asymptotic of $\mathcal{W}_\chi(f)$

$$\left| \mathcal{W}_\chi(f) - \int_{\mathbb{T}} f|\omega| \right| \leq \frac{C_f}{p^{1/4}},$$

where C_f is a constant that depends only on the function f. In Rudnick's lectures at MSRI, Berkeley 1999 [27], and ECM, Barcelona 2000 [28], he conjectured that a stronger bound should hold true, i.e.,

$$\left| \mathcal{W}_\chi(f) - \int_{\mathbb{T}} f|\omega| \right| \leq \frac{C_f}{p^{1/2}}. \tag{1}$$

A particular case (which implies (1)) of the above inequality is when $f = \xi$, where ξ is a non-trivial character. In this case, the integral $\int_{\mathbb{T}} \xi|\omega|$ vanishes and in addition it turns out that $C_\xi = 2 + o(1)$. Hence, we obtain the following simplified form of (1)

$$|\mathcal{W}_\chi(\xi)| \leq \frac{2 + o(1)}{\sqrt{p}}, \tag{2}$$

for sufficiently large p. These stronger bounds were proved in the paper [13]. It will be instructive to briefly recall the main ideas and techniques used in [13].

Geometric approach

The basic observation to be made is that the theory of quantum mechanics on the torus, in the case $\hbar = 1/p$, can be equivalently recast in the language of the representation theory of finite groups in characteristic p. We will endeavor to give a more precise explanation of this matter. Consider the quotient $\mathbb{F}_p$-vector space $V = \mathbb{T}^\vee / p\mathbb{T}^\vee$, where $\mathbb{T}^\vee \simeq \mathbb{Z}^2$ is the lattice of characters on $\mathbb{T}$. We denote by $H = H(V)$ the Heisenberg group associated to V. The group Sp is naturally identified with the group of linear symplectomorphisms of V. We have an action of Sp on H. The Stone–von Neumann theorem (see Theorem 5) states that there exists a unique irreducible representation $\pi : H \to GL(\mathcal{H})$, with a non-trivial character ψ of the center of H. As a consequence of its uniqueness, its isomorphism class is fixed by Sp. This is equivalent to saying that $\mathcal{H}$ is equipped with a compatible projective representation $\rho : Sp \to PGL(\mathcal{H})$, which in fact can be linearized to an honest representation. This representation is the celebrated Weil representation. Noting that Sp is the group of rational points of the algebraic group **Sp** (we use boldface letters to denote algebraic varieties), it is natural to ask whether there exists an algebro-geometric object that underlies the representation ρ. The answer to this question is positive. The construction is proposed in an unpublished letter of Deligne to Kazhdan [7], which appears now in [13, 16]. Briefly, the content of this letter is a construction of *representation sheaf* $\mathcal{K}_\rho$ on the algebraic variety **Sp**. We obtain, as a consequence, the following general principle:

Motivic principle. *All quantum mechanical quantities in the Hannay–Berry model are motivic in nature.*

By this we mean that every quantum-mechanical quantity $\mathcal{Q}$ is associated with a vector space $V_\mathcal{Q}$ (certain cohomology of a suitable ℓ-adic sheaf) endowed with a Frobenius action $\mathrm{Fr} : V_\mathcal{Q} \to V_\mathcal{Q}$ so that $\mathcal{Q} = \mathrm{Tr}(\mathrm{Fr}_{|V_\mathcal{Q}})$. In particular, it was shown in [13] that there exists a two-dimensional vector space V_χ, endowed with an action $\mathrm{Fr} : V_\chi \to V_\chi$, so that

$$\mathcal{W}_\chi(\xi) = \mathrm{Tr}(\mathrm{Fr}_{|V_\chi}). \tag{3}$$

This, combined with the purity condition that the eigenvalues of Fr are of absolute value $1/\sqrt{p}$, implies the estimate (2).

The higher-dimensional Hannay–Berry model

The higher-dimensional Hannay–Berry model is obtained as a quantization of a $2N$-dimensional symplectic torus $(\mathbb{T}, \omega)$ acted upon by the group $\Gamma \simeq Sp(2N, \mathbb{Z})$ of linear symplectic automorphisms. It was first constructed in [12], where, in particular, a quantization of the whole group of symmetries Γ was obtained. Consider a regular ergodic element $A \in \Gamma$, i.e., A generates an ergodic discrete dynamical system and it is regular in the sense that it has distinct eigenvalues over $\mathbb{C}$. It is natural to ask whether quantum unique ergodicity will hold true in this setting as well, as long as one takes into account the whole group of hidden (Hecke) symmetries? Interestingly, the answer to this question is NO! Several new results in this direction have been announced recently. In the case where the automorphism A is *non-generic*, meaning that it has an invariant Lagrangian (and more generally co-isotropic) sub-torus $\mathbb{T}_L \subset \mathbb{T}$, an interesting new phenomenon was revealed. There exists a sequence $\{\Psi_\hbar\}$ of Hecke eigenstates which are closely related to the physical phenomena of *localization*, known in the physics literature (cf. [20, 24]) as *scars*. We will call them *Hecke scars*. These states are localized in the sense that the associated Wigner distribution converges to the Haar measure μ on the invariant Lagrangian sub-torus

$$\mathcal{W}_{\Psi_\hbar}(f) \to \int_{\mathbb{T}_L} f d\mu, \text{ as } \hbar \to 0, \tag{4}$$

for every smooth observable f. These special kinds of Hecke eigenstates were first established in [10]. The semi-classical interpretation of the localization phenomena (4) was announced in [23].

The above phenomenon motivates the following definition:

Definition 1. *An element $A \in \Gamma$ is called generic if it is regular and admits no non-trivial invariant co-isotropic sub-tori.*

Remark 1. The collection of generic elements constitutes an open subscheme of Γ. In particular, a generic element need not be ergodic automorphism of $\mathbb{T}$. However, in the case where $\Gamma \simeq SL_2(\mathbb{Z})$ every ergodic (i.e., hyperbolic) element is generic. An example of a generic element which is not ergodic is given by the Weyl element $\mathrm{w} = \begin{pmatrix} 0 & 1 \\ -1 & 0 \end{pmatrix}$.

In these notes we will require the automorphism $A \in \Gamma$ to be generic. This case was first considered in [14], where using similar geometric techniques as in [13] the analogue of inequality (2) was obtained. For the sake of simplicity, let us assume that the automorphism A is *strongly generic*, i.e., it has no non-trivial invariant sub-tori.

Theorem 1 ([14]). *Let ξ be a non-trivial character of $\mathbb{T}$. The following bound holds*

$$|\mathcal{W}_\chi(\xi)| \leq \frac{[2+o(1)]^N}{\sqrt{p}^N}, \tag{5}$$

where p is a sufficiently large prime number.

In particular, using the bound (5), we obtain the following statement for general observable:

Corollary 1 (Hecke quantum unique ergodicity). *Consider an observable $f \in C^\infty(\mathbb{T})$ and a sufficiently large prime number p. Then*

$$\left| \mathcal{W}_\chi(f) - \int_{\mathbb{T}} f d\mu \right| \leq \frac{C_f}{\sqrt{p}^N},$$

where $\mu = |\omega|^N$ is the corresponding volume form and C_f is an explicit computable constant which depends only on the function f.

In these notes, using the self-reducibility property of the Weil representation, we improve the above estimates and obtain the following theorem:

Theorem 2 (Sharp bound). *Let ξ be a non-trivial character of $\mathbb{T}$. For sufficiently large prime number p we have*

$$|\mathcal{W}_\chi(\xi)| \leq \frac{[2+o(1)]^{r_p}}{\sqrt{p}^N}, \tag{6}$$

where the number r_p is an integer between 1 and N, that we will call the symplectic rank of T_A.

Remark 2. It will be shown (see Subsection 6.2) that the distribution of the symplectic rank r_p (6) in the set $\{1, \ldots, N\}$ is governed by the Chebotarev density theorem applied to a suitable Galois group. For example, in the case where $A \in Sp(4, \mathbb{Z})$ is strongly generic we have

$$\lim_{x\to\infty} \frac{\#\{r_p = r \mid p \leq x\}}{\pi(x)} = \tfrac{1}{2}, \qquad r = 1,\ 2,$$

where $\pi(x)$ denotes the number of primes up to x.

Remark 3. For the more general version of Theorem 2, one that holds in the general generic case (Definition 1), see Subsection 6.3.

In order to witness the improvement of (6) over (5), it would be instructive to consider the following extreme scenario. Assume that the Hecke torus T_A acts on $V \simeq \mathbb{F}_p^{2N}$ irreducibly. In this case it turns out that $r_p = 1$. Hence, (6) becomes

$$|\mathcal{W}_\chi(\xi)| \leq \frac{2+o(1)}{\sqrt{p}^N},$$

which constitutes a significant improvement over the coarse topological estimate (5). Let us elaborate on this. Recall the motivic interpretation (3) of the Wigner distribution. In [14] an analogous interpretation was given to the higher-dimensional Wigner distributions, realizing them as $\mathcal{W}_\chi(\xi) = \mathrm{Tr}(\mathrm{Fr}_{|V_\chi})$, where, by the purity condition, the eigenvalues of Fr are of absolute value $1/\sqrt{p}^N$. But, in this setting the dimension of V_χ is not 2, but 2^N, i.e., the Frobenius looks like

$$\mathrm{Fr} = \begin{pmatrix} \lambda_1 & * & * & * \\ & \cdot & * & * \\ & & \cdot & * \\ & & & \lambda_{2^N} \end{pmatrix}.$$

Hence, if we use only this amount of information, then the best estimate which can be obtained is (5). Therefore, in this respect the problem that we confront is showing cancellations between different eigenvalues, more precisely angles, of the Frobenius operator acting on a high-dimensional vector space, i.e., cancellations in the sum $\sum_{j=1}^{2^N} e^{i\theta_j}$, where the angles $0 \leq \theta_j < 2\pi$ are defined via $\lambda_j = e^{i\theta_j}/\sqrt{p}^N$. This problem is of a completely different nature, which is not accounted for by standard cohomological techniques (we thank R. Heath-Brown for pointing out to us [19] about the phenomenon of cancellations between Frobenius eigenvalues in the presence of high-dimensional cohomologies).

Remark 4. Choosing a realization $\mathcal{H} \simeq \mathbb{C}(\mathbb{F}_p^N)$, the matrix coefficient $\mathcal{W}_\chi(\xi)$ is equivalent to an exponential sum of the form

$$\langle \Psi | \pi(\xi) \Psi \rangle = \sum_{x \in \mathbb{F}_p^N} \Psi(x) e^{\frac{2\pi i}{p} \xi_+ x} \overline{\Psi}(x + \xi_-). \tag{7}$$

Here one encounters two problems. First, it is not so easy to describe the eigenstates Ψ. Second, the sum (7) is a high-dimensional exponential sum (over $\mathbb{F}_p$), which is known to be hard to analyze using standard techniques. The crucial point that we explain in these notes is that it can be realized, essentially, as a one-dimensional exponential sum over $\mathbb{F}_q$, where $q = p^N$.

1.3 Solution via self-reducibility

Let us explain the main idea underlying the proof of estimate (6). Let us assume for the sake of simplicity that the Hecke torus is completely inert, i.e., acts irreducibly on the vector space $V \simeq \mathbb{F}_p^{2N}$.

Representation theoretic interpretation of the Wigner distribution

The Hecke eigenstate Ψ is a vector in a representation space $\mathcal{H}$. The space $\mathcal{H}$ supports the Weil representation of the symplectic group $Sp \simeq Sp(2N, k)$, $k = \mathbb{F}_p$. The vector Ψ is completely characterized in representation theoretic terms, as being a character vector of the Hecke torus T_A. As a consequence, all quantities associated to Ψ, and in particular the Wigner distribution $\mathcal{W}_\chi$ are characterized in terms of the Weil representation. The main observation to be made is that the Hecke state Ψ can be characterized in terms of another Weil representation, this time of a group of much smaller dimension. In fact, it can be characterized, roughly, in terms of the Weil representation of $SL_2(K)$, $K = \mathbb{F}_{p^N}$.

Self-reducibility property

A fundamental notion in our study is that of a *symplectic module structure.* A symplectic module structure is a triple $(K, V, \overline{\omega})$, where K is a finite dimensional commutative algebra over k, equipped with an action on the vector space V, and $\overline{\omega}$ is a K-linear symplectic form satisfying the property $\mathrm{Tr}_{K/k}(\overline{\omega}) = \omega$. Let us assume for the sake of simplicity that K is a field. Let

$\overline{Sp} = Sp(V, \overline{\omega})$ be the group of K-linear symplectomorphisms with respect to the form $\overline{\omega}$. There exists a canonical embedding

$$\iota : \overline{Sp} \hookrightarrow Sp. \tag{8}$$

We will be mainly concerned with symplectic module structures which are associated to maximal tori in Sp. More precisely, it will be shown that associated to a maximal torus $T \subset Sp$ there exists a canonical symplectic module structure $(K, V, \overline{\omega})$ so that $T \subset \overline{Sp}$. The most extreme situation is when the torus $T \subset Sp$ is completely inert, i.e., acts irreducibly on the vector space V. In this particular case, the algebra K is in fact a field with $\dim_K V = 2$ which implies that $\overline{Sp} \simeq SL_2(K)$, i.e., using (8) we get $T \subset SL_2(K) \subset Sp$.

Let us denote by $(\rho, Sp, \mathcal{H})$ the Weil representation of Sp. The main observation now is (cf. [9]) the following:

Theorem 3 (Self-reducibility property). *The restricted representation* $(\overline{\rho} = \iota^* \rho, SL_2(K), \mathcal{H})$ *is the Weil representation of* $SL_2(K)$.

Applying the self-reducibility property to the Hecke torus T_A, it follows that the Hecke eigenstates Ψ can be characterized in terms of the Weil representation of $SL_2(K)$. Therefore, in this respect, Theorem 2 is reduced to the result obtained in [13].

1.4 Quantum unique ergodicity for statistical states

Let $A \in \Gamma$ be a generic linear symplectomorphism. As in harmonic analysis, one would like to use Theorem 2 concerning the Hecke eigenstates in order to extract information on the spectral theory of the operator $\rho(A)$ itself. For the sake of simplicity, let us assume that A is strongly generic, i.e., it acts on the torus $\mathbb{T}$ with no non-trivial invariant sub-tori. Next, a possible reformulation of the quantum unique ergodicity statement, one which is formulated for the automorphism A itself instead of the Hecke group of symmetries, is presented.

The element A acts via the Weil representation ρ on the space $\mathcal{H}$ and decomposes it into a direct sum of $\rho(A)$-eigenspaces

$$\mathcal{H} = \bigoplus_{\lambda \in Spec(\rho(A))} \mathcal{H}_\lambda. \tag{9}$$

Considering an $\rho(A)$-eigenstate Ψ and the corresponding projector P_Ψ one usually studies the Wigner distribution $\langle \Psi | \pi(\xi) \Psi \rangle = \mathrm{Tr}(\pi(\xi) P_\Psi)$ which, due to the fact that $rank(P_\Psi) = 1$, is sometimes called *pure state.* In the same way, we might think about an Hecke–Wigner distribution $\langle \Psi | \pi(\xi) \Psi \rangle = \mathrm{Tr}(\pi(\xi) P_\chi)$, attached to a T_A-eigenstate Ψ, as a *pure Hecke state.* Following von Neumann [33] we suggest the possibility of looking at the more general *statistical state*, defined by a non-negative, self-adjoint operator D, called the von Neumann density operator, normalized to have $\mathrm{Tr}(D) = 1$. For example, to the automorphism A we can attach the natural family of density operators $D_\lambda = \frac{1}{m_\lambda} P_\lambda$, where P_λ is the projector on the eigenspace $\mathcal{H}_\lambda$ (9), and $m_\lambda = \dim(\mathcal{H}_\lambda)$. Consequently, we obtain a family of statistical states

$$\mathcal{W}_\lambda(\cdot) = \mathrm{Tr}(\pi(\cdot)D_\lambda).$$

Theorem 4. *Let ξ be a non-trivial character of $\mathbb{T}$. For a sufficiently large prime number p, and every statistical state $\mathcal{W}_\lambda$, we have*

$$|\mathcal{W}_\lambda(\xi)| \leq \frac{(2+o(1))^{r_p}}{\sqrt{p}^N}, \tag{10}$$

where r_p is an explicit integer $1 \leq r_p \leq N$ which is determined by A.

Theorem 4 follows from the fact that the Hecke torus T_A acts on the spaces $\mathcal{H}_\lambda$, and hence, one can use the Hecke eigenstates, and the bound (6). In particular, using (10) we obtain for a general observable the following bound:

Corollary 2. *Consider an observable $f \in C^\infty(\mathbb{T})$ and a sufficiently large prime number p. Then*

$$\left| \mathcal{W}_\lambda(f) - \int_{\mathbb{T}} f d\mu \right| \leq \frac{C_f}{\sqrt{p}^N},$$

where $\mu = |\omega|^N$ is the corresponding volume form and C_f is an explicit computable constant which depends only on the function f.

1.5 Results

1. *Bounds of higher-dimensional exponential sums.* The main results of these notes are a sharp estimates of certain higher-dimensional exponential sums attached to tori in $Sp(2N, \mathbb{F}_q)$. This is the content of Theorems 12 and 14 and is obtained using the self-reducibility property of the Weil representation as stated in Theorems 9 and 10.
2. *Hecke quantum unique ergodicity theorem.* The main application of these notes is the proof of the *Hecke quantum unique ergodicity theorem*, i.e., Theorems 17 and 18, for generic linear symplectomorphism of the torus in any dimension. The proof of the theorem is a direct application of the sharp bound on the higher-dimensional exponential sums.
3. *Multiplicities formula.* Exact formula for the multiplicities, i.e., the dimensions of the character spaces for the action of maximal tori in the Weil representation are derived. This is obtained first for the $SL_2(\mathbb{F}_q)$ case in Theorem 8 using the character formula presented in Theorem 7. Then, as a direct application of the self-reducibility property, the formula is extended in Theorem 11 to the higher-dimensional cases.

In addition, a formulation of the quantum unique ergodicity statement for quantum chaos problems, close in spirit to the von Neumann idea about density operator, is suggested in Theorem 4. The statement includes only the quantum operator A rather than the whole Hecke group of symmetries [25]. The proof of the statement uses the Hecke operators as a harmonic analysis tool.

1.6 Structure of the notes

Apart from the introduction, the notes consist of five sections.

In Section 2 we give some preliminaries on representation theory which are used in the notes. In Subsection 2.3 we recall the invariant presentation of the Weil representation over finite fields [16], and we discuss applications to multiplicities. Section 3 constitutes the main technical part of this work. Here we develop the theory that underlies the self-reducibility property of the Weil representation. In particular, in Subsection 3.1 we introduce the notion of symplectic module structure. In Subsection 3.2 we prove the existence of symplectic module structure associated with a maximal torus in Sp. Finally, we establish the self-reducibility property of the Weil representation, i.e., Theorem 10, and apply this property to get information on multiplicities in Subsection 3.4. Section 4 is devoted to an application of the theory developed in previous sections to estimating higher-dimensional exponential sums which originate from the mathematical theory of quantum chaos. In Section 5 we describe the higher-dimensional Hannay–Berry model of quantum mechanics on the torus. Finally, in Section 6 we present the main application of these notes—the proof of the Hecke quantum unique ergodicity theorem for generic linear symplectomorphisms of the $2N$-dimensional torus.

Remark 5. Complete proofs for the statements appearing in these notes will be given elsewhere.

1.7 Acknowledgments

It is a pleasure to thank our teacher J. Bernstein for his interest and guidance in this project. We thank D. Kazhdan for several interesting discussions. We appreciate the support of the Technion, and in particular of M. Baruch. We thank D. Kelmer for sharing with us computer simulation data. Finally, we would like to thank O. Ceyhan and the organizers of the conference AGAQ, held in Istanbul during June 2006, and J. Wolf and the organizers of the conference Lie Groups, Lie Algebras and Their Representations, Berkeley, November 2006, for the invitation to present this work.

2 Preliminaries

In this section, we denote by $k = \mathbb{F}_q$ the finite field of q elements and odd characteristic.

2.1 The Heisenberg representation

Let (V, ω) be a $2N$-dimensional symplectic vector space over the finite field k. There exists a two-step nilpotent group $H = H(V, \omega)$ associated to the symplectic vector space (V, ω). The group H is called the *Heisenberg group.* It can be realized as the set $H = V \times k$, equipped with the multiplication rule

$$(v, z) \cdot (v', z') = (v + v', z + z' + \tfrac{1}{2}\omega(v, v')).$$

The center of H is $Z(H) = \{(0, z) : z \in k\}$. Fix a non-trivial central character $\psi : Z(H) \longrightarrow \mathbb{C}^{\times}$. We have the following fundamental theorem:

Theorem 5 (Stone–von Neumann). *There exists a unique (up to isomorphism) irreducible representation* $(\pi, H, \mathcal{H})$ *with central character* ψ, *i.e.,* $\pi(z) = \psi(z) Id_{\mathcal{H}}$ *for every* $z \in Z(H)$.

We call the representation π appearing in Theorem 5, the *Heisenberg representation* associated with the central character ψ.

Remark 6. The representation π, although it is unique, admits a multitude of different models (realizations). In fact, this is one of its most interesting and powerful attributes. In particular, to any Lagrangian splitting $V = L' \oplus L$, there exists a model $(\pi_{L',L}, H, \mathbb{C}(L))$, where $\mathbb{C}(L)$ denotes the space of complex valued functions on L. In this model, we have the following actions:

- $\pi_{L',L}(l')[f](x) = \psi(\omega(l', x)) f(x)$;
- $\pi_{L',L}(l)[f](x) = f(x + l)$;
- $\pi_{L',L}(z)[f](x) = \psi(z) f(x)$,
  where $l' \in L'$, $x, l \in L$, and $z \in Z(H)$.

The above model is called the Schrödinger realization associated with the splitting $V = L' \oplus L$.

2.2 The Weyl transform

Given a linear operator $A : \mathcal{H} \to \mathcal{H}$ we can associate to it a function on the group H defined as follows

$$W(A)(h) = \frac{1}{\dim \mathcal{H}} \mathrm{Tr}(A\pi(h^{-1})). \tag{11}$$

The transform $W : \mathrm{End}(\mathcal{H}) \to \mathbb{C}(H)$ is called the *Weyl transform [21, 35].* The Weyl transform admits a left inverse $\pi : \mathbb{C}(H) \to \mathrm{End}(\mathcal{H})$ given by the extended action $\pi(K) = \sum_{h \in H} K(h)\pi(h)$.

2.3 The Weil representation

Let $Sp = Sp(V, \omega)$ denote the group of linear symplectic automorphisms of V. The group Sp acts by group automorphisms on the Heisenberg group through its tautological action on the vector space V. A direct consequence of Theorem 5 is the existence of a projective representation $\widetilde{\rho} : Sp \rightarrow PGL(\mathcal{H})$. The classical construction of $\widetilde{\rho}$ out of the Heisenberg representation π is due to Weil [34]. Considering the Heisenberg representation π and an element $g \in Sp$, one can define a new representation π^g acting on the same Hilbert space via $\pi^g(h) = \pi(g(h))$. Clearly both π and π^g have central character ψ; hence, by Theorem 5, they are isomorphic. Since the space $\mathsf{Hom}_H(\pi, \pi^g)$ is one-dimensional, choosing for every $g \in Sp$ a non-zero representative $\widetilde{\rho}(g) \in \mathsf{Hom}_H(\pi, \pi^g)$ gives the required projective representation. In more concrete terms, the projective representation $\widetilde{\rho}$ is characterized by the formula

$$\widetilde{\rho}(g)\, \pi(h)\, \widetilde{\rho}\left(g^{-1}\right) = \pi(g(h)), \tag{12}$$

for every $g \in Sp$ and $h \in H$. It is a peculiar phenomenon of the finite field setting that the projective representation $\widetilde{\rho}$ can be linearized into an honest representation. This linearization is unique, except in the case the finite field is $\mathbb{F}_3$ and $\dim V = 2$ (for the canonical choice in the latter case see [17]).

Theorem 6. *There exists a canonical unitary representation*

$$\rho : Sp \longrightarrow GL(\mathcal{H}),$$

satisfying the formula (12).

Invariant presentation of the Weil representation

An elegant description of the Weil representation can be obtained [16] using the Weyl transform (see Subsection 2.2). Given an element $g \in Sp$, the operator $\rho(g)$ can be written as $\rho(g) = \pi(K_g)$, where K_g is the Weyl transform $K_g = W(\rho(g))$. The homomorphism property of ρ is manifested as

$$K_g * K_h = K_{gh} \quad \text{for every} \quad g, h \in Sp,$$

where $*$ denotes (properly normalized) group theoretic convolution on H. Finally, the function K can be explicitly described on an appropriate subset of Sp [16]. Let $U \subset Sp$ denote the subset consisting of all elements $g \in Sp$ such that $g - I$ is invertible. For every $g \in U$ and $v \in V$ we have

$$K_g(v) = \nu(g)\psi(\tfrac{1}{4}\omega(\kappa(g)v, v)), \tag{13}$$

where $\kappa(g) = \frac{g+I}{g-I}$ is the Cayley transform [21, 36], and

$$\nu(g) = (G/q)^{2N}\sigma(\det(g - I)),$$

with σ the unique quadratic character of the multiplicative group $\mathbb{F}_q^\times$, and $G = \sum\limits_{z \in Z(H)} \psi(z^2)$ the quadratic Gauss sum.

2.4 The Heisenberg–Weil representation

Let J denote the semi-direct product $J = Sp \ltimes H$. The group J is sometimes referred to as the *Jacobi* group. The compatible pair (π, ρ) is equivalent to a single representation $\tau : J \longrightarrow GL(\mathcal{H})$ of the Jacobi group defined by the formula $\tau(g, h) = \rho(g)\pi(h)$. It is an easy exercise to verify that the Egorov identity (12) implies the multiplicativity of the map τ.

In these notes, we would like to adopt the name *Heisenberg–Weil representation* when referring to the representation τ.

2.5 Character formulas

The invariant presentation (11) and formula (13) imply [16] a formula for the character of the $2N$-dimensional Heisenberg–Weil representation over a finite field (cf. [9, 22]).

Theorem 7 (Character formulas [16]). *The character* ch_ρ *of the Weil representation, when restricted to the subset* U*, is given by*

$$ch_\rho(g) = \sigma((-1)^N \det(g - I)), \tag{14}$$

and the character ch_τ *of the Heisenberg–Weil representation, when restricted to the subset* $U \times H$*, is given by*

$$\mathrm{ch}_\tau(g, v, z) = \mathrm{ch}_\rho(g)\psi(\tfrac{1}{4}\omega(\kappa(g)v, v) + z). \tag{15}$$

2.6 Application to multiplicities

We would like to apply the formula (15) to the study of the multiplicities arising from actions of tori via the Weil representation (cf. [1, 9, 32]). Let us start with the two-dimensional case (see Theorem 11 for the general case). Let $T \subset Sp \simeq SL_2(\mathbb{F}_q)$ be a maximal torus. The torus T acts semisimply on $\mathcal{H}$, decomposing it into a direct sum of character spaces $\mathcal{H} = \bigoplus_{\chi:T\to\mathbb{C}^\times} \mathcal{H}_\chi$.

As a consequence of having the explicit formula (14), we obtain a simple description for the multiplicities $m_\chi = \dim \mathcal{H}_\chi$. Denote by $\sigma : T \to \mathbb{C}^\times$ the unique quadratic character of T.

Theorem 8 (Multiplicities formula). *We have* $m_\chi = 1$ *for any character* $\chi \neq \sigma$*. Moreover,* $m_\sigma = 2$ *or* 0*, depending on whether the torus* T *is split or inert, respectively.*

What about the multiplicities for action of tori in the Weil representation of higher-dimensional symplectic groups? This problem can be answered (see Theorem 11) using the self-reducibility property of the Weil representation.

3 Self-reducibility of the Weil representation

In this section, unless stated otherwise, the field k is an *arbitrary field* of characteristic different from two.

3.1 Symplectic module structures

Let K be a finite-dimensional commutative algebra over the field k. Let $\mathrm{Tr} : K \to k$ be the trace map, associating to an element $x \in K$ the trace of the k-linear operator $m_x : K \to K$ obtained by left multiplication by the element x. Consider a symplectic vector space (V, ω) over k.

Definition 2. *A symplectic K-module structure on (V, ω) is an action $K \otimes_k V \to V$, and a K-linear symplectic form $\overline{\omega} : V \times V \to K$ such that*

$$Tr \circ \overline{\omega} = \omega. \tag{16}$$

Given a symplectic module structure $(K, V, \overline{\omega})$ on a symplectic vector space (V, ω), we denote by $\overline{Sp} = Sp(V, \overline{\omega})$ the group of K-linear symplectomorphisms with respect to the form $\overline{\omega}$. The compatibility condition (16) gives a natural embedding

$$\iota : \overline{Sp} \hookrightarrow Sp. \tag{17}$$

3.2 Symplectic module structure associated with a maximal torus

Let $T \subset Sp$ be a maximal torus.

A particular case

In order to simplify the presentation, let us assume first that T acts irreducibly on the vector space V, i.e., there exists no non-trivial T-invariant subspaces. Let $A = Z(T, \mathrm{End}(V))$, be the centralizer of T in the algebra of all linear endomorphisms. Clearly (due to the assumption of irreducibility) A is a division algebra. Moreover, we have

Claim. The algebra A is commutative.

In particular, this claim implies that A is a field extension of k. Let us now describe a special quadratic element in the Galois group $\mathrm{Gal}(A/k)$. Denote by $(\cdot)^t : \mathrm{End}(V) \to \mathrm{End}(V)$ the symplectic transpose characterized by the property

$$\omega(Rv, u) = \omega(v, R^t u),$$

for all $v, u \in V$, and every $R \in \mathrm{End}(V)$. It can be easily verified that $(\cdot)^t$ preserves A, leaving the subfield k fixed, hence, it defines an element $\Theta \in \mathrm{Gal}(A/k)$, satisfying $\Theta^2 = \mathrm{Id}$. Denote by $K = A^\Theta$ the subfield of A consisting of elements fixed by Θ. We have the following proposition:

Proposition 1 (Hilbert's Theorem 90). *We have* $\dim_K V = 2$.

Corollary 3. *We have* $\dim_K A = 2$.

As a corollary, we have the following description of T. Denote by $N_{A/K} : A \to K$ the standard norm map.

Corollary 4. *We have* $T = S(A) = \{a \in A : N_{A/K}(a) = 1\}$

The symplectic form ω can be lifted to a K-linear symplectic form $\overline{\omega}$, which is invariant under the action of the torus T. This is the content of the following proposition:

Proposition 2 (Existence of canonical symplectic module structure). *There exists a canonical T-invariant K-linear symplectic form $\overline{\omega} : V \times V \to K$ satisfying the property $Tr \circ \overline{\omega} = \omega$.*

Concluding, we obtained a T-invariant symplectic K-module structure on V.

Let $\overline{Sp} = Sp(V, \overline{\omega})$ denote the group of K-linear symplectomorphisms with respect to the symplectic form $\overline{\omega}$. We have (17) a natural embedding $\overline{Sp} \subset Sp$. The elements of T commute with the action of K, and preserve the symplectic form $\overline{\omega}$ (Proposition 2); hence, we can consider T as a subgroup of $\overline{Sp}$. By Proposition 1 we can identify $\overline{Sp} \simeq SL_2(K)$, and using (17) we obtain

$$T \subset SL_2(K) \subset Sp. \tag{18}$$

To conclude we see that T consists of the K-rational points of a maximal torus $\mathbf{T} \subset \mathbf{SL}_2$ (in this case T consists of the rational points of an inert torus).

General case

Here, we drop the assumption that T acts irreducibly on V. By the same argument as before, the algebra $A = Z(T, \mathrm{End}(V))$ is commutative, yet, it may no longer be a field. The symplectic transpose $(\cdot)^t$ preserves the algebra A, and induces an involution $\Theta : A \to A$. Let $K = A^{\Theta}$ be the subalgebra consisting of elements $a \in A$ fixed by Θ. Following the same argument as in the proof of Proposition 1, we can show that V is a free K-module of rank 2. Following the same arguments as in the proof of Proposition 2, we can show that there exists a canonical symplectic form $\overline{\omega} : V \times V \to K$, which is K-linear and invariant under the action of the torus T. Concluding, associated to a maximal torus T there exists a T-invariant symplectic K-module structure

$$(K, V, \overline{\omega}). \tag{19}$$

Denote by $\overline{Sp} = Sp(V, \overline{\omega})$ the group of K-linear symplectomorphisms with respect to the form $\overline{\omega}$. We have a natural embedding

$$\iota_S : \overline{Sp} \hookrightarrow Sp \tag{20}$$

and we can consider T as a subgroup of $\overline{Sp}$. Finally, we have $\overline{Sp} \simeq SL_2(K)$, and T consists of the K-rational points of a maximal torus $\mathbf{T} \subset \mathbf{SL}_2$. In particular, the relation (18) holds also in this case: $T \subset SL_2(K) \subset Sp$.

We shall now proceed to give a finer description of all objects discussed so far. The main technical result is summarized in the following lemma:

Lemma 1 (Symplectic decomposition). *We have a canonical decomposition*

$$(V, \omega) = \bigoplus_{\alpha \in \Xi} (V_\alpha, \omega_\alpha), \tag{21}$$

into (T, A)-invariant symplectic subspaces. In addition, we have the following associated canonical decompositions

1. $T = \prod T_\alpha$, where T_α consists of elements $t \in T$ such that $t_{|V_\beta} = \mathrm{Id}$ for every $\beta \neq \alpha$.
2. $A = \bigoplus A_\alpha$, where A_α consists of elements $a \in A$ such that $a_{|V_\beta} = \mathrm{Id}$ for every $\beta \neq \alpha$. Moreover, each sub-algebra A_α is preserved under the involution Θ.
3. $K = \bigoplus K_\alpha$, where $K_\alpha = A_\alpha^\Theta$. Moreover, K_α is a field and $\dim_{K_\alpha} V_\alpha = 2$.
4. $\overline{\omega} = \bigoplus \overline{\omega}_\alpha$, where $\overline{\omega}_\alpha : V_\alpha \times V_\alpha \to K_\alpha$ is a K_α-linear T_α-invariant symplectic form satisfying $\mathrm{Tr} \circ \overline{\omega}_\alpha = \omega_\alpha$.

Definition 3. *We will call the set Ξ (21) the symplectic type of T and the number $|\Xi|$ the symplectic rank of T.*

Using the results of Lemma 1, we have an isomorphism

$$\overline{Sp} \simeq \prod \overline{Sp}_\alpha, \tag{22}$$

where $\overline{Sp}_\alpha = Sp(V_\alpha, \overline{\omega}_\alpha)$ denotes the group of K_α-linear symplectomorphisms with respect to the form $\overline{\omega}_\alpha$. Moreover, for every $\alpha \in \Xi$ we have $T_\alpha \subset \overline{Sp}_\alpha$. In particular, under the identifications $\overline{Sp}_\alpha \simeq SL_2(K_\alpha)$, there exist the following sequence of inclusions

$$T = \prod T_\alpha \subset \prod SL_2(K_\alpha) = SL_2(K) \subset Sp, \tag{23}$$

and for every $\alpha \in \Xi$ the torus T_α coincides with the K_α-rational points of a maximal torus $\mathbf{T}_\alpha \subset \mathbf{SL}_2$.

3.3 Self-reducibility of the Weil representation

In this subsection we assume that the field k is a finite field of odd characteristic (although, the results continue to hold true also for local fields of characteristic $\neq 2$, i.e., with the appropriate modification, replacing the group

Sp with its double cover $\widetilde{Sp}$). Let $(\tau, J, \mathcal{H})$ be the Heisenberg–Weil representation associated with a central character $\psi : Z(J) = Z(H) \to \mathbb{C}^{\times}$. Recall that $J = Sp \ltimes H$, and τ is obtained as a semi-direct product, $\tau = \rho \ltimes \pi$, of the Weil representation ρ and the Heisenberg representation π. Let $T \subset Sp$ be a maximal torus.

A particular case

For clarity of presentation, assume first that T acts irreducibly on V. Using the results of the previous section, there exists a symplectic module structure $(K, V, \overline{\omega})$ (in this case K/k is a field extension of degree $[K : k] = N$). The group $\overline{Sp} = Sp(V, \overline{\omega})$ is imbedded as a subgroup $\iota_S : \overline{Sp} \hookrightarrow Sp$. Our goal is to describe the restriction

$$(\overline{\rho} = \iota_S^* \rho, \overline{Sp}, \mathcal{H}). \tag{24}$$

Define an auxiliary Heisenberg group

$$\overline{H} = V \times K, \tag{25}$$

with the multiplication given by $(v, z) \cdot (v', z') = (v + v', z + z' + \frac{1}{2}\overline{\omega}(v, v'))$. There exists homomorphism

$$\iota_H : \overline{H} \to H, \tag{26}$$

given by $(v, z) \mapsto (v, \mathrm{Tr}(z))$. Consider the pullback $(\overline{\pi} = \iota_H^* \pi, \overline{H}, \mathcal{H})$. We have

Proposition 3. *The representation $(\overline{\pi} = \iota_H^* \pi, \overline{H}, \mathcal{H})$ is the Heisenberg representation associated with the central character $\overline{\psi} = \psi \circ Tr$.*

The group $\overline{Sp}$ acts by automorphisms on the group $\overline{H}$ through its tautological action on the V-coordinate. This action is compatible with the action of Sp on H, i.e., we have $\iota_H(g \cdot h) = \iota_S(g) \cdot \iota_H(h)$ for every $g \in \overline{Sp}$, and $h \in \overline{H}$. The description of the representation $\overline{\rho}$ (24) now follows easily (cf. [9]).

Theorem 9 (Self-reducibility property (particular case)). *The representation $(\overline{\rho}, \overline{Sp}, \mathcal{H})$ is the Weil representation associated with the Heisenberg representation $(\overline{\pi}, \overline{H}, \mathcal{H})$.*

Remark 7. We can summarize the result in a slightly more elegant manner using the Jacobi groups. Let $J = Sp \ltimes H$ and $\overline{J} = \overline{Sp} \ltimes \overline{H}$ be the Jacobi groups associated with the symplectic spaces (V, ω) and $(V, \overline{\omega})$ respectively. We have a homomorphism $\iota : \overline{J} \to J$, given by $\iota(g, h) = (\iota_S(g), \iota_H(h))$. Let $(\tau, J, \mathcal{H})$ be the Heisenberg–Weil representation of J associated with a character ψ of the center $Z(J)$ (note that $Z(J) = Z(H)$), then the pullback $(\iota^* \tau, \overline{J}, \mathcal{H})$ is the Heisenberg–Weil representation of $\overline{J}$, associated with the character $\overline{\psi} = \psi \circ \mathrm{Tr}$ of the center $Z(\overline{J})$.

The general case

Here, we drop the assumption that T acts irreducibly on V. Let $(K, V, \overline{\omega})$ be the associated symplectic module structure (19). Using the results of Subsection 3.2, we have decompositions

$$(V, \omega) = \bigoplus_{\alpha \in \Xi} (V_\alpha, \omega_\alpha), \qquad (V, \overline{\omega}) = \bigoplus_{\alpha \in \Xi} (V_\alpha, \overline{\omega}_\alpha), \tag{27}$$

where $\overline{\omega}_\alpha : V_\alpha \times V_\alpha \to K_\alpha$. Let (cf. 25) $\overline{H} = V \times K$, be the Heisenberg group associated with $(V, \overline{\omega})$. There exists (cf. (26) a homomorphism $\iota_H : \overline{H} \to H$. Let us describe the pullback $\overline{\pi} = \iota_H^* \pi$ of the Heisenberg representation. First, we note that the decomposition (27) induces a corresponding decomposition of the Heisenberg group, $\overline{H} = \prod \overline{H}_\alpha$, where $\overline{H}_\alpha$ is the Heisenberg group associated with $(V_\alpha, \overline{\omega}_\alpha)$. We have the following proposition

Proposition 4. *There exists an isomorphism*

$$(\overline{\pi}, \overline{H}, \mathcal{H}) \simeq (\bigotimes \overline{\pi}_\alpha, \prod \overline{H}_\alpha, \bigotimes \mathcal{H}_\alpha),$$

where $(\overline{\pi}_\alpha, \overline{H}_\alpha, \mathcal{H}_\alpha)$ is the Heisenberg representation of $\overline{H}_\alpha$ associated with the central character $\overline{\psi}_\alpha = \psi \circ Tr_{K_\alpha/k}$.

Let $\iota_S : \overline{Sp} \hookrightarrow Sp$, be the embedding (20). Our next goal is to describe the restriction $\overline{\rho} = \iota_S^* \rho$. Recall that we have a decomposition $\overline{Sp} = \prod \overline{Sp}_\alpha$ (see (22)). In terms of this decomposition we have (cf. [9])

Theorem 10 (Self reducibility property—general case). *There exists an isomorphism*

$$(\overline{\rho}, \overline{Sp}, \mathcal{H}) \simeq (\bigotimes \overline{\rho}_\alpha, \prod \overline{Sp}_\alpha, \bigotimes \mathcal{H}_\alpha),$$

where $(\overline{\rho}_\alpha, \overline{Sp}_\alpha, \mathcal{H}_\alpha)$ is the Weil representation associated with the Heisenberg representation $\overline{\pi}_\alpha$.

Remark 8. As before, we can state an equivalent result using the Jacobi groups $J = Sp \ltimes H$ and $\overline{J} = \overline{Sp} \ltimes \overline{H}$. We have a decomposition $\overline{J} = \prod \overline{J}_\alpha$, where $\overline{J}_\alpha = \overline{Sp}_\alpha \ltimes \overline{H}_\alpha$. Let τ be the Heisenberg–Weil representation of J associated with a character ψ of the center $Z(J)$ (note that $Z(J) = Z(H)$). Then the pullback $\overline{\tau} = \iota^* \tau$ is isomorphic to $\bigotimes \overline{\tau}_\alpha$, where $\overline{\tau}_\alpha$ is the Heisenberg–Weil representation of $\overline{J}_\alpha$, associated with the character $\overline{\psi}_\alpha = \psi \circ \mathrm{Tr}_{K_\alpha/k}$ of the center $Z(\overline{J}_\alpha)$.

3.4 Application to multiplicities

Let us specialize to the case where the filed k is a finite field of odd characteristic. Let $T \subset Sp$ be a maximal torus. The torus T acts, via the Weil representation ρ, on the space $\mathcal{H}$, decomposing it into a direct sum of T-character spaces $\mathcal{H} = \bigoplus_{\chi:T\to\mathbb{C}^\times} \mathcal{H}_\chi$. Consider the problem of determining the multiplicities $m_\chi = \dim(\mathcal{H}_\chi)$. Using Lemma 1, we have (see (23)) a canonical decomposition of T

$$T = \prod T_\alpha, \tag{28}$$

where each of the tori T_α coincides with a maximal torus inside $\overline{Sp} \simeq SL_2(K_\alpha)$, for some field extension $K_\alpha \supset k$. In particular, by (28) we have a decomposition

$$\mathcal{H}_\chi = \bigotimes_{\chi_\alpha:T_\alpha\to\mathbb{C}^\times} \mathcal{H}_{\chi_\alpha}, \tag{29}$$

where $\chi = \prod \chi_\alpha : \prod T_\alpha \to \mathbb{C}^\times$. Hence, by Theorem 10, and the result about the multiplicities in the two-dimensional case (see Theorem (8)), we can compute the integer m_χ as follows. Denote by σ_α the quadratic character of T_α (note that by Theorem (8) the quadratic character σ_α cannot appear in the decomposition (29) if the torus T_α is inert).

Theorem 11. *We have*

$$m_\chi = 2^l,$$

where $l = \#\{\alpha : \chi_\alpha = \sigma_\alpha\}$.

4 Bounds on Higher-Dimensional Exponential Sums

In this section we present an application of the self-reducibility technique to bound higher-dimensional exponential sums attached to tori in $Sp = Sp(V, \omega)$, where (V, ω) is a $2N$-dimensional symplectic vector space over the finite field $\mathbb{F}_p$, $p \neq 2$. These exponential sums originated from the theory of quantum chaos (see Sections 5 and 6). Let $(\tau, J, \mathcal{H})$ be the Heisenberg–Weil representation associated with a central character $\psi : Z(J) = Z(H) \to \mathbb{C}^\times$. Recall that $J = Sp \ltimes H$, and τ is obtained as a semi-direct product $\tau = \rho \ltimes \pi$ of the Weil representation ρ and the Heisenberg representation π. Consider a maximal torus $T \subset Sp$. The torus T acts semisimply on $\mathcal{H}$, decomposing it into a direct sum of character spaces $\mathcal{H} = \bigoplus_{\chi:T\to\mathbb{C}^\times} \mathcal{H}_\chi$. We shall study common eigenstates $\Psi \in \mathcal{H}_\chi$. In particular, we will be interested in estimating matrix coefficients of the form $\langle \Psi | \pi(\xi) \Psi \rangle$ where $\xi \in V$ is not contained in any proper T-invariant subspace. It will be convenient to assume first that the torus T is completely inert (i.e., acts irreducibly on V). In this case one can show (see Theorem 11) that $\dim \mathcal{H}_\chi = 1$ for every χ. Below we sketch a proof of the following estimate.

Theorem 12. *For $\xi \in V$ which is not contained in any proper T-invariant subspace, we have*

$$|\langle \Psi | \pi(\xi) \Psi \rangle| \leq \frac{2 + o(1)}{\sqrt{p}^N}.$$

Let us explain way it is not easy to get such a bound by a direct calculation. Choosing a Schrödinger realization (see Remark 6), we can identify $\mathcal{H} = \mathbb{C}(\mathbb{F}_p^N)$. Under this identification, the matrix coefficient is equivalent to a sum

$$\langle \Psi | \pi(\xi) \Psi \rangle = \sum_{x \in \mathbb{F}_p^N} \Psi(x) e^{\frac{2\pi i}{p} \xi_+ x} \overline{\Psi}(x + \xi_-). \tag{30}$$

In this respect two problems are encountered. First, it is not easy to describe the eigenstates Ψ. Second, the sum (30) is a high-dimensional exponential sum, which is known to be hard to analyze using standard techniques.

Interestingly enough, representation theory suggests a remedy for both problems. Our strategy will be to interpret the matrix coefficient $\langle \Psi | \pi(\xi) \Psi \rangle$ in representation theoretic terms, and then to show, using the self-reducibility technique, that (30) is equivalent to a 1-dimensional sum over $\mathbb{F}_q$, $q = p^N$.

Representation theory and dimensional reduction of (30)

The torus T acts irreducibly on the vector space V. Invoking the result of Section 3.2, there exists a canonical symplectic module structure $(K, V, \overline{\omega})$ associated to T. Recall that in this particular case the algebra K is in fact a field, and $\dim_K V = 2$ (in our case $K = \mathbb{F}_q$, where $q = p^N$). Let $\overline{J} = \overline{Sp} \ltimes \overline{H}$ be the Jacobi group associated to the (two-dimensional) symplectic vector space $(V, \overline{\omega})$ over K. There exists a natural homomorphism $\iota : \overline{J} \to J$. Invoking the results of Section 3.3, the pullback $\overline{\tau} = \iota^* \tau$ is the Heisenberg–Weil representation of $\overline{J}$, i.e., $\overline{\tau} = \overline{\rho} \ltimes \overline{\pi}$.

Let $\Psi \in \mathcal{H}_\chi$. Denote by P_χ the orthogonal projector on the vector space $\mathcal{H}_\chi$. We can write P_χ in terms of the Weil representation $\overline{\rho}$

$$P_\chi = \frac{1}{|T|} \sum_{B \in T} \chi^{-1}(B) \overline{\rho}(B). \tag{31}$$

Since $\dim \mathcal{H}_\chi = 1$ (Theorem 11) we realize that

$$\langle \Psi | \pi(\xi) \Psi \rangle = \mathrm{Tr}(P_\chi \pi(\xi)). \tag{32}$$

Substituting (31) in (32), we can write

$$\langle \Psi | \pi(\xi) \Psi \rangle = \frac{1}{|T|} \sum_{B \in T} \chi^{-1}(B) \mathrm{Tr}(\overline{\rho}(B) \pi(\xi)).$$

Noting that $\mathrm{Tr}(\overline{\rho}(B)\pi(\xi))$ is nothing other than the character $\mathrm{ch}_{\overline{\tau}}(B \cdot \xi)$ of the Heisenberg–Weil representation $\overline{\tau}$. and that $|T| = p^N + 1$, we deduce that it is enough to prove that

$$\left| \sum_{B \in T} \chi^{-1}(B) \mathrm{ch}_{\overline{\tau}}(B \cdot \xi) \right| \leq 2\sqrt{q}, \tag{33}$$

where $q = p^N$. Now, note that the left-hand side of (33) is a one-dimensional exponential sum over $\mathbb{F}_q$, which is defined completely in terms of the two-dimensional Heisenberg–Weil representation $\overline{\tau}$. Estimate (33) is then a particular case of the following theorem, proved in [13].

Theorem 13. *Let (V, ω) be a two-dimensional symplectic vector space over a finite field $k = \mathbb{F}_q$, and $(\tau, J, \mathcal{H})$ be the corresponding Heisenberg–Weil representation. Let $T \subset Sp$ be a maximal torus. We have the following estimate*

$$\left| \sum_{B \in T} \chi(B) ch_{\tau}(B \cdot \xi) \right| \leq 2\sqrt{q}, \tag{34}$$

where χ is a character of T, and $0 \neq \xi \in V$ is not an eigenvector of T.

4.1 General case

In this subsection we state and prove the analogue of Theorem 12, where we drop the assumption of T being completely inert. In what follows, we use the results of Subsections 3.2 and 3.3.

Let $(K, V, \overline{\omega})$ be the symplectic module structure associated with the torus T. The algebra K is no longer a field, but decomposes into a direct sum of fields $K = \bigoplus_{\alpha \in \Xi} K_\alpha$. We have canonical decompositions

$$(V, \omega) = \bigoplus (V_\alpha, \omega_\alpha), \qquad (V, \overline{\omega}) = \bigoplus (V_\alpha, \overline{\omega}_\alpha).$$

Recall that V_α is a two-dimensional vector space over the field K_α. The Jacobi group $\overline{J}$ decomposes into $\overline{J} = \prod \overline{J}_\alpha$, where $\overline{J}_\alpha = \overline{Sp}_\alpha \ltimes \overline{H}_\alpha$ is the Jacobi group associated to $(V_\alpha, \overline{\omega}_\alpha)$. The pullback $(\overline{\tau} = \iota^* \tau, \overline{J}, \mathcal{H})$ decomposes into a tensor product $(\bigotimes \overline{\tau}_\alpha, \prod \overline{J}_\alpha, \bigotimes \mathcal{H}_\alpha)$, where $\overline{\tau}_\alpha$ is the Heisenberg–Weil representation of $\overline{J}_\alpha$. The torus T decomposes into $T = \prod T_\alpha$, where T_α is a maximal torus in $\overline{Sp}_\alpha$. Consequently, the character $\chi : T \to \mathbb{C}^\times$ decomposes into a product $\chi = \Pi \chi_\alpha : \Pi T_\alpha \to \mathbb{C}^\times$, and the space $\mathcal{H}_\chi$ decomposes into a tensor product

$$\mathcal{H}_\chi = \bigotimes \mathcal{H}_{\chi_\alpha}. \tag{35}$$

It follows from the above decomposition that it is enough to estimate matrix coefficients with respect to *pure tensor* eigenstates, i.e., eigenstates Ψ of the form $\Psi = \bigotimes \Psi_\alpha$, where $\Psi_\alpha \in \mathcal{H}_{\chi_\alpha}$. For a vector of the form $\xi = \bigoplus \xi_\alpha$, we have

$$\left\langle \bigotimes \Psi_\alpha | \pi(\xi) \bigotimes \Psi_\alpha \right\rangle = \prod \left\langle \Psi_\alpha | \pi(\xi_\alpha) \Psi_\alpha \right\rangle. \tag{36}$$

Hence, we need to estimate the matrix coefficients $\langle \Psi_\alpha | \pi(\xi_\alpha) \Psi_\alpha \rangle$, but these are defined in terms of the two-dimensional Heisenberg–Weil representation $\overline{\tau}_\alpha$. In addition, we recall the assumption that the vector $\xi \in V$ is not contained in any proper T-invariant subspace. This condition in turn implies that no summand ξ_α is an eigenvector of T_α. Hence, we can use Lemma 13, obtaining

$$|\langle \Psi_\alpha | \pi(\xi_\alpha) \Psi_\alpha \rangle| \leq 2/\sqrt{p}^{[K_\alpha:\mathbb{F}_p]}. \tag{37}$$

Consequently, using (36) and (37) we obtain

$$\left|\left\langle \bigotimes \Psi_\alpha | \pi(\xi) \bigotimes \Psi_\alpha \right\rangle\right| \leq 2^{|\Xi|}/\sqrt{p}^{\sum [K_\alpha:\mathbb{F}_p]} = 2^{|\Xi|}/\sqrt{p}^{[K:\mathbb{F}_p]} = 2^{|\Xi|}/\sqrt{p}^{N}.$$

Recall that the number $r_p = |\Xi|$ is called the symplectic rank of the torus T. The main application of the self-reducibility property, presented in these notes, is summarized in the following theorem.

Theorem 14. *Let (V, ω) be a $2N$-dimensional vector space over the finite field $\mathbb{F}_p$, and $(\tau, J, \mathcal{H})$ the corresponding Heisenberg–Weil representation. Let $\Psi \in \mathcal{H}_\chi$ be a unit χ-eigenstate with respect to a maximal torus $T \subset Sp$. We have the following estimate:*

$$|\langle \Psi | \pi(\xi) \Psi \rangle| \leq \frac{m_\chi \cdot (2 + o(1))^{r_p}}{\sqrt{p}^N},$$

where $1 \leq r_p \leq N$ is the symplectic rank of T, $m_\chi = \dim \mathcal{H}_\chi$, and $\xi \in V$ is not contained in any T-invariant subspace.

5 The Hannay–Berry model

We shall proceed to describe the higher-dimensional Hannay–Berry model of quantum mechanics on toral phase spaces. This model plays an important role in the mathematical theory of quantum chaos as it serves as a model where general phenomena, which are otherwise treated only on a heuristic basis, can be rigorously proven.

5.1 The classical phase space

Our classical phase space is the $2N$-dimensional symplectic torus $(\mathbb{T}, \omega)$. We denote by Γ the group of linear symplectic automorphisms of $\mathbb{T}$. Note that $\Gamma \simeq Sp(2N, \mathbb{Z})$. On the torus $\mathbb{T}$ we consider an algebra of complex functions (observables) $\mathcal{A} = \mathcal{F}(\mathbb{T})$. We denote by $\Lambda \simeq \mathbb{Z}^{2N}$ the lattice of characters (exponents) of $\mathbb{T}$. The form ω induces a skew-symmetric form on Λ, which we denote also by ω, and we assume it takes integral values on Λ and is normalized so that $\int_{\mathbb{T}} |\omega|^N = 1$.

5.2 The classical mechanical system

We take our classical mechanical system to be of a very simple nature. Let $A \in \Gamma$ be a generic element (see Definition 1), i.e., A is regular and admits no invariant co-isotropic sub-tori. The last condition can be equivalently restated in dual terms, namely, requiring that A admits no invariant isotropic subvectorspaces in $\Lambda_{\mathbb{Q}} = \Lambda \otimes_{\mathbb{Z}} \mathbb{Q}$. The element A generates, via its action as an automorphism $A : \mathbb{T} \longrightarrow \mathbb{T}$, a discrete time dynamical system.

5.3 Quantization

Before we employ the formal model, it is worthwhile to discuss the general phenomenological principles of quantization which are common to all models. Principally, quantization is a protocol by which one associates a Hilbert space $\mathcal{H}$ to the classical phase space, which in our case is the torus $\mathbb{T}$; In addition, the protocol gives a rule

$$f \rightsquigarrow Op(f) : \mathcal{H} \to \mathcal{H},$$

by which one associates an operator on the Hilbert space to every classical observable, i.e., a function $f \in \mathcal{F}(\mathbb{T})$. This rule should send a real function into a self-adjoint operator. In addition, in the presence of classical symmetries which in our case are given by the group Γ, the Hilbert space $\mathcal{H}$ should support a (projective unitary) representation $\Gamma \to PGL(\mathcal{H})$,

$$\gamma \mapsto U(\gamma) : \mathcal{H} \to \mathcal{H},$$

which is compatible with the quantization rule $Op(\cdot)$.

More precisely, quantization is not a single protocol, but a one-parameter family of protocols, parameterized by a parameter $\hbar$ called the Planck constant. Accepting these general principles, one searches for a formal model by which to quantize. In this work we employ a model called the *non-commutative torus* model.

5.4 The non-commutative torus model

Denote by $\mathcal{A}$ the algebra of trigonometric polynomials on $\mathbb{T}$, i.e., $\mathcal{A}$ consists of functions f which are a finite linear combinations of characters. We shall construct a one-parametric deformation of $\mathcal{A}$ called the non-commutative torus [6, 29].

Let $\hbar = 1/p$, where p is an odd prime number, and consider the additive character $\psi : \mathbb{F}_p \longrightarrow \mathbb{C}^\times$, $\psi(t) = e^{\frac{2\pi i t}{p}}$. We give here the following presentation of the algebra $\mathcal{A}_\hbar$. Let $\mathcal{A}_\hbar$ be the free non-commutative $\mathbb{C}$-algebra generated by the symbols $s(\xi)$, $\xi \in \Lambda$, and the relations

$$s(\xi)s(\eta) = \psi(\tfrac{1}{2}\omega(\xi,\eta))s(\xi+\eta). \tag{38}$$

Here we consider ω as a map $\omega : \Lambda \times \Lambda \longrightarrow \mathbb{F}_p$.

Note that $\mathcal{A}_\hbar$ satisfies the following properties:

- As a vector space $\mathcal{A}_\hbar$ is equipped with a natural basis $s(\xi)$, $\xi \in \Lambda$. Hence we can identify the vector space $\mathcal{A}_\hbar$ with the vector space $\mathcal{A}$ for each value of $\hbar$,
$$\mathcal{A}_\hbar \simeq \mathcal{A}. \tag{39}$$
- Substituting $\hbar = 0$ we have $\mathcal{A}_0 = \mathcal{A}$. Hence, we see that indeed $\mathcal{A}_\hbar$ is a deformation of the algebra of (algebraic) functions on $\mathbb{T}$.
- The group Γ acts by automorphisms on the algebra $\mathcal{A}_\hbar$, via $\gamma \cdot s(\xi) = s(\gamma\xi)$, where $\gamma \in \Gamma$ and $\xi \in \Lambda$. This action induces an action of Γ on the category of representations of $\mathcal{A}_\hbar$, taking a representation π and sending it to the representation π^γ, where $\pi^\gamma(f) = \pi(\gamma f)$, $f \in \mathcal{A}_\hbar$.

Using the identification (39) , we can describe a choice for the quantization of the functions. We just need to pick a representation of the quantum algebra $\mathcal{A}_\hbar$. But what representation to pick? It turns out that, we have a canonical choice. All the irreducible algebraic representations of $\mathcal{A}_\hbar$ are classified [12] and each of them is of dimension p^N. We have

Theorem 15 (Invariant representation [12]). *Let $\hbar = 1/p$ where p is a prime number. There exists a unique (up to isomorphism) irreducible representation $\pi : \mathcal{A}_\hbar \to End(\mathcal{H}_\hbar)$ which is fixed by the action of Γ. Namely, π^γ is isomorphic to π for every $\gamma \in \Gamma$.*

Let $(\pi, \mathcal{A}_\hbar, \mathcal{H})$ be a representative of the special representation defined in Theorem 15. For every element $\gamma \in \Gamma$ we have an isomorphism $\widetilde{\rho}(\gamma) : \mathcal{H} \to \mathcal{H}$ intertwining the representations π and π^γ , namely, it satisfies $\widetilde{\rho}(\gamma)\pi(f)\widetilde{\rho}(\gamma)^{-1} = \pi(\gamma f)$, for every $f \in \mathcal{A}_\hbar$ and $\gamma \in \Gamma$. The isomorphism $\widetilde{\rho}(\gamma)$ is not unique but unique up to a scalar (this is a consequence of Schur's lemma and the fact that π and π^γ are irreducible representations). It is easy to realize that the collection $\{\widetilde{\rho}(\gamma)\}$ constitutes a projective representation $\widetilde{\rho} : \Gamma \to PGL(\mathcal{H})$. Let $\hbar = 1/p$ where p is an odd prime $\neq 3$. We have the following linearization theorem (cf. [11, 13])

Theorem 16 (Linearization). *The projective representation $\widetilde{\rho}$ can be linearized uniquely to an honest representation $\rho : \Gamma \to GL(\mathcal{H})$ that factors through the finite quotient group $Sp \simeq Sp(2N, \mathbb{F}_p)$.*

Remark 9. The representation $\rho : Sp \to GL(\mathcal{H})$ is the celebrated Weil representation, here obtained via quantization of the torus.

5.5 The quantum dynamical system

Recall that we started with a classical dynamic on $\mathbb{T}$, generated by a generic (i.e., regular with no non-trivial invariant co-isotropic sub-tori) element $A \in \Gamma$. Using the Weil representation, we can associate to A the unitary operator

$\rho(A) : \mathcal{H} \to \mathcal{H}$, which constitutes the generator of discrete time quantum dynamics. We would like to study the $\rho(A)$-eigenstates

$$\rho(A)\Psi = \lambda\Psi,$$

which satisfy additional symmetries. This we do in the next section.

6 The Hecke quantum unique ergodicity theorem

It turns out that the operator $\rho(A)$ has degeneracies namely, its eigenspaces might be extremely large. This is manifested in the existence of a group of hidden symmetries commuting with $\rho(A)$ (note that classically the group of linear symplectomorphisms of $\mathbb{T}$ that commute with A, i.e., $\mathbf{T}_A(\mathbb{Z})$, does not contribute much to the harmonic analysis of $\rho(A)$). These symmetries can be computed. Indeed, let $T_A = Z(A, Sp)$, be the centralizer of the element A in the group Sp. Clearly T_A contains the cyclic group $\langle A \rangle$ generated by the element A, but it often happens that T_A contains additional elements. The assumption that A is regular (i.e., has distinct eigenvalues) implies that for sufficiently large p the group T_A consists of the $\mathbb{F}_p$-rational points of a maximal torus $\mathbf{T}_A \subset \mathbf{Sp}$, i.e., $T_A = \mathbf{T}_A(\mathbb{F}_p)$ (more precisely, p large enough so that it does not divides the discriminant of A). The group T_A is called the *Hecke* torus. It acts semisimply on $\mathcal{H}$, decomposing it into a direct sum of character spaces $\mathcal{H} = \bigoplus_{\chi : T_A \to \mathbb{C}^\times} \mathcal{H}_\chi$. We shall study common eigenstates $\Psi \in \mathcal{H}_\chi$, which we call *Hecke eigenstates* and will be assumed to be normalized so that $\|\Psi\|_{\mathcal{H}} = 1$. In particular, we will be interested in estimating matrix coefficients of the form $\langle \Psi | \pi(f) \Psi \rangle$, where $f \in \mathcal{A}$ is a classical observable on the torus $\mathbb{T}$ (see Subsection 5.4). We will call these matrix coefficients *Hecke–Wigner distributions*. It will be convenient for us to start with the following case.

6.1 The strongly generic case

Let us assume first that the automorphism A acts on $\mathbb{T}$ with no invariant sub-tori. In dual terms, this means that the element A acts irreducibly on the $\mathbb{Q}$-vector space $\Lambda_{\mathbb{Q}} = \Lambda \otimes_{\mathbb{Z}} \mathbb{Q}$.

We denote by r_p the symplectic rank of T_A, i.e., $r_p = |\Xi|$ where $\Xi = \Xi(T_A)$ is the symplectic type of T_A (see Definition 3). By definition we have $1 \leq r_p \leq N$ (for example, we get the two extreme cases: $d_p = 1$ when the torus T_A acts irreducibly on $V \simeq \mathbb{F}_p^{2N}$, and $d_p = N$ when T_A splits). We have

Theorem 17. *Consider a non-trivial exponent $0 \neq \xi \in \Lambda$ and a sufficiently large prime number p. Then for every normalized Hecke eigenstate $\Psi \in \mathcal{H}_\chi$ the following bound holds:*

$$|\langle \Psi | \pi(\xi) \Psi \rangle| \leq \frac{m_\chi \cdot 2^{r_p}}{\sqrt{p}^N}, \tag{40}$$

where $m_\chi = \dim(\mathcal{H}_\chi)$.

The lattice Λ constitutes a basis for $\mathcal{A}$, hence, using the bound (40) we obtain

Corollary 5 (Hecke quantum unique ergodicity—strongly generic case). *Consider an observable* $f \in \mathcal{A}$ *and a sufficiently large prime number* p. *For every normalized Hecke eigenstate* Ψ *we have*

$$\left| \langle \Psi | \pi(f) \Psi \rangle - \int_{\mathbb{T}} f d\mu \right| \leq \frac{C_f}{\sqrt{p}^N},$$

where $\mu = |\omega|^N$ *is the corresponding volume form and* C_f *is an explicit computable constant which depends only on the function* f.

Remark 10. In Subsection 6.2 we will elaborate on the distribution of the symplectic rank r_p (40) and in Subsection 6.3 the more general statements where $A \in \Gamma$ is any generic element (see Definition 1) will be stated and proved.

Proof of Theorem 17

The proof is by reduction to the bound on the Hecke–Wigner distributions obtained in Section 4, namely reduction to Theorem 14. Our first goal is to interpret the Hecke–Wigner distribution $\langle \Psi | \pi(\xi) \Psi \rangle$ in terms of the Heisenberg–Weil representation.

Step 1. *Replacing the non-commutative torus by the finite Heisenberg group.* Note that the Hilbert space $\mathcal{H}$ is a representation space of both the algebra $\mathcal{A}_\hbar$ and the group Sp. We will show next that the representation $(\pi, \mathcal{A}_\hbar, \mathcal{H})$ is *equivalent* to the Heisenberg representation of some finite Heisenberg group. The representation π is determined by its restriction to the lattice Λ. However, the restriction

$$\pi_{|\Lambda} : \Lambda \to GL(\mathcal{H}),$$

is not multiplicative and in fact constitutes (see Formula (38)) a projective representation of the lattice given by

$$\pi(\xi)\pi(\eta) = \psi(\tfrac{1}{2}\omega(\xi, \eta))\pi(\xi + \eta). \tag{41}$$

It is evident from (41) that the map $\pi_{|\Lambda}$ factors through the quotient $\mathbb{F}_p$-vector space V

$$\Lambda \to V = \Lambda / p\Lambda \to GL(\mathcal{H}). \tag{42}$$

The vector space V is equipped with a symplectic structure ω obtained via specialization of the corresponding form on Λ. Let $H = H(V, \omega)$ be the Heisenberg group associated with (V, ω). Recall that as a set $H = V \times \mathbb{F}_p$ and the multiplication is given by

$$(v, z) \cdot (v', z') = (v + v', z + z' + \tfrac{1}{2}\omega(v, v')). \tag{43}$$

From formula (41), the factorization (42), and the multiplication rule (43) we learn that the map $\pi : V \to GL(\mathcal{H})$, given by (42), lifts to an honest representation of the Heisenberg group $\pi : H \to GL(\mathcal{H})$. Finally, the pair (ρ, π), where ρ is the Weil representation obtained using quantization of the torus (see Theorem 16) glues into a single representation $\tau = \rho \ltimes \pi$ of the Jacobi group $J = Sp \ltimes H$, which is of course nothing other than the Heisenberg–Weil representation

$$\tau : J \to GL(\mathcal{H}). \tag{44}$$

Having the Heisenberg–Weil representation at our disposal we proceed to

Step 2. *Reformulation.* Let $\mathbf{V}$ and $\mathbf{T}_A$ be the algebraic group scheme defined over $\mathbb{Z}$ so that $\Lambda = \mathbf{V}(\mathbb{Z})$ and for every prime p we have $V = \mathbf{V}(\mathbb{F}_\mathbf{p})$ and $T_A = \mathbf{T}_A(\mathbb{F}_p)$. In this setting for every prime number p we can consider the lattice element $\xi \in \Lambda$ as a vector in the $\mathbb{F}_p$-vector space V.

Let $(\tau, J, \mathcal{H})$ be the Heisenberg–Weil representation (44) and consider a normalized Hecke eigenstate $\Psi \in \mathcal{H}_\chi$. We need to verify that for a sufficiently large prime number p we have

$$|\langle \Psi | \pi(\xi) \Psi \rangle| \leq \frac{m_\chi \cdot 2^{r_p}}{\sqrt{p}^N}, \tag{45}$$

where m_χ denotes the multiplicity $m_\chi = \dim \mathcal{H}_\chi$ and r_p is the symplectic rank of T_A. This verification is what we do next.

Step 3. *Verification.* We need to show that we meet the conditions of Theorem 14. What is left to check is that for sufficiently large prime number p the vector $\xi \in V$ is not contained in any T_A-invariant subspace of V. Let us denote by O_ξ the orbit $O_\xi = T_A \cdot \xi$. We need to show that for sufficiently large p we have

$$\mathrm{Span}_{\mathbb{F}_p}\{O_\xi\} = V. \tag{46}$$

The condition (46) is satisfied since it holds globally. In more details, our assumption on A guarantees that it holds for the corresponding objects over the field of rational numbers $\mathbb{Q}$, i.e., $\mathrm{Span}_{\mathbb{Q}}\{\mathbf{T}_A(\mathbb{Q}) \cdot \xi\} = \mathbf{V}(\mathbb{Q})$. Hence (46) holds for a sufficiently large prime number p.

6.2 The distribution of the symplectic rank

We would like to compute the asymptotic distribution of the symplectic rank r_p (45) in the set $\{1, \ldots, N\}$, i.e.,

$$\delta(r) = \lim_{x\to\infty} \frac{\#\{r_p = r\ ;\ p \le x\}}{\pi(x)}, \tag{47}$$

where $\pi(x)$ denotes the number of prime numbers up to x.

We fix an algebraic closure $\overline{\mathbb{Q}}$ of the field $\mathbb{Q}$, and denote by G the Galois group $G = \mathrm{Gal}(\overline{\mathbb{Q}}/\mathbb{Q})$. Consider the vector space $\mathbf{V} = \mathbf{V}(\overline{\mathbb{Q}})$. By extension of scalars the symplectic form ω on $\mathbf{V}(\mathbb{Q})$ induces a $\overline{\mathbb{Q}}$-linear symplectic form on $\mathbf{V}$, which we will also denote by ω. Let $\mathbf{T}$ denote the algebraic torus $\mathbf{T} = \mathbf{T}_A(\overline{\mathbb{Q}})$. The action of $\mathbf{T}$ on $\mathbf{V}$ is completely reducible, decomposing it into one-dimensional character spaces $\mathbf{V} = \bigoplus_{\chi\in\mathfrak{X}} \mathbf{V}_\chi$.

Let Θ be the restriction of the symplectic transpose $(\cdot)^t : \mathrm{End}(\mathbf{V}) \to \mathrm{End}(\mathbf{V})$ to $\mathbf{T}$. The involution Θ acts on the set of characters $\mathfrak{X}$ by $\chi \mapsto \Theta(\chi) = \chi^{-1}$ and this action is compatible with the action of the Galois group G on $\mathfrak{X}$ by conjugation $\chi \mapsto g\chi g^{-1}$, where $\chi \in \mathfrak{X}$ and $g \in G$. This means (recall that A is strongly generic) that we have a transitive action of G on the set $\mathfrak{X}/\Theta$. Consider the kernel $\mathrm{K} = \ker(G \to \mathrm{Aut}(\mathfrak{X}/\Theta))$, and the corresponding finite Galois group $Q = G/\mathrm{K}$. Considering Q as a subgroup of $\mathrm{Aut}(\mathfrak{X}/\Theta)$ we define the cycle number $c(C)$ of a conjugacy class $C \subset Q$ to be the number of irreducible cycles that compose a representative of C. By a direct application of the Chebotarev theorem [5] we get

Proposition 5 (Chebotarev's theorem). *The distribution δ (47) obeys*

$$\delta(r) = \frac{|C_r|}{|Q|},$$

where $C_r = \bigcup_{\substack{C\subset Q\\ c(C)=r}} C$.

6.3 The general generic case

Let us now treat the more general case where the automorphism A acts on $\mathbb{T}$ in a generic way (Definition 1). In dual terms, this means that the torus $\mathbf{T}(\mathbb{Q}) = \mathbf{T}_A(\mathbb{Q})$ acts on the symplectic vector space $\mathbf{V}(\mathbb{Q}) = \Lambda \otimes_{\mathbb{Z}} \mathbb{Q}$ decomposing it into an orthogonal symplectic direct sum

$$(\mathbf{V}(\mathbb{Q}), \omega) = \bigoplus_{\alpha\in\Xi} (\mathbf{V}_\alpha(\mathbb{Q}), \omega_\alpha), \tag{48}$$

with an irreducible action of $\mathbf{T}(\mathbb{Q})$ on each of the spaces $\mathbf{V}_\alpha(\mathbb{Q})$. For an exponent $\xi \in \Lambda$ define its support with respect to the decomposition (48) by $S_\xi = \mathrm{Supp}(\xi) = \{\alpha;\ P_\alpha\xi \neq 0\}$, where $P_\alpha : \mathbf{V}(\mathbb{Q}) \to \mathbf{V}(\mathbb{Q})$ is the projector onto the space $\mathbf{V}_\alpha(\mathbb{Q})$ and denote by d_ξ the dimension $d_\xi = \sum_{\alpha\in S_\xi} \dim \mathbf{V}_\alpha(\mathbb{Q})$.

The decomposition (48) induces a decomposition of the torus $\mathbf{T}(\mathbb{Q})$ into a product of completely inert tori

$$\mathbf{T}(\mathbb{Q}) = \prod_{\alpha\in\Xi} \mathbf{T}_\alpha(\mathbb{Q}). \tag{49}$$

Consider now a sufficiently large prime number p and specialize all the objects involved to the finite filed $\mathbb{F}_p$. The Hecke torus $T = \mathbf{T}(\mathbb{F}_p)$ acts on the quantum Hilbert space $\mathcal{H}$ decomposing it into an orthogonal direct sum $\mathcal{H} = \bigoplus_{\chi:T\to\mathbb{C}^\times} \mathcal{H}_\chi$. The decomposition (49) induces decompositions on the level of groups of points $T = \prod_{\alpha\in\Xi} T_\alpha$, where $T_\alpha = \mathbf{T}_\alpha(\mathbb{F}_p)$, on the level of characters $\chi = \underset{\alpha}{\Pi}\chi : \prod_\alpha T_\alpha \to \mathbb{C}^\times$, and on the level of character spaces $\mathcal{H}_\chi = \bigotimes_\alpha \mathcal{H}_{\chi_\alpha}$.

For each torus T_α we denote by $r_{p,\alpha} = r_p(T_\alpha)$ its symplectic rank (see Definition 3) and we consider the integer $|S_\xi| \leq r_{p,\xi} \leq d_\xi$ given by $r_{p,\xi} = \underset{\alpha\in S_\xi}{\Pi} r_{p,\alpha}$.

Let us denote by m_{χ_ξ} the dimension $m_{\chi_\xi} = \sum_{\alpha\in S_\xi} \dim \mathcal{H}_{\chi_\alpha}$. Finally, we can state the theorem for the generic case. We have

Theorem 18 (Hecke quantum unique ergodicity—generic case). *Consider a non-trivial exponent $0 \neq \xi \in \Lambda$ and a sufficiently large prime number p. Then for every normalized Hecke eigenstate $\Psi \in \mathcal{H}_\chi$ the following bound holds:*

$$|\langle \Psi|\pi(\xi)\Psi\rangle| \leq \frac{m_{\chi_\xi}\cdot 2^{r_{p,\xi}}}{\sqrt{p}^{d_\xi}}. \tag{50}$$

Considering the decomposition (48) we denote by d the dimension $d = \min_\alpha \mathbf{V}_\alpha(\mathbb{Q})$. Since the lattice Λ constitutes a basis for the algebra $\mathcal{A}$ of observables on $\mathbb{T}$, then using the bound (50) we obtain

Corollary 6. *Consider an observable $f \in \mathcal{A}$ and a sufficiently large prime number p. Then for every normalized Hecke eigenstate Ψ we have*

$$\left| \langle \Psi|\pi(f)\Psi\rangle - \int_{\mathbb{T}} f d\mu \right| \leq \frac{C_f}{\sqrt{p}^d},$$

where $\mu = |\omega|^N$ is the corresponding volume form and C_f is an explicit computable constant which depends only on the function f.

The proof of Theorem 18 is a straightforward application of Theorem 17. Indeed, considering the decomposition (48) of the torus $\mathbf{T}(\mathbb{Q})$ to a product of completely inert tori $\mathbf{T}_\alpha(\mathbb{Q})$, we may apply the theory developed for the strongly generic case in Subsection (6.1) to each of the tori $\mathbf{T}_\alpha(\mathbb{Q})$ to deduce Theorem 18.

Remark 11. As explained in Subsection (6.2) the distribution of the symplectic rank $r_{p,\xi}$ is determined by the Chebotarev theorem applied to (now a product of) suitable finite Galois groups Q_α attached to the tori $\mathbf{T}_\alpha$, $\alpha \in S_\xi$ (49).

Remark 12. The corresponding quantum unique ergodicity theorem for statistical states of generic automorphism A of $\mathbb{T}$ (see Theorem 4) follows directly from Theorem 18.

References

1. Adams J. and Moy A., Unipotent representations and reductive dual pairs over finite fields. *Trans. Amer. Math. Soc. 340, no. 1, (1993), 309–321*
2. Berry M.V., Regular and irregular semiclassical wavefunctions. *J. Phys. A 10, no. 12* (1977), 2083–2091.
3. Berry M.V., Semiclassical Mechanics of regular and irregular motion. *Les Houches Lecture Series Session XXXVI, eds. G. Iooss, R.H.G. Helleman and R. Stora, North Holland, Amsterdam* (1983), 171–271.
4. Bouzouina A. and De Biévre S., Equipartition of the eigenstates of quantized ergodic maps on the torus. *Commun. Math. Phys. 178* (1996), 83–105.
5. Chebotarev N.G., Determination of the density of the set of primes corresponding to a given class of permutations, *Izv. Akad. Nauk., 17* (1923), 205–230.
6. Connes A., Algebres et geometrie differentielle. *C. R. Acad. Sci. Paris Ser. A-B 290 (1980), no. 13, A599–A604.*
7. Deligne P., Metaplectique. *A letter to Kazhdan* (1982).
8. Faure F., Nonnenmacher S. and De Bievre S. Scarred eigenstates for quantum cat maps of minimal periods. *Commun. Math. Phys. 239* (2003), 449–492.
9. Gérardin P., Weil representations associated to finite fields. *J. Algebra 46*, no. 1, (1977), 54–101.
10. Gurevich S., Weil representation, Deligne's sheaf, and proof of the Kurlberg-Rudnick conjecture. *Ph.D. Thesis* (submitted: March 2005)
11. Gurevich S. and Hadani R., On Hannay-Berry equivariant quantization of the Torus. *arXiv:math-ph/0312039* (2001).
12. Gurevich S. and Hadani R., On Hannay-Berry equivariant quantization of higher-dimensional tori. *arXiv:math-ph/0403036* (2002).
13. Gurevich S. and Hadani R., Proof of the Kurlrberg-Rudnick rate conjecture. Comptes Rendus Mathematique, Volume 342, Issue 1, (2006), Pages 69–72.
14. Gurevich S. and Hadani R., Hecke quantum unique ergodicity for the higher-dimensional Hannay-Berry models. *arXiv:math-ph/0403036* (2004).
15. Gurevich S. and Hadani R., Heisenberg realizations, eigenfunctions and proof of the Kurlberg-Rudnick Supremum Conjecture. *arXiv:math-ph/0511036* (2005).
16. Gurevich S. and Hadani R., The geometric Weil representation. *Selecta Mathematica, New Series, Birkhäuser Basel* (Accepted: Dec. 2006).
17. Gurevich S. and Hadani R., Quantization of symplectic vector spaces over finite fields. *arXiv:0705.4556* (2005).
18. Hannay J.H. and Berry M.V., Quantization of linear maps on the torus—Fresnel diffraction by a periodic grating. *Physica D1* (1980), 267–291.
19. Heath-Brown R., Private communication. Montreal, July 2005.
20. Heller J.E., Bound-state eigenstates of classically chaotic Hamiltonian systems: scars of periodic orbits. *Phys. Rev. Lett. 53* (1984), 1515–1518.
21. Howe R., Invariant theory and duality for classical groups over finite fields with applications to their singular representation theory, preprint, Yale University.

22. Howe R., On the character of Weil's representation. *Trans. Amer. Math. Soc. 177* (1973), 287–298.
23. Kelmer D., Arithmetic Quantum Unique Ergodicity for Symplectic Linear Maps of the Multidimensional Torus. *math-ph/0510079.*
24. Kaplan L. and Heller E.J. Measuring scars of periodic orbits. *Phys. Rev.* E 59 (1999).
25. Kurlberg P. and Rudnick Z., Hecke theory and equidistribution for the quantization of linear maps of the torus. *Duke Math. Jour. 103* (2000), 47–78.
26. Marklof J., Arithmetic quantum chaos. Encyclopedia of Mathematical Physics, eds. J.-P. Francoise, G.L. Naber and Tsou S.T. Oxford: Elsevier, Volume 1 (2006), 212–220.
27. Rudnick Z., The quantized cat map and quantum ergodicity. *Lecture at the MSRI conference—Random Matrices and their Applications, Berkeley, June 7–11, 1999.*
28. Rudnick Z., On quantum unique ergodicity for linear maps of the torus. *European Congress of Mathematics, Vol. II (Barcelona, 2000), Progr. Math., 202, Birkhäuser, Basel.* (2001), 429–437.
29. Rieffel M.A., Non-commutative tori—a case study of non-commutative differentiable manifolds, *Contemporary Math. 105* (1990), 191–211.
30. Sarnak P., Arithmetic quantum chaos. *The Schur lectures, 1992, Tel Aviv. Israel Math. Conf. Proc., 8, Bar-Ilan Univ., Ramat Gan* (1995) 183–236.
31. Shnirelman A. I., Ergodic properties of eigenstates. *Uspehi Mat. Nauk 29, no. 6(180)* (1974), 181–182.
32. Srinivasan B., Weil representations of finite classical groups. *Invent. Math. 51, no. 2, 143–153* (1979).
33. von Neumann J., Mathematical foundations of quantum mechanics. *Princeton University Press* (1955).
34. Weil A., Sur certains groupes d'operateurs unitaires. *Acta Math. 111* (1964) 143–211.
35. Weyl H., Gruppentheorie und Quantenmechanik. *1st edition Hirzel, Leipzig* (1928).
36. Weyl H., The Classical Groups. Their Invariants and Representations. *Princeton University Press, Princeton, N.J.* (1939).

Notes on Canonical Quantization of Symplectic Vector Spaces over Finite Fields

Shamgar Gurevich* and Ronny Hadani†

* Department of Mathematics, University of California, Berkeley, CA 94720, USA
shamgar@math.berkeley.edu
† Department of Mathematics, University of Chicago, IL 60637, USA
hadani@math.uchicago.edu

Summary. In these notes we construct a quantization functor, associating a Hilbert space $\mathcal{H}(V)$ to a finite dimensional symplectic vector space V over a finite field $\mathbb{F}_q$. As a result, we obtain a canonical model for the Weil representation of the symplectic group $Sp(V)$. The main technical result is a proof of a stronger form of the Stone–von Neumann theorem for the Heisenberg group over $\mathbb{F}_q$. Our result answers, for the case of the Heisenberg group, a question of Kazhdan about the possible existence of a canonical Hilbert space attached to a coadjoint orbit of a general unipotent group over $\mathbb{F}_q$.

Key words: Quantization functor, Weil representation, Quantization of Lagrangian correspondences, Geometric intertwining operators.

AMS codes: 11F27, 53D50.

1 Introduction

Quantization is a fundamental procedure in mathematics and in physics. Although it is widely used in both contexts, its precise nature remains to some extent unclear. From the physical side, quantization is the procedure by which one associates to a classical mechanical system its quantum counterpart. From the mathematical side, it seems that quantization is a way to construct interesting Hilbert spaces out of symplectic manifolds, suggesting a method for constructing representations of the corresponding groups of symplectomorphisms [14, 16].

Probably, one of the principal manifestation of quantization in mathematics appears in the form of the Weil representation [19, 20, 22] of the metaplectic group

$$\rho : Mp(2n, \mathbb{R}) \to U\left(L^2\left(\mathbb{R}^n\right)\right),$$

Ö. Ceyhan et al. (eds.), *Arithmetic and Geometry Around Quantization*,
Progress in Mathematics 279, DOI: 10.1007/978-0-8176-4831-2_8,

where $Mp(2n,\mathbb{R})$ is a double cover of the linear symplectic group $Sp(2n,\mathbb{R})$. The general ideology [23, 24] suggests that the Weil representation appears through a quantization of the standard symplectic vector space $(\mathbb{R}^{2n},\omega)$. This means that there should exist a quantization functor $\mathcal{H}$, associating to a symplectic manifold (M,ω) an Hilbert space $\mathcal{H}(M)$, such that when applied to $(\mathbb{R}^{2n},\omega)$ it yields the Weil representation in the form of

$$\mathcal{H} : Sp(2n,\mathbb{R}) \to U\left(\mathcal{H}\left(\mathbb{R}^{2n},\omega\right)\right).$$

As stated, this ideology is too naive since it does not account for the metaplectic cover.

1.1 Main results

In these notes, we show that the quantization ideology can be made precise when applied in the setting of symplectic vector spaces over the finite field $\mathbb{F}_q$, where q is odd. Specifically, we construct a quantization functor $\mathcal{H} : \mathsf{Symp} \to \mathsf{Hilb}$, where Symp denotes the (groupoid) category whose objects are finite dimensional symplectic vector spaces over $\mathbb{F}_q$ and morphisms are linear isomorphisms of symplectic vector spaces and Hilb denotes the category of finite dimensional Hilbert spaces.

As a consequence, for a fixed symplectic vector space $V \in \mathsf{Symp}$, we obtain, by functoriality, a homomorphism $\mathcal{H} : Sp(V) \to U(\mathcal{H}(V))$, which we refer to as the canonical model of the Weil representation of the symplectic group $Sp(V)$.

Properties of the quantization functor

In addition, we show that the functor $\mathcal{H}$ satisfies the following basic properties (cf. [24]):

- **Compatibility with Cartesian products.** The functor $\mathcal{H}$ is a monoidal functor: Given $V_1, V_2 \in \mathsf{Symp}$, we have a natural isomorphism

$$\mathcal{H}(V_1 \times V_2) \simeq \mathcal{H}(V_1) \otimes \mathcal{H}(V_2).$$

- **Compatibility with duality.** Given $V = (V,\omega) \in \mathsf{Symp}$, its symplectic dual is $\overline{V} = (V,-\omega)$. There exists a natural non-degenerate pairing

$$\langle\cdot,\cdot\rangle_V : \mathcal{H}\left(\overline{V}\right) \times \mathcal{H}(V) \to \mathbb{C}.$$

- **Compatibility with linear symplectic reduction**. Given $V \in \mathsf{Symp}$, $I \subset V$ an isotropic subspace in V and $o_I \in \wedge^{\mathrm{top}} I$ a non-zero vector, there exists a natural isomorphism

$$\mathcal{H}(V)^I \simeq \mathcal{H}\left(I^{\perp}/I\right), \tag{1}$$

where $\mathcal{H}(V)^I$ stands for the subspace of I-invariant vectors in $\mathcal{H}(V)$ (an operation which will be made precise in the sequel) and $I^{\perp}/I \in \mathsf{Symp}$ is the symplectic reduction of V with respect to I [3]. (A pair (I, o_I), where $o_I \in \wedge^{\mathrm{top}} I$ is non-zero vector, is called an *oriented isotropic subspace*).

Quantization of oriented Lagrangian subspaces. A particular situation is when $I = L$ is a Lagrangian subspace. In this situation, $L^{\perp}/L = 0$ and (1) yields an isomorphism $\mathcal{H}(V)^L \simeq \mathcal{H}(0) = \mathbb{C}$, which associates to $1 \in \mathbb{C}$ a vector $v_{L^\circ} \in \mathcal{H}(V)$. This means that we establish a mechanism which associates to every oriented Lagrangian subspace in V a well defined vector in $\mathcal{H}(V)$

$$L^\circ \longmapsto v_{L^\circ} \in \mathcal{H}(V).$$

Interestingly, to the best of our knowledge (cf. [9]), this kind of structure, which exists in the setting of the Weil representation of the group $Sp(V)$, was not observed before.

Quantization of oriented Lagrangian correspondences. It is also interesting to consider simultaneously the compatibility of $\mathcal{H}$ with Cartesian product, duality, and linear symplectic reduction. The first and second properties imply that $\mathcal{H}\left(\overline{V}_1 \times V_2\right)$ is naturally isomorphic to the vector space $\mathrm{Hom}\left(\mathcal{H}(V_1), \mathcal{H}(V_2)\right)$. The third property implies that every oriented Lagrangian L° in $\overline{V}_1 \times V_2$ (i.e., oriented canonical relation from V_1 to V_2 (cf. [23, 24])) can be quantized into a well defined operator

$$L^\circ \longmapsto A_{L^\circ} \in \mathrm{Hom}\left(\mathcal{H}(V_1), \mathcal{H}(V_2)\right).$$

In this regard, a particular kind of oriented Lagrangian in $\overline{V} \times V$ is the graph Γ_g of a symplectic linear map $g : V \to V$, $g \in Sp(V)$. The orientation is automatic in this case—it is induced from $\omega^{\wedge - n}$, $\dim(V) = 2n$, through the isomorphism $p_V : \Gamma_g \to V$, where $p_V : \overline{V} \times V \to V$ is the projection on the V-coordinate.

A further and more detailed study of these properties will appear in a subsequent work.

The strong Stone–von Neumann theorem

The main technical result of these notes is a proof ([10, 11] unpublished) of a stronger form of the Stone–von Neumann theorem for the Heisenberg group over $\mathbb{F}_q$. In this regard we describe an algebro-geometric object (an ℓ-adic perverse Weil sheaf $\mathcal{K}$), which, in particular, implies the strong Stone–von Neumann theorem. The construction of the sheaf $\mathcal{K}$ is one of the main contributions of this work.

Finally, we note that our result answers, for the case of the Heisenberg group, a question of Kazhdan [13] about the possible existence of a canonical Hilbert space attached to a coadjoint orbit of a general unipotent group over $\mathbb{F}_q$.

We devote the rest of the introduction to an intuitive explanation of the main ideas and results of these notes.

1.2 Quantization of symplectic vectors spaces over finite fields

Let (V, ω) be a $2n$-dimensional symplectic vector space over the finite field $\mathbb{F}_q$, assuming q is odd. The vector space V considered as an abelian group admits a non-trivial central extension H, called the *Heisenberg group*, which can be presented as $H = V \times \mathbb{F}_q$ with center $Z = Z(H) = \{(0,z) : z \in \mathbb{F}_q\}$. The group $Sp = Sp(V)$ acts on H by group automorphisms via its tautological action on the V-coordinate.

The celebrated Stone–von Neumann theorem [18, 21] asserts that given a non-trivial central character $\psi : Z \to \mathbb{C}^{\times}$, there exists a unique (up to isomorphism) irreducible representation $\pi : H \to GL(\mathcal{H})$ such that the center acts by ψ, i.e., $\pi_{|Z} = \psi \cdot \mathrm{Id}_{\mathcal{H}}$. The representation π is called the *Heisenberg representation.*

Choosing a Lagrangian subvector space $L \in \mathrm{Lag}(V)$ (the set $\mathrm{Lag}(V)$ is called the *Lagrangian Grassmanian*) we can define a model $(\pi_L, H, \mathcal{H}_L)$ of the Heisenberg representation, where $\mathcal{H}_L$ consists of functions $f : H \to \mathbb{C}$ satisfying $f(z \cdot l \cdot h) = \psi(z) f(h)$ for every $l \in L$, $z \in Z$ and the action π_L is given by right translation. The problematic issue in this construction is that there is no preferred choice of a Lagrangian subspace $L \in V$ and consequently none of the spaces $\mathcal{H}_L$ admit an action of the group Sp. In fact, an element $g \in Sp$ induces an isomorphism

$$g : \mathcal{H}_L \to \mathcal{H}_{gL}, \tag{2}$$

for every $L \in \mathrm{Lag}(V)$.

The strong Stone–von Neumann theorem

The strategy that we will employ is: "If you can not choose a preferred Lagrangian subspace then work with all of them simultaneously".

We can think of the system of models $\{\mathcal{H}_L\}$ as a vector bundle $\mathfrak{H}$ on Lag with fibers $\mathfrak{H}_{|L} = \mathcal{H}_L$, the condition (2) means that $\mathfrak{H}$ is equipped with an Sp-equivariant structure and what we seek is a canonical trivialization of $\mathfrak{H}$. More formally, we seek for a canonical system of intertwining morphisms $F_{M,L} \in \mathrm{Hom}_H(\mathcal{H}_L, \mathcal{H}_M)$, for every $L, M \in \mathrm{Lag}(V)$. The existence of such a system is the content of the strong Stone–von Neumann theorem.

Theorem 1 (Strong Stone–von Neumann theorem). *There exists a canonical system of intertwining morphisms $\{F_{M,L} \in \mathrm{Hom}_H(\mathcal{H}_L, \mathcal{H}_M)\}$ satisfying the multiplicativity property $F_{N,M} \circ F_{M,L} = F_{N,L}$, for every $N, M, L \in$ Lag(V).*

Remark 1 (Important remark). It is important to note here that the precise statement involves the finer notion of an oriented Lagrangian subspace [1,17], but for the sake of the introduction we will ignore this technical nuance.

The Hilbert space $\mathcal{H}(V)$ consists of systems of vectors $(v_L \in \mathcal{H}_L)_{L \in \mathrm{Lag}}$ such that $F_{M,L}(v_L) = v_M$, for every $L, M \in \mathrm{Lag}(V)$. The vector space $\mathcal{H}(V)$ can be thought of as the space of horizontal sections of $\mathfrak{H}$.

As it turns out, the symplectic group Sp naturally acts on $\mathcal{H}(V)$. We denote this representation by $(\rho_V, Sp, \mathcal{H}(V))$, and refer to it as the *canonical model* of the Weil representation. We proceed to explain the main underlying idea behind the construction of the system $\{F_{M,L}\}$.

1.3 Canonical system of intertwining morphisms

The construction will be close in spirit to the procedure of "analytic continuation". We consider the subset $U \subset \mathrm{Lag}(V)^2$, consisting of pairs $(L, M) \in \mathrm{Lag}(V)^2$ which are in general position, that is $L \cap M = 0$. The basic idea is that for a pair $(L, M) \in U$, $F_{M,L}$ can be given by an explicit formula—**ansatz**. The main statement is that this formula admits a unique multiplicative extension to the set of all pairs. The extension is constructed using algebraic geometry.

Extension to singular pairs

It will be convenient to work in the setting of *kernels*. In more detail, every intertwining morphism $F \in \mathrm{Hom}_H(\mathcal{H}_L, \mathcal{H}_M)$ can be presented by a kernel function $K \in \mathbb{C}(H, \psi)$ satisfying $K(m \cdot h \cdot l) = K(h)$, for every $m \in M$ and $l \in L$ (we denote by $\mathbb{C}(H, \psi)$ the subspace of functions $f \in \mathbb{C}(H)$ which are ψ-equivariant with respect to the center, that is $f(z \cdot h) = \psi(z) f(h)$, for every $z \in Z$). Moreover, this presentation is unique when $(M, L) \in U$; hence, in this case, we have a unique kernel $K_{M,L}$ representing our given $F_{M,L}$. If we denote by O the set $U \times H$, we see that the collection $\{K_{M,L} : (M, L) \in U\}$ forms a function $K_O \in \mathbb{C}(O)$ given by $K_O(M, L) = K_{M,L}$ for every $(M, L) \in U$. The problem is how to (correctly) extend the function K_O to the set $X = \mathrm{Lag}(V)^2 \times H$. In order to do that, we invoke the procedure of geometrization, which we briefly explain below.

Geometrization

A general ideology due to Grothendieck is that any meaningful set-theoretic object is governed by a more fundamental algebro-geometric one. The procedure by which one translates from the set theoretic setting to algebraic geometry is called *geometrization*, which is a formal procedure by which sets are replaced by algebraic varieties and functions are replaced by certain sheaf-theoretic objects.

The precise setting consists of a set $X = \mathbf{X}(\mathbb{F}_q)$ of rational points of an algebraic variety $\mathbf{X}$, defined over $\mathbb{F}_q$ and a complex valued function $f \in \mathbb{C}(X)$ governed by an ℓ-adic Weil sheaf $\mathcal{F}$.

The variety $\mathbf{X}$ is a space equipped with an automorphism $Fr : \mathbf{X} \to \mathbf{X}$ (called Frobenius), such that the set X is naturally identified with the set of fixed points $X = \mathbf{X}^{Fr}$.

The sheaf $\mathcal{F}$ can be considered as a vector bundle on the variety $\mathbf{X}$, equipped with an endomorphism $\theta : \mathcal{F} \to \mathcal{F}$ which lifts Fr.

The procedure by which f is obtained from $\mathcal{F}$ is called Grothendieck's *sheaf-to-function correspondence* and it can be described, roughly, as follows. Given a point $x \in X$, the endomorphism θ restricts to an endomorphism $\theta_x : \mathcal{F}_{|x} \to \mathcal{F}_{|x}$ of the fiber $\mathcal{F}_{|x}$. The value of f on the point x is defined to be

$$f(x) = \mathrm{Tr}(\theta_x : \mathcal{F}_{|x} \to \mathcal{F}_{|x}).$$

The function defined by this procedure is denoted by $f = f^{\mathcal{F}}$.

Solution to the extension problem

Our extension problem fits nicely into the geometrization setting: The sets O, X are sets of rational points of corresponding algebraic varieties $\mathbf{O}, \mathbf{X}$, the imbedding $j : O \hookrightarrow X$ is induced from an open imbedding $j : \mathbf{O} \hookrightarrow \mathbf{X}$ and, finally, the function K_O comes from a Weil sheaf $\mathcal{K}_{\mathbf{O}}$ on the variety $\mathbf{O}$.

The extension problem is solved as follows: First extend the sheaf $\mathcal{K}_{\mathbf{O}}$ to a sheaf $\mathcal{K}$ on the variety $\mathbf{X}$ and then take the corresponding function $K = f^{\mathcal{K}}$, which establishes the desired extension. The reasoning behind this strategy is that in the realm of sheaves there exist several functorial operations of extension, probably the most interesting one is called *perverse extension* [2]. The sheaf $\mathcal{K}$ is defined as the perverse extension of $\mathcal{K}_{\mathbf{O}}$.

1.4 Structure of the notes

Apart from the introduction, the notes consists of three sections.

In Section 2, all basic constructions are introduced and main statements are formulated. We begin with the definition of the Heisenberg group and the Heisenberg representation. Next, we introduce the canonical system of intertwining morphisms between different models of the Heisenberg representation and formulate the strong Stone von–Neumann theorem (Theorem 3). We proceed to explain how to present an intertwining morphism by a kernel function, and we reformulate the strong Stone von–Neumann theorem in the setting of kernels (Theorem 4). Using Theorem 3, we construct a quantization functor $\mathcal{H}$. We finish this section by showing that $\mathcal{H}$ is a monoidal functor and that it is compatible with duality and the operation of linear symplectic reduction. In section 3, we construct a sheaf theoretic counterpart for the canonical system of intertwining morphisms (Theorem 5). This sheaf is then used to prove Theorem 4. Finally, in Section 4 we sketch the proof of Theorem 5. Complete proofs for the statements appearing in these notes will appear elsewhere.

1.5 Acknowledgments

We would like to thank our scientific advisor J. Bernstein for his interest and guidance, and for his idea about the notion of oriented Lagrangian subspace. It is a pleasure to thank D. Kazhdan for sharing with us his thoughts about the possible existence of canonical Hilbert spaces. We thank A. Weinstein for teaching us some of his ideas concerning quantization, and for the opportunity to present this work in the symplectic geometry seminar, Berkeley, February 2007. We acknowledge O. Gabber for his remark concerning the characterization of the Weil representation. We would like to acknowledge M. Vergne for her encouragement. Finally, we would like to thank O. Ceyhan and the organizers of the conference AGAQ, held in Istanbul during June 2006, for the invitation to present this work.

2 Quantization of symplectic vector spaces over finite fields

2.1 The Heisenberg group

Let (V,ω) be a $2n$-dimensional symplectic vector space over the finite field $\mathbb{F}_q$. Considering V as an abelian group, it admits a non-trivial central extension called the *Heisenberg* group. Concretely, the group $H = H(V)$ can be presented as the set $H = V \times \mathbb{F}_q$ with the multiplication given by

$$(v,z)\cdot(v',z') = \Big(v+v', z+z'+\frac{1}{2}\omega(v,v')\Big).$$

The center of H is $Z = Z(H) = \{(0,z) : z \in \mathbb{F}_q\}$. The symplectic group $Sp = Sp(V)$ acts by automorphism of H through its tautological action on the V-coordinate.

2.2 The Heisenberg representation

One of the most important attributes of the group H is that it admits, principally, a unique irreducible representation. We will call this property *The Stone–von Neumann property* (S-vN for short). The precise statement goes as follows. Let $\psi : Z \to \mathbb{C}^\times$ be a non-trivial character of the center. For example we can take $\psi(z) = e^{\frac{2\pi i}{p}\operatorname{tr}(z)}$. It is not hard to show

Theorem 2 (Stone–von Neumann property). *There exists a unique (up to isomorphism) irreducible unitary representation $(\pi, H, \mathcal{H})$ with the center acting by ψ, i.e., $\pi_{|Z} = \psi \cdot Id_{\mathcal{H}}$.*

The representation π which appears in the above theorem will be called the *Heisenberg representation.*

2.3 The strong Stone–von Neumann property

Although the representation π is unique, it admits a multitude of different models (realizations); in fact this is one of its most interesting and powerful attributes. These models appear in families. In this work we will be interested in a particular family of such models which are associated with Lagrangian subspaces in V.

Let us denote by $\text{Lag} = \text{Lag}(V)$ the set of Lagrangian subspaces in V. Let $\mathbb{C}(H, \psi)$ denote the subspace of functions $f \in \mathbb{C}(H)$, satisfying the equivariance property $f(z \cdot h) = \psi(z) f(h)$, for every $z \in Z$.

Given a Lagrangian subspace $L \in \text{Lag}$, we can construct a model $(\pi_L, H, \mathcal{H}_L)$ of the Heisenberg representation: The vector space $\mathcal{H}_L$ consists of functions $f \in \mathbb{C}(H, \psi)$ satisfying $f(l \cdot h) = f(h)$, for every $l \in L$ and the Heisenberg action is given by right translation $(\pi_L(h) \triangleright f)(h') = f(h' \cdot h)$, for $f \in \mathcal{H}_L$.

Definition 1. *An* <u>*oriented Lagrangian*</u> L° *is a pair* $L^\circ = (L, o_L)$, *where* L *is a Lagrangian subspace in* V *and* o_L *is a non-zero vector in* $\bigwedge^{top} L$.

Let $\text{Lag}^\circ = \text{Lag}^\circ(V)$ denote the set of oriented Lagrangian subspaces in V. We associate to each oriented Lagrangian subspace L°, a model $(\pi_{L^\circ}, H, \mathcal{H}_{L^\circ})$ of the Heisenberg representation simply by forgetting the orientation, taking $\mathcal{H}_{L^\circ} = \mathcal{H}_L$ and $\pi_{L^\circ} = \pi_L$. Sometimes, we will use a more informative notation $\mathcal{H}_{L^\circ} = \mathcal{H}_{L^\circ}(V)$ or $\mathcal{H}_{L^\circ} = \mathcal{H}_{L^\circ}(V, \psi)$.

Canonical system of intertwining morphisms

Given a pair $(M^\circ, L^\circ) \in \text{Lag}^{\circ 2}$, the models $\mathcal{H}_{L^\circ}$ and $\mathcal{H}_{M^\circ}$ are isomorphic as representations of H by Theorem 2, moreover, since the Heisenberg representation is irreducible, the vector space $\text{Hom}_H(\mathcal{H}_{L^\circ}, \mathcal{H}_M^\circ)$ of intertwining morphisms is one-dimensional. Roughly, the strong Stone–von Neumann property asserts the existence of a distinguished element $F_{M^\circ, L^\circ} \in \text{Hom}_H(\mathcal{H}_{L^\circ}, \mathcal{H}_M^\circ)$, for every pair $(M^\circ, L^\circ) \in \text{Lag}^{\circ 2}$. The precise statement involves the following definition:

Definition 2. *A system* $\{F_{M^\circ, L^\circ} \in \text{Hom}_H(\mathcal{H}_{L^\circ}, \mathcal{H}_M^\circ) : (M^\circ, L^\circ) \in Lag^{\circ 2}\}$ *of intertwining morphisms is called* <u>*multiplicative*</u> *if for every triple* $(N^\circ, M^\circ, L^\circ) \in Lag^{\circ 3}$ *the following equation holds*

$$F_{N^\circ, L^\circ} = F_{N^\circ, M^\circ} \circ F_{M^\circ, L^\circ}.$$

We proceed as follows. Let $U \subset \text{Lag}^{\circ 2}$ denote the set of pairs $(M^\circ, L^\circ) \in \text{Lag}^{\circ 2}$ which are in general position, i.e., $L \cap M = 0$. For $(M^\circ, L^\circ) \in U$, we define F_{M°, L° by the following explicit formula:

$$F_{M^\circ, L^\circ} = C_{M^\circ, L^\circ} \cdot \widetilde{F}_{M,L}, \tag{3}$$

where $\widetilde{F}_{M,L} : \mathcal{H}_{L^\circ} \to \mathcal{H}_{M^\circ}$ is the averaging morphism

$$\widetilde{F}_{M,L}[f](h) = \sum_{m \in M} f(m \cdot h),$$

for every $f \in \mathcal{H}_{L^\circ}$ and C_{M°,L° is a normalization constant given by

$$C_{M^\circ,L^\circ} = (G_1/q)^n \cdot \sigma\left((-1)^{\binom{n}{2}} \omega_\wedge (o_L, o_M)\right),$$

where $n = \frac{\dim(V)}{2}$, σ is the unique quadratic character (also called the Legendre character) of the multiplicative group $G_m = \mathbb{F}_q^\times$, G_1 is the one-dimensional Gauss sum

$$G_1 = \sum_{z \in \mathbb{F}_q} \psi\left(\frac{1}{2} z^2\right),$$

and $\omega_\wedge$ is the pairing $\omega_\wedge : \bigwedge^{\text{top}} L \bigotimes \bigwedge^{\text{top}} M \to \mathbb{F}_q$ induced by the symplectic form.

Theorem 3 (The strong Stone–von Neumann property). *There exists a unique system $\{F_{M^\circ,L^\circ}\}$ of intertwining morphisms satisfying*

1. *Restriction. For every pair $(M^\circ, L^\circ) \in U$, F_{M°,L° is given by (3).*
2. *Multiplicativity. For every triple $(N^\circ, M^\circ, L^\circ) \in Lag^{\circ 3}$,*

$$F_{N^\circ,L^\circ} = F_{N^\circ,M^\circ} \circ F_{M^\circ,L^\circ}.$$

Theorem 3 will follow from Theorem 4 below.

Granting the existence and uniqueness of the system $\{F_{M^\circ,L^\circ}\}$, we can write F_{M°,L° in a closed form, for a general pair $(M^\circ, L^\circ) \in \text{Lag}^{\circ 2}$. In order to do that we need to fix some additional terminology.

Let $I = M \cap L$. We have canonical tensor product decompositions

$$\bigwedge^{\text{top}} M = \bigwedge^{\text{top}} I \bigotimes \bigwedge^{\text{top}} M/I,$$
$$\bigwedge^{\text{top}} L = \bigwedge^{\text{top}} I \bigotimes \bigwedge^{\text{top}} L/I.$$

In terms of the above decompositions, the orientation can be written in the form $o_M = \iota_M \otimes o_{M/I}$, $o_L = \iota_L \otimes o_{L/I}$. Using the same notations as before, we denote by $\widetilde{F}_{M,L} : \mathcal{H}_{L^\circ} \to \mathcal{H}_{M^\circ}$ the averaging morphism

$$\widetilde{F}_{M,L}[f](h) = \sum_{\overline{m} \in M/I} f(m \cdot h),$$

for $f \in \mathcal{H}_{L^\circ}$ and by C_{M°,L° the normalization constant

$$C_{M^\circ,L^\circ} = (G_1)^k \cdot \sigma\left((-1)^{\binom{k}{2}} \frac{\iota_M}{\iota_L} \cdot \omega_\wedge (o_{L/I}, o_{M/I})\right),$$

where $k = \dfrac{\dim(I^\perp/I)}{2}$.

Proposition 1. *For every* $(M^\circ, L^\circ) \in Lag^{\circ 2}$

$$F_{M^\circ,L^\circ} = C_{M^\circ,L^\circ} \cdot \widetilde{F}_{M,L}.$$

2.4 Kernel presentation of an intertwining morphism

An explicit way to present an intertwining morphism is via a kernel function. Fix a pair $(M^\circ, L^\circ) \in \mathrm{Lag}^{\circ 2}$ and let $\mathbb{C}(M\backslash H/L, \psi)$ denote the subspace of functions $f \in \mathbb{C}(H, \psi)$ satisfying the equivariance property $f(m \cdot h \cdot l) = f(h)$ for every $m \in M$ and $l \in L$. Given a function $K \in \mathbb{C}(M\backslash H/L, \psi)$, we can associate to it an intertwining morphism $I[K] \in \mathrm{Hom}_H(\mathcal{H}_{L^\circ}, \mathcal{H}^\circ_M)$ defined by

$$I[K](f) = K * f = m_!(K \boxtimes_{Z \cdot M} f),$$

for every $f \in \mathcal{H}_{L^\circ}$. Here, $K \boxtimes_{Z \cdot L} f$ denotes the function $K \boxtimes f \in \mathbb{C}(H \times H)$, factored to the quotient $H \times_{Z \cdot L} H$ and $m_!$ denotes the operation of summation along the fibers of the multiplication mapping $m : H \times H \to H$. The function K is called an *intertwining kernel.* The procedure just described defines a linear transform

$$I : \mathbb{C}(M\backslash H/L, \psi) \longrightarrow \mathrm{Hom}_H(\mathcal{H}_{L^\circ}, \mathcal{H}_{M^\circ}).$$

An easy verification reveals that I is surjective, but it is injective only when M, L are in general position.

Fix a triple $(N^\circ, M^\circ, L^\circ) \in \mathrm{Lag}^{\circ 3}$. Given kernels $K_1 \in \mathbb{C}(N\backslash H/M, \psi)$ and $K_2 \in \mathbb{C}(M\backslash H/L, \psi)$, their convolution $K_1 * K_2 = m_!(K_1 \boxtimes_{Z \cdot M} K_2)$ lies in $\mathbb{C}(N\backslash H/L, \psi)$. The transform I sends convolution of kernels to composition of operators

$$I[K_1 * K_2] = I[K_1] \circ I[K_2].$$

Canonical system of intertwining kernels

Below, we formulate a slightly stronger version of Theorem 3, in the setting of kernels.

Definition 3. *A system* $\{K_{M^\circ,L^\circ} \in \mathbb{C}(M\backslash H/L, \psi) : (M^\circ, L^\circ) \in Lag^{\circ 2}\}$ *of kernels is called* <u>*multiplicative*</u> *if for every triple* $(N^\circ, M^\circ, L^\circ) \in Lag^{\circ 3}$ *the following equation holds*

$$K_{N^\circ,L^\circ} = K_{N^\circ,M^\circ} * K_{M^\circ,L^\circ}$$

A multiplicative system of kernels $\{K_{M^\circ,L^\circ}\}$ can be equivalently thought of as a single function $K \in \mathbb{C}\left(\mathrm{Lag}^{\circ 2} \times H\right)$, $K(M^\circ, L^\circ) = K_{M^\circ,L^\circ}$, satisfying the following multiplicativity relation on $\mathrm{Lag}^{\circ 3} \times H$

$$p_{12}^* K * p_{23}^* K = p_{13}^* K, \tag{4}$$

where $p_{ij}((L_1^\circ, L_2^\circ, L_3^\circ), h) = ((L_i^\circ, L_j^\circ), h)$ are the projections on the i, j copies of Lag° and the left-hand side of (4) means fiberwise convolution, namely $p_{12}^* K * p_{23}^* K(L_1^\circ, L_2^\circ, L_3^\circ) = K(L_1^\circ, L_2^\circ) * K(L_2^\circ, L_3^\circ)$. To simplify notations, we will sometimes suppress the projections p_{ij} from (4) obtaining a much cleaner formula

$$K * K = K.$$

We proceed along lines similar to Section 2.3. For every $(M^\circ, L^\circ) \in U$, there exists a unique kernel $K_{M^\circ,L^\circ} \in \mathbb{C}(M\backslash H/L, \psi)$ such that $F_{M^\circ,L^\circ} = I[K_{M^\circ,L^\circ}]$, which is given by the following explicit formula

$$K_{M^\circ,L^\circ} = C_{M^\circ,L^\circ} \cdot \widetilde{K}_{M^\circ,L^\circ}, \tag{5}$$

where $\widetilde{K}_{M^\circ,L^\circ} = (\iota^{-1})^* \psi$, $\iota = \iota_{M^\circ,L^\circ}$ is the isomorphism given by the composition $Z \hookrightarrow H \twoheadrightarrow M\backslash H/L$. The system $\{K_{M^\circ,L^\circ} : (M^\circ, L^\circ) \in U\}$ yields a well defined function $K_U \in \mathbb{C}(U \times H)$.

Theorem 4 (Canonical system of kernels). *There exists a unique function $K \in \mathbb{C}(Lag^{\circ 2} \times H)$ satisfying*

1. *Restriction.* $K_{|U} = K_U$.
2. *Multiplicativity.* $K * K = K$.

We note that the proof of the uniqueness part in Theorem 4 is easy, it follows from the fact that for every pair $N^\circ, L^\circ \in \mathrm{Lag}^\circ$ one can find a third $M^\circ \in \mathrm{Lag}^\circ$ such that the pairs N°, M° and M°, L° are in general position. Therefore, by the multiplicativity property (Property 2), $K_{N^\circ,L^\circ} = K_{N^\circ,M^\circ} * K_{M^\circ,L^\circ}$. The proof of the existence part will be algebro-geometric (see Section 3). Finally, we note that Theorem 3 follows from Theorem 4 by taking $F_{M^\circ,L^\circ} = I[K_{M^\circ,L^\circ}]$.

2.5 The canonical vector space

Let us denote by Symp the category whose objects are symplectic vector spaces over $\mathbb{F}_q$ and morphisms are linear isomorphisms of symplectic vector spaces. Using the canonical system of intertwining morphisms $\{F_{M^\circ,L^\circ}\}$ we can associate, in a functorial manner, a vector space $\mathcal{H}(V)$ to a symplectic vector space $V \in$ Symp. The construction proceeds as follows.

Let $\Gamma(V)$ denote the total vector space

$$\Gamma(V) = \bigoplus_{L^\circ \in \mathrm{Lag}^\circ(V)} \mathcal{H}_{L^\circ},$$

Define $\mathcal{H}(V)$ to be the subvector space of $\Gamma(V)$ consisting of sequences $(v_{L^\circ} \in \mathcal{H}_{L^\circ} : L^\circ \in \mathrm{Lag}^\circ)$ satisfying $F_{M^\circ,L^\circ}(v_{L^\circ}) = v_{M^\circ}$ for every $(M^\circ, L^\circ) \in \mathrm{Lag}^{\circ 2}(V)$. We will call the vector space $\mathcal{H}(V)$ the *canonical vector space* associated with V. Sometimes we will use the more informative notation $\mathcal{H}(V) = \mathcal{H}(V, \psi)$.

Proposition 2 (Functoriality). *The rule $V \mapsto \mathcal{H}(V)$ establishes a contravariant (quantization) functor*

$$\mathcal{H} : \mathsf{Symp} \longrightarrow \mathsf{Vect},$$

where Vect *denote the category of finite dimensional complex vector spaces.*

Considering a fixed symplectic vector space V, we obtain as a consequence a representation $(\rho_V, Sp(V), \mathcal{H}(V))$, with $\rho_V(g) = \mathcal{H}(g^{-1})$, for every $g \in Sp(V)$. The representation ρ_V is isomorphic to the Weil representation and we call it the *canonical model* of the Weil representation.

Remark 2. The canonical model ρ_V can be viewed from another perspective: We begin with the total vector space Γ and make the following two observations. First observation is that the symplectic group Sp acts naturally on Γ, the action is of a geometric nature, i.e., induced from the diagonal action on $\mathrm{Lag}^{\circ} \times H$. Second observation is that the system $\{F_{M^\circ, L^\circ}\}$ defines an Sp-invariant idempotent (total Fourier transform) $F : \Gamma \to \Gamma$ given by

$$F(v_{L^\circ}) = \frac{1}{\#(\mathrm{Lag}^\circ)} \bigoplus_{M^\circ \in \mathrm{Lag}^\circ} F_{M^\circ, L^\circ}(v_{L^\circ}),$$

for every $L^\circ \in \mathrm{Lag}^\circ$ and $v_{L^\circ} \in \mathcal{H}_{L^\circ}$. The situation is summarized in the following diagram:

$$Sp \circlearrowright \Gamma \circlearrowleft F.$$

The canonical model is given by the image of F, that is, $\mathcal{H}(V) = F\Gamma$. The nice thing about this point of view is that it shows a clear distinction between operators associated with action of the symplectic group and operators associated with intertwining morphisms. Finally, we remark that one can also consider the Sp-invariant idempotent $F^{\perp} = \mathrm{Id} - F$ and the associated representation $\left(\rho_V^{\perp}, Sp, \mathcal{H}(V)^{\perp}\right)$, with $\mathcal{H}(V)^{\perp} = F^{\perp}\Gamma$. The meaning of this representation is unclear.

Compatibility with Cartesian products

The category Symp admits a monoidal structure given by Cartesian product of symplectic vector spaces. The category Vect admits the standard monoidal structure given by tensor product. With respect to these monoidal structures, the functor $\mathcal{H}$ is a monoidal functor.

Proposition 3. *There exists a natural isomorphism*

$$\alpha_{V_1 \times V_2} : \mathcal{H}(V_1 \times V_2) \rightarrow \mathcal{H}(V_1) \otimes \mathcal{H}(V_2),$$

where $V_1, V_2 \in \mathsf{Symp}$.

As a result, we obtain the following compatibility condition between the canonical models of the Weil representation

$$\alpha_{V_1 \times V_2} : (\rho_{V_1 \times V_2})_{|Sp(V_1)\times Sp(V_2)} \longrightarrow \rho_{V_1} \otimes \rho_{V_2}. \tag{6}$$

Remark 3 ([5]). Condition (6) has an interesting consequence in case the ground field is $\mathbb{F}_3$. In this case, the group $Sp(V)$ is not perfect when $\dim(V) = 2$, therefore, a priori, the Weil representation is not uniquely defined in this particular situation. However, since the group $Sp(V)$ becomes perfect when $\dim(V) > 2$, the canonical model gives a natural choice for the Weil representation in the *singular* dimension, $\dim(V) = 2$.

Compatibility with symplectic duality

Let $V = (V, \omega) \in \mathsf{Symp}$ and let us denote by $\overline{V} = (V, -\omega)$ the symplectic dual of V.

Proposition 4. *There exists a natural non-degenerate pairing*

$$\langle \cdot, \cdot \rangle_V : \mathcal{H}(\overline{V}, \psi) \times \mathcal{H}(V, \psi) \to \mathbb{C},$$

where $V \in \mathsf{Symp}$.

Compatibility with symplectic reduction

Let $V \in \mathsf{Symp}$ and let I be an isotropic subspace in V considered as an abelian subgroup in $H(V)$. On the one hand, we can associate to I the subspace $\mathcal{H}(V)^I$ of I-invariant vectors. On the other hand, we can form the symplectic reduction $I^{\perp}/I$ and consider the vector space $\mathcal{H}(I^{\perp}/I)$ (note that since I is isotropic then $I \subset I^{\perp}$ and $I^{\perp}/I$ is equipped with a natural symplectic structure). Roughly, we claim that the vector spaces $\mathcal{H}(I^{\perp}/I)$ and $\mathcal{H}(V)^I$ are naturally isomorphic. The precise statement involves the following definition.

Definition 4. *An oriented isotropic subspace in V is a pair $I^{\circ} = (I, o_I)$, where $I \subset V$ is an isotropic subspace and o_I is a non-trivial vector in $\bigwedge^{top} I$.*

Proposition 5. *There exists a natural isomorphism*

$$\alpha_{(I^{\circ},V)} : \mathcal{H}(V)^I \to \mathcal{H}(I^{\perp}/I),$$

where, $V \in \mathsf{Symp}$ and I° an oriented isotropic subspace in V. The naturality condition is $\mathcal{H}(f_I) \circ \alpha_{(J^{\circ},U)} = \alpha_{(I^{\circ},V)} \circ \mathcal{H}(f)$, for every $f \in \mathrm{Mor}_{\mathsf{Symp}}(V, U)$ such that $f(I^{\circ}) = J^{\circ}$ and $f_I \in \mathrm{Mor}_{\mathsf{Symp}}(I^{\perp}/I, J^{\perp}/J)$ is the induced morphism.

As a result we obtain another compatibility condition between the canonical models of the Weil representation. In order to see this, fix $V \in \mathsf{Symp}$ and let I° be an oriented isotropic subspace in V. Let $P \subset Sp(V)$ be the subgroup of elements $g \in Sp(V)$ such that $g(I^{\circ}) = I^{\circ}$. The isomorphism $\alpha_{(I^{\circ},V)}$ establishes the following isomorphism:

$$\alpha_{(I^{\circ},V)} : (\rho_V)_{|P} \longrightarrow \rho_{I^{\perp}/I} \circ \pi, \tag{7}$$

where $\pi : P \rightarrow Sp\left(I^{\perp}/I\right)$ is the canonical homomorphism.

3 Geometric intertwining morphisms

In this section we are going to prove Theorem 4, by constructing a geometric counterpart to the set-theoretic system of intertwining kernels. This will be achieved using geometrization.

3.1 Preliminaries from algebraic geometry

We denote by k an algebraic closure of $\mathbb{F}_q$. Next we have to take some space to recall notions and notations from algebraic geometry and the theory of ℓ-adic sheaves.

Varieties

In the sequel, we are going to translate back and forth between algebraic varieties defined over the finite field $\mathbb{F}_q$ and their corresponding sets of rational points. In order to prevent confusion between the two, we use bold-face letters to denote a variety $\mathbf{X}$ and normal letters X to denote its corresponding set of rational points $X = \mathbf{X}(\mathbb{F}_q)$. For us, a variety $\mathbf{X}$ over the finite field is a quasi-projective algebraic variety, such that the defining equations are given by homogeneous polynomials with coefficients in the finite field $\mathbb{F}_q$. In this situation, there exists a (geometric) *Frobenius* endomorphism $Fr : \mathbf{X} \rightarrow \mathbf{X}$, which is a morphism of algebraic varieties. We denote by X the set of points fixed by Fr, i.e.,

$$X = \mathbf{X}(\mathbb{F}_q) = \mathbf{X}^{Fr} = \{x \in \mathbf{X} : Fr(x) = x\}.$$

The category of algebraic varieties over $\mathbb{F}_q$ will be denoted by $\mathsf{Var}_{\mathbb{F}_q}$.

Sheaves

Let $\mathsf{D}^b(\mathbf{X})$ denote the bounded derived category of constructible ℓ-adic sheaves on $\mathbf{X}$ [2,4]. We denote by $\mathsf{Perv}(\mathbf{X})$ the Abelian category of perverse sheaves on the variety $\mathbf{X}$, i.e., the heart with respect to the autodual perverse t-structure in $\mathsf{D}^b(\mathbf{X})$. An object $\mathcal{F} \in \mathsf{D}^b(\mathbf{X})$ is called n-perverse if $\mathcal{F}[n] \in \mathsf{Perv}(\mathbf{X})$. Finally, we recall the notion of a Weil structure (Frobenius structure) [4]. A Weil structure associated to an object $\mathcal{F} \in \mathsf{D}^b(\mathbf{X})$ is an isomorphism

$$\theta : Fr^*\mathcal{F} \longrightarrow \mathcal{F}.$$

A pair $(\mathcal{F}, \theta)$ is called a Weil object. By an abuse of notation we often denote θ also by Fr. We choose once an identification $\overline{\mathbb{Q}}_\ell \simeq \mathbb{C}$, hence all sheaves are considered over the complex numbers.

Remark 4. All the results in this section make perfect sense over the field $\overline{\mathbb{Q}}_\ell$, in this respect the identification of $\overline{\mathbb{Q}}_\ell$ with $\mathbb{C}$ is redundant. The reason it is specified is in order to relate our results with the standard constructions of the Weil representation [7, 12].

Given a Weil object $(\mathcal{F}, Fr^*\mathcal{F} \simeq \mathcal{F})$ one can associate to it a function $f^{\mathcal{F}} : X \to \mathbb{C}$ to $\mathcal{F}$ as follows

$$f^{\mathcal{F}}(x) = \sum_i (-1)^i \mathrm{Tr}(Fr_{|H^i(\mathcal{F}_x)}).$$

This procedure is called *Grothendieck's sheaf-to-function correspondence.* Another common notation for the function $f^{\mathcal{F}}$ is $\chi_{Fr}(\mathcal{F})$, which is called the *Euler characteristic* of the sheaf $\mathcal{F}$.

3.2 Canonical system of geometric intertwining kernels

We shall now start the geometrization procedure.

Replacing sets by varieties

The first step we take is to replace all sets involved by their geometric counterparts, i.e., algebraic varieties. The symplectic space (V, ω) is naturally identified as the set $V = \mathbf{V}(\mathbb{F}_q)$, where $\mathbf{V}$ is a $2n$-dimensional symplectic vector space in $\mathsf{Var}_{\mathbb{F}_q}$. The Heisenberg group H is naturally identified as the set $H = \mathbf{H}(\mathbb{F}_q)$, where $\mathbf{H} = \mathbf{V} \times \mathbb{A}^1$ is the corresponding group variety. Finally, $\mathrm{Lag}^\circ = \mathbf{Lag}^\circ(\mathbb{F}_q)$, where $\mathbf{Lag}^\circ$ is the variety of oriented Lagrangians in $\mathbf{V}$.

Replacing functions by sheaves

The second step is to replace functions by their sheaf-theoretic counterparts [6]. The additive character $\psi : \mathbb{F}_q \longrightarrow \mathbb{C}^\times$ is associated via the sheaf-to-function correspondence to the Artin–Schreier sheaf $\mathcal{L}_\psi$ living on $\mathbb{A}^1$, i.e., we have $f^{\mathcal{L}_\psi} = \psi$. The Legendre character σ on $\mathbb{F}_q^\times \simeq \mathbb{G}_m(\mathbb{F}_q)$ is associated to the Kummer sheaf $\mathcal{L}_\sigma$ on $\mathbb{G}_m$. The one-dimensional Gauss sum G_1 is associated with the Weil object

$$\mathcal{G}_1 = \int_{\mathbb{A}^1} \mathcal{L}_{\psi(z^2)} \in \mathsf{D}^b(\mathbf{pt}),$$

where, for the rest of these notes, $\int = \int_!$ denotes integration with compact support [2]. Grothendieck's Lefschetz trace formula [8] implies that, indeed, $f^{\mathcal{G}_1} = G_1$. In fact, there exists a quasi-isomorphism $\mathcal{G}_1 \longrightarrow H^1(\mathcal{G}_1)[-1]$ and $\dim H^1(\mathcal{G}_1) = 1$, hence, $\mathcal{G}_1$ can be thought of as a one-dimensional vector space, equipped with a Frobenius operator, sitting at cohomological degree 1.

Our main objective, in this section, is to construct a multiplicative system of kernels $K : \mathrm{Lag}^{\circ 2} \times H \longrightarrow \mathbb{C}$ extending the subsystem K_U (see 2.4). The extension appears as a direct consequence of the following geometrization theorem:

Theorem 5 (Geometric kernel sheaf). *There exists a geometrically irreducible* $[\dim(\mathbf{Lag}^{\circ 2}) + \mathbf{n} + \mathbf{1}]$*-perverse Weil sheaf* $\mathcal{K}$ *on* $\mathbf{Lag}^{\circ 2} \times \mathbf{H}$ *of pure weight* $w(\mathcal{K}) = 0$*, satisfying the following properties:*

1. *Multiplicativity property. There exists an isomorphism*
$$\mathcal{K} \simeq \mathcal{K} * \mathcal{K}.$$
2. *Function property. We have* $f^{\mathcal{K}}_{|U} = K_U$.

For a proof, see Section 4.

Proof of Theorem 4

Let $K = f^{\mathcal{K}}$. Invoking Theorem 5, we obtain that K is multiplicative (Property 1) and extends K_U (Property 2). Hence, we see that K satisfies the conditions of Theorem 4. The nice thing about this construction is that it uses geometry and, in particular, the notion of perverse extension which has no counterpart in the set-function theoretic setting.

4 Proof of the geometric kernel sheaf theorem

Section 4 is devoted to sketching the proof of Theorem 5.

4.1 Construction

The construction of the sheaf $\mathcal{K}$ is based on formula (5). Let $\mathbf{U} \subset \mathbf{Lag}^{\circ 2}$ be the open subvariety consisting of pairs $(M^\circ, L^\circ) \in \mathbf{Lag}^{\circ 2}$ in general position. The construction proceeds as follows:

- *Non-normalized kernel.* On the variety $\mathbf{U} \times \mathbf{H}$ define the sheaf
$$\widetilde{\mathcal{K}}_{\mathbf{U}}(M^\circ, L^\circ) = \left(\iota^{-1}\right)^* \mathcal{L}_\psi,$$
where $\iota = \iota_{M^\circ, L^\circ}$ is the composition $\mathbf{Z} \hookrightarrow \mathbf{H} \twoheadrightarrow \mathbf{M} \backslash \mathbf{H} / \mathbf{L}$.

- *Normalization coefficient.* On the open subvariety $\mathbf{U} \times \mathbf{H}$ define the sheaf

$$\mathcal{C}(M^\circ, L^\circ) = \mathcal{G}_1^{\otimes n} \otimes \mathcal{L}_\sigma\left((-1)^{\binom{n}{2}} \omega_\wedge (o_L, o_M)\right)[2n](n). \qquad (8)$$

- *Normalized kernels.* On the open subvariety $\mathbf{U} \times \mathbf{H}$ define the sheaf

$$\mathcal{K}_\mathbf{U} = \mathcal{C} \otimes \widetilde{\mathcal{K}}_\mathbf{U}.$$

Finally, take

$$\mathcal{K} = j_{!*}\mathcal{K}_\mathbf{U}, \qquad (9)$$

where $j : \mathbf{U} \times \mathbf{H} \hookrightarrow \mathbf{Lag}^{\circ 2} \times \mathbf{H}$ is the open imbedding, and $j_{!*}$ is the functor of perverse extension [2] (in our setting, $j_{!*}$ might better be called irreducible extension, since the sheaves we consider are not perverse but perverse up to a cohomological shift). It follows directly from the construction that the sheaf $\mathcal{K}$ is irreducible $[\dim(\mathbf{Lag}^{\circ 2}) + n + 1]$-perverse of pure weight 0.

The function property (Property 2) is clear from the construction. We are left to prove the multiplicativity property (Property 2).

4.2 Proof of the multiplicativity property

We need to show that

$$p_{13}^*\mathcal{K} \simeq p_{12}^*\mathcal{K} * p_{23}^*\mathcal{K}, \qquad (10)$$

where $p_{ij} : \mathbf{Lag}^{\circ 3} \times \mathbf{H} \to \mathbf{Lag}^{\circ 2} \times \mathbf{H}$ are the projectors on the i, j copies of $\mathbf{Lag}^\circ$. We will need the following notations. Let $\mathbf{U}_3 \subset \mathbf{Lag}^{\circ 3}$ denote the open subvariety consisting of triples $(L_1^\circ, L_2^\circ, L_3^\circ)$ which are in general position pairwisely. Let $n_k = \dim(\mathbf{Lag}^{\circ k}) + n + 1$.

Lemma 1. *There exists, on $\mathbf{U}_3 \times \mathbf{H}$, an isomorphism*

$$p_{13}^*\mathcal{K} \simeq p_{12}^*\mathcal{K} * p_{23}^*\mathcal{K}.$$

Let $\mathbf{V}_3 \subset \mathbf{Lag}^{\circ 2}$ be the open subvariety consisting of triples $(L_1^\circ, L_2^\circ, L_3^\circ) \in \mathbf{Lag}^{\circ 4}$ such that L_1°, L_2° and L_2°, L_3° are in general position. Lemma 1 admits a slightly stronger form.

Lemma 2. *There exists, on $\mathbf{V}_3 \times \mathbf{H}$, an isomorphism*

$$p_{13}^*\mathcal{K} \simeq p_{12}^*\mathcal{K} * p_{23}^*\mathcal{K}.$$

We can now finish the proof of (10). Lemma 1 implies that the sheaves $p_{13}^*\mathcal{K}$ and $p_{12}^*\mathcal{K} * p_{23}^*\mathcal{K}$ are isomorphic on the open subvariety $\mathbf{U}_3 \times \mathbf{H}$. The sheaf $p_{13}^*\mathcal{K}$ is irreducible $[n_3]$-perverse as a pullback by a smooth, surjective with connected fibers morphism, of an irreducible $[n_2]$-perverse sheaf on $\mathbf{Lag}^{\circ 2} \times \mathbf{H}$. Hence, it is enough to show that the sheaf $p_{12}^*\mathcal{K} * p_{23}^*\mathcal{K}$ irreducible $[n_3]$-perverse. Let $\mathbf{V}_4 \subset \mathbf{Lag}^{\circ 4}$ be the open subvariety consisting of quadruples $(L_1^\circ, L_2^\circ, L_3^\circ, L_4^\circ) \in \mathbf{Lag}^{\circ 4}$ such that the pairs L_1°, L_2° and L_2°, L_3° are in general position. Consider the projection $p_{134} : \mathbf{V}_4 \times \mathbf{H} \rightarrow \mathbf{Lag}^{\circ 3} \times \mathbf{H}$, it is clearly smooth and surjective, with connected fibers. It is enough to show that the pull-back $p_{134}^* (p_{12}^* \mathcal{K} * p_{23}^* \mathcal{K})$ is irreducible $[n_4]$-perverse. Using Lemma 2 and also invoking some direct diagram chasing one obtains

$$p_{123}^* (p_{12}^* \mathcal{K} * p_{23}^* \mathcal{K}) \simeq p_{12}^* \mathcal{K} * p_{23}^* \mathcal{K} * p_{34}^* \mathcal{K}. \tag{11}$$

The right-hand side of (11) is principally a subsequent application of a properly normalized, Fourier transforms on $p_{34}^* \mathcal{K}$, hence by the Katz–Laumon theorem [15] it is irreducible $[n_4]$-perverse.

Let us summarize. We showed that both sheaves $p_{13}^* \mathcal{K}$ and $p_{12}^* \mathcal{K} * p_{23}^* \mathcal{K}$ are irreducible $[n_3]$-perverse and are isomorphic on an open subvariety. This implies that they must be isomorphic. This concludes the proof of the multiplicativity property.

References

1. Bernstein J., *Private communication, Max-Planck Institute, Bonn, Germany* (August, 2004).
2. Beilinson A., Bernstein J. and Deligne P. Faisceaux pervers. *Analysis and topology on singular spaces, I, Asterisque, 100, Soc. Math. France, Paris* (1982), 5–171.
3. Bates S, and Weinstein A., Lectures on the geometry of quantization. *Berkeley Mathematics Lecture Notes, 8. American Mathematical Society, Providence, RI* (1997).
4. Deligne P., La conjecture de Weil II. *Publ. Math. I.H.E.S 52* (1981), 313–428.
5. Gabber O., Private communication (2006).
6. Gaitsgory D., Informal introduction to geometric Langlands. *An introduction to the Langlands program, Jerusalem 2001, Birkhäuser, Boston, MA* (2003) 269–281.
7. Gérardin P., Weil representations associated to finite fields. *J. Algebra 46*, no. 1, (1977), 54–101.
8. Grothendieck A., Formule de Lefschetz et rationalité des fonctions *L. Seminaire Bourbaki, Vol. 9, Exp. No. 279* (1964).
9. Guillemin V. and Sternberg S., Some problems in integral geometry and some related problems in microlocal analysis. *Amer. J. Math. 101* (1979) 915-955.
10. Gurevich S. and Hadani R., Heisenberg Realizations, Eigenfunctions and Proof of the Kurlberg-Rudnick Supremum Conjecture. *arXiv:math-ph/0511036* (2005).
11. Hadani R., The Geometric Weil Representation and some applications, *Ph.D. Thesis, Tel-Aviv University* (June, 2006).
12. Howe R., On the character of Weil's representation. *Trans. Amer. Math. Soc. 177* (1973), 287–298.
13. Kazhdan D., *Private communication, Hebrew University, Jerusalem, Israel* (March, 2004).
14. Kirillov, A. A., Unitary representations of nilpotent Lie groups. *Uspehi Mat. Nauk 17* (1962) 67–110.

15. Katz N. M. and Laumon G., Transformation de Fouriér et majoration de sommes exponentièlles. *Inst. Hautes Etudes Sci. Publ. Math. No. 62* (1985), 361–418.
16. Kostant B., Quantization and Unitary Representations. *Lecture Notes in Math. 170, Springer-Verlag* (1970), 87–207.
17. Lion G. and Vergne M., The Weil representation, Maslov index and theta series. *Progress in Mathematics, 6. Birkhäuser, Boston, Mass* (1980).
18. Neumann J. von., Die Eindeutigkeit der Schrödingerschen Operationen. *Math. Ann. 104. (*1931), 570–578.
19. Segal I. E., Transforms for operators and symplectic automorphisms over a locally compact group. *Math. Scand., vol. 13* (1963) 31–43.
20. Shale D., Linear symmetries of free boson fields. *Trans. Amer. Math. Soc. 103* (1962), 149–167.
21. Stone M. H., Linear transformations in Hilbert space, III: operational methods and group theory. Proc. Nat. Acad. Sci. U.S.A. 16 (1930) 172–175.
22. Weil A., Sur certains groupes d'operateurs unitaires. *Acta Math. 111* (1964) 143–211.
23. Weinstein A., Symplectic geometry. *Bull. Amer. Math. Soc. (N.S.) 5, no. 1 (*1981) 1–13.
24. Weinstein A., The symplectic "category". *Differential geometric methods in mathematical physics. Lecture Notes in Math., 905, Springer, Berlin-New York* (1982) 45–51.

Noncommutative Geometry in the Framework of Differential Graded Categories

Snigdhayan Mahanta

Dept. of Math., University of Toronto
smahanta@math.toronto.edu

Mathematics Subject Classification (2000). Primary 18E30; Secondary 14A22

Summary. In this survey we discuss a framework of noncommutative geometry with differential graded categories as models for spaces. We outline a construction of the category of noncommutative spaces and also include a discussion on noncommutative motives. We propose a motivic measure with values in a motivic ring. This enables us to introduce certain zeta functions of noncommutative spaces.

1 Introduction

This expository note is meant to be a brief introduction to noncommutative geometry in a differential graded (DG) framework, i.e., such categories playing the role of spaces. Keeping in mind the significance of certain correspondences in classical geometry, we are naturally led to include them as some sort of generalised morphisms of spaces. In this sense this geometry is *motivic*. However, although the construction of the category of noncommutative spaces will follow closely that of motives, the resulting category of noncommutative spaces will not even be additive (only semiadditive). Passing on to the homotopy categories of the DG categories (considered as spaces) one recovers most of the results known at the level of triangulated categories. However, in this setting one does not run into some unpleasant technical problems which one would otherwise have to deal with at the level of triangulated categories. The theory seems to be a blend between algebraic topology (or homotopical geometry) and algebraic geometry. There is a possibility of recasting many different models of noncommutative geometry in this general setting. We also include some pointers to some other areas of mathematics, which are well adapted to be seen in this context.

Ö. Ceyhan et al. (eds.), *Arithmetic and Geometry Around Quantization*,
Progress in Mathematics 279, DOI: 10.1007/978-0-8176-4831-2_9,

It must also be emphasized that this is noncommutative geometry 'at a large scale' (after Ginzburg [28]) and, therefore, there are some natural new phenomena which are not quite compatible with the classical picture. One such instance is the isomorphism between certain classical spaces, e.g., an abelian variety and its dual, which need not be isomorphic as classical spaces (varieties). For noncommutative geometry 'at a small scale', i.e., viewed as a deformation of classical geometry one may look at, e.g., [32],[43].

The outline of some of the constructions presented here can be found in a recent preprint of Kontsevich [39]. In fact, this is a simplified version of the proposed one in ibid. Readers should also refer to the articles of J. Lurie [45, 46] and Toën–Vezzosi [65, 66], who seem to have developed a geometry based on a functor of points approach from simplicial algebras (equivalent to connective differential graded algebras in characteristic 0) to simplicial sets. Noncommutative geometry in their parlance is homotopical algebraic geometry (HAG) or derived algebraic geometry (DAG). The author would like to thank D. Ben-Zvi for providing quite convincing arguments to dispel any idea that DAG or HAG can subsume noncommutative geometry. The material presented here is mostly modelled on the ICM talk of Keller [37], though there are some minor deviations.

Acknowledgments: The author would like to thank Matilde Marcolli and Yu. I. Manin for generously sharing their thoughts and providing useful comments. A special word of thanks goes out to B. Keller for suggesting several changes and corrections based on an initial draft. The author would also like to thank B. Toën for patiently answering many questions and the anonymous referee for numerous meticulous remarks. The author is grateful to the Max-Planck-Institut für Mathematik, Bonn and the Fields Institute, Toronto for their hospitality, while this work was carried out and to the organizers of the AGAQ, Istanbul 2006 conference, which introduced the author to some of these recent interesting developments.

2 Noncommutative geometry in a DG framework

For a long time it was felt that the language of triangulated categories is deficient for many purposes in geometry. The language of DG categories seems to have resolved most of the technical and aesthetic problems. We first prepare the readers for the seemingly abstruse definition of the category of noncommutative spaces. We propose a theory over an algebraically closed field of characteristic zero and by the *Lefschetz Principle* there is no harm in assuming our ground field k to be actually $\mathbb{C}$. This reduces a lot of technical difficulties. For brevity, we denote most of the functors by their underived notation, for instance, $\otimes^{\mathbb{L}}$ is written simply as $\otimes$.

2.1 Motivation

The traditional way of doing geometry with the emphasis on spaces is deficient in many physical situations. Most notably, due to Heisenberg's Uncertainty Principle one is forced to consider polynomial algebras with noncommuting variables, like Weyl algebras. One has to do away with the notion of points of a space quite naturally. However, one has perfectly well-defined algebras, albeit noncommutative, with which one can work. One very successful approach from this point of view is that of Connes [15]. It has many applications and a large part of classical (differential) geometry can be subsumed in this setting. One might also want to take a closer look at the key features of classical (algebraic) geometry and try to generalise them.

From spaces to categories; from functions to sheaves: It is quite common in mathematics to study an object via its representations (in an appropriate sense). It is neat to assemble all representations into a category and study it. In this manner, from groups one is led to the study of Tannakian categories, from algebras to that of certain triangulated categories and so on. This process is roughly some sort of categorification.

We have already done away with the traditional notion of a space and its points. For the time being it is described by its functions. The topology of a space allows us to define functions locally and glue them (if possible to a global one). A better way of keeping track of such information is using the language of the sheaf of local sections or functions on a space. Every classical space comes hand in hand with its structure sheaf of admissible functions, e.g., continuous, smooth, holomorphic, algebraic, etc. according to the structure of the underlying space. The representations of the structure sheaf, which for us are nothing but quasicoherent sheaves, determine the space. In this manner one replaces the notion of a space by its category of quasicoherent sheaves, an idea that goes back to Gabriel, Grothendieck, Manin, and Serre.

The category of quasicoherent sheaves is a Grothendieck category when the scheme is quasicompact and quasiseparated [62]. There are many approaches towards developing a theory by treating abelian categories (or some modifications thereof, like Grothendieck categories) as the category of quasicoherent sheaves on noncommutative spaces, e.g., [1, 55, 67].

Remark 1 *There is another point of view inspired by the Geometric Langlands programme and the details can be found, for instance, in [25]. The guiding principle here is a generalisation of Grothendieck's faisceaux-fonctions correspondence. The faisceaux-fonctions correspondence appears naturally in the context of étale ℓ-adic sheaves. Associated to any complex of étale ℓ-adic sheaves $\mathscr{K}^\bullet$ over a variety V defined over a finite field $\mathbb{F}_q$ is a function $f^{\mathscr{K}^\bullet} : V \longrightarrow \mathbb{C}$ given by*

$$f^{\mathscr{K}^\bullet}(x) = \sum (-1)^i \mathrm{Tr}(\mathrm{Fr}_{\bar{x}} \mid \mathrm{H}^i(\mathscr{K}^\bullet)_{\bar{x}}).$$

Here $x \in V(\mathbb{F}_q)$ and $\bar{x}$ denotes a geometric point of V over x. Of course, one has to fix an identification $\overline{\mathbb{Q}}_\ell \xrightarrow{\sim} \mathbb{C}$. According to Grothendieck all interesting functions appear in this manner and extrapolating this idea we regard constructible sheaves as the only source of interesting functions over $\mathbb{C}$.

The lack of Verdier Duality, which is a generalisation of Poincaré Duality and hence an important feature, makes the naïve category of constructible ℓ-adic sheaves undesirable. Instead one works with the category of so-called perverse sheaves. It is known that the derived category of coherent sheaves also admits a dualising complex imitating Grothendieck–Serre duality in place of Verdier duality (see Proposition 1 [5]). They are objects which live in a bigger derived category. Via a version of the Riemann–Hilbert correspondence over $\mathbb{C}$ the category of perverse sheaves (of middle perversity) is equivalent to the category of regular holonomic $\mathscr{D}$-modules. More precisely, let X be a complex manifold, $D^b_{rh}(\mathscr{D}_X)$ denote the bounded derived category of complexes of $\mathscr{D}_X$-modules with regular holonomic cohomologies and $D^b_c(\mathbb{C}_X)$ denote the bounded derived category of sheaves of complex vector spaces with constructible cohomologies. Then Kashiwara proved in [33] $\mathscr{R}\mathscr{H}om_{\mathscr{D}_X}(-,\mathcal{O}_X) : D^b_{rh}(\mathscr{D}_X) \xrightarrow{\sim} D^b_c(\mathbb{C}_X)^{\mathrm{op}}$ is an equivalence of triangulated categories. Under this equivalence the standard t-structure on $D^b_{rh}(\mathscr{D}_X)$, whose heart is the abelian category of regular holonomic $\mathscr{D}$-modules on X, is mapped to the heart of the t-structure of middle perversity on $D^b_c(\mathbb{C}_X)$. The heart of this t-structure is the category of perverse sheaves (of middle perversity), which can be regarded as another generalisation of functions. As opposed to a quasicoherent sheaf, the model for a function in this setting is a $\mathscr{D}$-module, which is roughly a quasicoherent sheaf with a flat connection. A quasicoherent sheaf (resp. a $\mathscr{D}$-module) corresponds to a polynomial (resp. a constructible locally constant) function.

The passage to derived categories: In the category of smooth schemes any morphism $f : X \to Y$ gives rise to two canonical functors on the category of sheaves, viz., pull-back f^* and push-forward f_*. One should naturally expect any generalisation of classical geometry to allow such operations. We see that restricting to abelian categories is not enough as functors like push-forwards are not exact. The natural framework for such functors to exist is that of derived categories or abstract triangulated categories. Besides, if one chooses to work with perverse sheaves as substitutes for functions one has to view them as elements of an abelian category sitting inside a bigger derived category.

Adding correspondences to morphisms: Denoting by `Var` the category of complex algebraic varieties, `Top` that of *nice* topological spaces (here *nice* should imply all properties typical of the complex points of a complex algebraic variety) one has a tensor functor `Var` $\to$ `Top` associating to a complex algebraic variety its underlying space with the analytic topology. The tensor structure on the two categories is given by direct product. To a topological space in `Top` one can associate its singular cochain complex which is also a tensor functor to D_{ab}, the category of complexes of finitely generated abelian groups whose cohomology is bounded. According to Beĭlinson and Vologodsky [3] the basic objective of the theory of motives is to fill in a commutative diagram

$$\begin{array}{ccc} \mathtt{Var} & \longrightarrow & \mathcal{D}_{\mathcal{M}} \\ \downarrow & & \downarrow \\ \mathtt{Top} & \longrightarrow & D_{ab} \end{array}$$

where $\mathcal{D}_{\mathcal{M}}$ is the rigid tensor triangulated category of motives. The upper horizontal arrow should be faithful and defined purely geometrically and the right vertical arrow should respect the tensor structures. In order to construct the upper horizontal arrow one first needs to enrich Var to include correspondences (modulo some equivalence relation). This endows Var with an additive structure.

Triangulated structure is not enough: The goal is to construct a rigid tensor category of *motivic* noncommutative spaces which allows basic operations like pull-back, push-forward and finite correspondences (as morphisms). In the classical setting, we have a construction of $\mathcal{D}_{\mathcal{M}}$ as a triangulated category due to Voevodsky (see e.g., [26]). However, one would like to extract the *right* category of motives inside it (possibly as an abelian rigid tensor category). One basic operation is direct product, which endows Var with the tensor structure. It should also survive in $\mathcal{D}_{\mathcal{M}}$. The tensor product of two triangulated categories unfortunately does not carry a natural triangulated structure. Also one runs into trouble in trying to define inner Hom's. This is where the framework of DG (differential graded) categories comes in handy.

2.2 Overview of DG categories

Before we are able to spell out the definition of the category of noncommutative spaces we need some preparation on DG categories, which will be quite concise. For details we refer the readers to e.g., [23],[37],[63]. They can be defined over k, where k is not necessarily a field. However, as mentioned before, we set $k = \mathbb{C}$ and, unless otherwise stated, all our categories are assumed to be k-linear.

A category $\mathcal{C}$ is called a DG category if for all $X, Y \in \mathrm{Obj}(\mathcal{C})$ $\mathrm{Hom}(X,Y)$ has the structure of a complex of k-linear spaces (in other words, a DG vector space) and the composition maps are associative k-linear maps of DG vector spaces. In particular, $\mathrm{Hom}(X,X)$ is a DG algebra with a unit.

Example 1 *Given any k-linear category $\mathcal{M}$ it is possible to construct a DG category $\mathcal{C}_{dg}(\mathcal{M})$ with complexes $(M^\bullet, d_M)$ over $\mathcal{M}$ as objects and setting $\mathrm{Hom}(M^\bullet, N^\bullet) = \oplus_n \mathrm{Hom}(M^\bullet, N^\bullet)_n$, where $\mathrm{Hom}(M^\bullet, N^\bullet)_n$ denotes the component of morphisms of degree n, i.e., $f_n : M^\bullet \longrightarrow N^\bullet[n]$ and whose differential is the graded commutator*

$$d(f) = d_M \circ f_n - (-1)^n f_n \circ d_N.$$

Let DGcat stand for the category of all small DG categories. The morphisms in this category are *DG functors*, i.e., $F : \mathcal{C} \to \mathcal{C}'$ such that for all $X, Y \in \mathrm{Obj}(\mathcal{C})$

$$F(X,Y) : \mathrm{Hom}(X,Y) \longrightarrow \mathrm{Hom}(FX,FY)$$

is a morphism of DG vector spaces compatible with the compositions and the units.

The tensor structure: The tensor product of two DG categories $\mathcal{C}$ and $\mathcal{D}$ can be defined in the obvious manner, viz., the objects of $\mathcal{C} \otimes \mathcal{D}$ are written as $X \otimes Y$, $X \in \mathrm{Obj}(\mathcal{C})$, $Y \in \mathrm{Obj}(\mathcal{D})$ and one sets

$$\mathrm{Hom}_{\mathcal{C}\otimes\mathcal{D}}(X \otimes Y, X' \otimes Y') = \mathrm{Hom}_{\mathcal{C}}(X, X') \otimes \mathrm{Hom}_{\mathcal{D}}(Y, Y')$$

with natural compositions and units.

The category of DG functors $\mathcal{H}om(\mathcal{C}, \mathcal{D})$ between two DG categories $\mathcal{C}$ and $\mathcal{D}$ with natural transformations as morphisms is once again a DG category. With respect to the above-mentioned tensor product DGcat becomes a symmetric tensor category with an inner $\mathcal{H}om$ functor given by

$$\mathrm{Hom}(\mathcal{B} \otimes \mathcal{C}, \mathcal{D}) = \mathrm{Hom}(\mathcal{B}, \mathcal{H}om(\mathcal{C}, \mathcal{D})).$$

However, in the category of noncommutative spaces (to be defined shortly), this notion of the inner Hom functor needs to be modified.

The derived category of a DG category: The standard reference for the construction is [34]. We recall some basic facts here. Let $\mathcal{C}$ be a small DG category. A right DG $\mathcal{C}$-module is by definition a DG functor $M : \mathcal{C}^{\mathrm{op}} \to \mathcal{C}_{dg}(k)$, where $\mathcal{C}_{dg}(k)$ denotes the DG category of complexes of k-linear spaces. Note that the composition of morphisms in the opposite category is defined by the Koszul sign rule: the composition of f and g in $\mathcal{C}^{\mathrm{op}}$ is equal to the morphism $(-1)^{|f||g|} gf$ in $\mathcal{C}$. Every object X of $\mathcal{C}$ defines canonically what is called a *free* right module $X^{\wedge} := \mathrm{Hom}(-, X)$. A morphism of DG modules $f : L \to M$ is by definition a morphism (natural transform) of DG functors such that $fX : LX \to MX$ is a morphism of complexes for all $X \in \mathrm{Obj}(\mathcal{C})$. We call such an f a quasiisomorphism if fX is a quasiisomorphism for all X, i.e., fX induces isomorphism on cohomologies.

Definition 2.1 *The derived category $D(\mathcal{C})$ of $\mathcal{C}$ is defined to be the localisation of the category of right DG $\mathcal{C}$-modules with respect to the class of quasi-isomorphisms.*

Remark 2 *With the translation induced by the shift of complexes and triangles coming from short exact sequences of complexes, $D(\mathcal{C})$ becomes a triangulated category. The Yoneda functor $X \mapsto X^\wedge$ induces an embedding $\mathrm{H}^0(\mathcal{C}) \to D(\mathcal{C})$. Here $\mathrm{H}^0(\mathcal{C})$ stands for the zeroth cohomology category whose objects are the same as $\mathcal{C}$ but the morphisms are replaced by the zeroth cohomology, i.e., $\mathrm{Hom}_{\mathrm{H}^0(\mathcal{C})}(X,Y) = \mathrm{H}^0\mathrm{Hom}_{\mathcal{C}}(X,Y)$. It is also called the homotopy DG category as it produces the homotopy category of complexes over any k-linear category $\mathcal{M}$ when specialised to $\mathcal{C}_{dg}(\mathcal{M})$.*

Definition 2.2 *The triangulated subcategory of $D\mathcal{C}$ generated by the free DG $\mathcal{C}$-modules $X^\wedge$ under translations in both directions, extensions and passage to direct factors is called the* **perfect** *derived category and denoted by* $\mathrm{per}(\mathcal{C})$. *A DG category $\mathcal{C}$ is said to be* **pretriangulated** *if the above-mentioned Yoneda functor induces an equivalence $\mathrm{H}^0(\mathcal{C}) \to \mathrm{per}(\mathcal{C})$. Our definition of a pretriangulated category is slightly stronger than [37], in that, in our definition the homotopy category of such a category is automatically idempotent complete.*

Remark 3 *A pretriangulated category does not have a triangulated structure. Rather it is a DG category, which is equivalent to the notion of an enhanced triangulated category in the sense of Bondal–Kapranov [11], whose homotopy category is Karoubian.*

Definition 2.3 *A DG functor $F : \mathcal{C} \to \mathcal{D}$ is called a Morita equivalence if it induces an equivalence $F^* : D(\mathcal{D}) \to D(\mathcal{C})$.*

2.3 The category of noncommutative spaces

The definition provided below is a culmination of the works of several people spanning over two decades, including Bondal, Drinfeld, Keller, Kontsevich, Lurie, Orlov, Quillen, and Toën, amongst others. This list of names is very far from complete and it only reflects the author's ignorance of the history behind this development.

Definition 2.4 *The category of noncommutative spaces* `NCS` *is the localisation of* `DGcat` *with respect to Morita equivalences.*

Thanks to Tabuada we know that `DGcat` has a cofibrantly generated Quillen model category structure, where the weak equivalences are the Morita equivalences and the fibrant objects are pretriangulated DG categories. It seems that there was a slight inaccuracy in the proof of the above statement that appeared in [58], which has now been corrected in [57]. This enables us to conclude that each object of `NCS` has a fibrant replacement, which is a pretriangulated DG category. The tensor product of `DGcat` induces one on `NCS` after replacing any object by its cofibrant model since the tensor product by a cofibrant DG module preserves weak equivalences. The category `NCS` can be regarded as an enhancement of the category of all small idempotent complete triangulated categories.

Remark 4 *We have deliberately included correspondences in the category of noncommutative spaces. These spaces are somewhat motivic in nature and it is expected to be a feature of this geometry. We do not want to treat* NCS *as a 2-category.*

However, the inner $\mathscr{H}om$ functor cannot be derived from DGcat. Thanks to Toën [64] (also cf. [36]) one knows that there does exist an inner $\mathscr{H}om$ functor given by

$$\mathscr{H}om(\mathcal{C}, \mathcal{D}) = \text{cat. of } A_\infty\text{-functors } \mathcal{C} \to \mathcal{D}. \tag{1}$$

Here $\mathcal{D}$ needs to be a pretriangulated DG category which is no restriction since we know that in NCS every object has a canonical pretriangulated replacement. The DG structure of $\mathcal{D}$ endows $\mathscr{H}om(\mathcal{C}, \mathcal{D})$ with a DG structure as well. We will not be able to discuss A_∞-categories and A_∞-functors here. Let us mention that a DG category is a special case of an A_∞-category and we refer the readers to, e.g., [36] for a highly readable survey of the same.

Remark 5 *The* Hom *sets in* NCS *are commutative monoids and it is possible to talk about exact sequences in* NCS *(see Definition 3.3 below).*

Definition 2.5 (Kontsevich)

- *A noncommutative space (DG category)* $\mathcal{C}$ *is called smooth if the bimodule given by the DG bifunctor* $(X, Y) \mapsto \mathrm{Hom}_{\mathcal{C}}(X, Y)$ *is in* $\mathrm{per}(\mathcal{C}^{\mathrm{op}} \otimes \mathcal{C})$.
- *It is called smooth and proper if if is isomorphic in* NCS *to a DG algebra whose homology is of finite total dimension.*

Remark 6 *There is a notion of affinity in this context which just says that a variety is D-affine (or derived affine, e.g., [6] for an analogous notion in the setting of* $\mathscr{D}$*-modules) if its triangulated category of quasicoherent sheaves is equivalent to the derived category of modules over some (possibly DG) algebra. A theorem of Bondal–Van den Bergh [10] (see also [56]) asserts that if* X *is a quasicompact and quasiseparated scheme, then* $D_{\mathrm{QCoh}}(X)$ *is equivalent to* $D(\Lambda)$ *for a suitable DG algebra* Λ *with bounded cohomology. Note that in this theorem* $D_{\mathrm{QCoh}}(X)$ *denotes the honest derived category of complexes of* $\mathcal{O}_X$*-modules with quasicoherent cohomologies and* $D(\Lambda)$ *likewise. As a consequence we deduce that in the DG setting every proper variety is D-affine.*

Viewing classical geometry in this setting: We define the *DG category of quasicoherent sheaves* on an honest scheme X as

$$\begin{aligned}\mathcal{C}_{dg}(X) &:= \mathcal{C}_{dg}(\mathrm{QCoh}(X))\\ &= \text{DG category of fibrant unbounded complexes over } \mathrm{QCoh}(X),\end{aligned}$$

which is how we view classical schemes in this framework. It is also known that $\mathrm{H}^0\mathcal{C}_{dg}(X) \xrightarrow{\sim} D_{\mathrm{QCoh}}(X)$. As mentioned above there are reconstruction theorems available from QCoh(X) (without any further assumption [27, 55]) and from $D_{\mathrm{QCoh}}(X)$ (only if the canonical or the anticanonical bundle is ample [9] or with the knowledge of the tensor and the triangulated structure [2]). They glaringly exclude abelian varieties or (weak) Calabi–Yau varieties, however, for abelian varieties we do have an understanding of the derived category and its autoequivalences [51].

Remark 7 *Those who prefer regular holonomic $\mathscr{D}$-modules as substitutes for functions can perform the above operation after replacing* QCoh(X) *by the category of regular holonomic $\mathscr{D}$-modules.*

Since we have enhanced the morphisms between our spaces by incorporating certain right perfect correspondences, we have also increased the chance of objects becoming isomorphic. Due to Mukai [50] we know that an abelian variety is derived equivalent to its dual precisely via a *correspondence-like* morphism, which is a Fourier–Mukai transform. Roughly, given any two smooth projective varieties X and Y and an object in $\mathscr{E} \in D^b(X \times Y)$ one constructs an exact Fourier–Mukai transform (also sometimes called an integral transform) $\Phi^{\mathscr{E}}_{X\to Y} : D^b(X) \longrightarrow D^b(Y)$ as follows:

$$\Phi^{\mathscr{E}}_{X\to Y}(-) = \pi_{Y*}\left(\mathscr{E} \otimes \pi_X^*(-)\right),$$

where π_X (resp. ϕ_Y) denotes the projection $X \times Y \to X$ (resp. $X \times Y \to Y$). Here all functors are assumed to be appropriately derived. The object $\mathscr{E}$ is called the *kernel* of the Fourier–Mukai transform. In the case of the equivalence between an abelian variety A and its dual $\hat{A}$ the kernel is given by the Poincaré sheaf $\mathcal{P}$. Given a divisorial correspondence in $X \times Y$ one can consider the corresponding line bundle on $X \times Y$ and use that as the kernel of a Fourier–Mukai transform. Conversely, given a kernel $\mathscr{E} \in D^b(X \times Y)$ of a Fourier–Mukai transform one obtains a cycle (correspondence modulo an equivalence relation) in $X \times Y$ by applying the Chern character to $\mathscr{E}$.

2.4 DG categories up to quasiequivalences

We gave a direct method of constructing the category NCS. There is an intermediate notion which one might also want to consider. We call a DG functor $F : \mathcal{C} \longrightarrow \mathcal{D}$ a *quasiequivalence* if the induced maps $\mathrm{Hom}_{\mathcal{C}}(X,Y) \longrightarrow \mathrm{Hom}_{\mathcal{D}}(FX,FY)$ are quasi-isomorphisms for all $X, Y \in \mathrm{Obj}(\mathcal{C})$ and the induced functor $\mathrm{H}^0(F) : \mathrm{H}^0(\mathcal{C}) \longrightarrow \mathrm{H}^0(\mathcal{D})$ is an equivalence. The category

DGcat admits a cofibrantly generated Quillen model category structure whose weak equivalences are quasi-isomorphisms [59]. Let us denote the homotopy category with respect to this model structure Hqe. Being quasiequivalent is stronger than being Morita equivalent. Therefore, the category of DG categories up to quasiequivalence is bigger (has more non-isomorphic objects) than NCS. There is a canonical localisation functor Hqe $\longrightarrow$ NCS inverting the Morita equivalences which are not quasiequivalences, which admits a section functor $\mathcal{A} \longrightarrow \mathrm{per}_{dg}(\mathcal{A})$, i.e., a right adjoint to the canonical localisation functor.

Let us explain the construction of $\mathrm{per}_{dg}(\mathcal{A})$ briefly. For a DG category $\mathcal{A}$ a right $\mathcal{A}$-module, i.e., a DG functor from $\mathcal{A}^{\mathrm{op}}$ to the DG category of complexes over k is called *semifree* if it admits a countable filtration such that the subquotients are free DG modules (up to shifts), i.e., modules formed by arbitrary sums of copies of $\mathrm{Hom}(-, X)$ for some $X \in \mathrm{Obj}(\mathcal{A})$, possibly with shifts. Let us denote the category of semifree modules over $\mathcal{A}$ by $\mathrm{SF}(\mathcal{A})$. The inclusion functor $\mathrm{SF}(\mathcal{A}) \to \mathcal{A}^{\mathrm{op}}$-modules induces an equivalence of triangulated categories between $\mathrm{H}^0(\mathrm{SF}(\mathcal{A}))$ and the derived category of $\mathcal{A}$ [23]. The category $\mathrm{per}_{dg}(\mathcal{A})$ is defined as the full DG subcategory of $\mathrm{SF}(\mathcal{A})$ consisting of objects which become isomorphic to an object in $\mathrm{per}(\mathcal{A})$ after passing on to the zeroth cohomology category. Roughly speaking, $\mathrm{per}_{dg}(\mathcal{A})$ is a DG version of $\mathrm{per}(\mathcal{A})$, i.e., $\mathrm{H}^0(\mathrm{per}_{dg}(\mathcal{A})) = \mathrm{per}(\mathcal{A})$.

In fact, the category NCS is equivalent to the full subcategory of Hqe consisting of the pretriangulated (or Morita fibrant) DG categories.

3 On noncommutative motives

We begin by reviewing the classical notion of pure motives corresponding to smooth and projective varieties.

3.1 Pure motives at a glance

The main steps involved in the construction of effective pure motives from Var are linearisation, pseudo-abelianisation and finally inversion of the Lefschetz motive, extending the tensor structure of Var given by the fibre product over k. Letting $\sim$ stand for any adequate relation, e.g., rational, algebraic, homological or numerical, we define $A^i(X)$ to be the abelian group of algebraic cycles of codimension i in X modulo $\sim$. We define an additive tensor category of correspondences, denoted by $\mathtt{Corr}_\sim$, keeping as objects those of Var and setting

$$\mathtt{Corr}_\sim(X, Y) = \bigoplus_j A^{\dim j}(X \times Y_j),$$

where each Y_j is an irreducible component of Y.

Definition 3.1 *An additive category $\mathcal{D}$ is called pseudo-abelian if for any projector (idempotent) $p \in \mathrm{Hom}(X,X)$, $X \in \mathrm{Obj}(\mathcal{D})$ there exists a kernel* $\ker p$.

There is a canonical *pseudo-abelian completion* $\overline{\mathcal{D}}$ of any additive category $\mathcal{D}$. The objects of $\overline{\mathcal{D}}$ are pairs (X,p), where $X \in \mathrm{Obj}(\mathcal{D})$ and $p \in \mathrm{Hom}_{\mathcal{D}}(X,X)$ is an arbitrary projector. Define Hom sets as

$$\mathrm{Hom}_{\overline{\mathcal{D}}}((X,p),(Y,q)) = \frac{\{f \in \mathrm{Hom}_{\mathcal{D}}(X,Y) \ \text{ such that } fp = qf\}}{\{\text{subgroup of } f \ \text{ such that } fp = qf = 0\}}$$

We can apply this machinery to construct the pseudo-abelianisation of $\mathtt{Corr}_{\sim}$. In the resulting category the motive of $\mathbb{P}^n$ decomposes as $\mathbb{P}^n = \mathrm{pt} \oplus \mathbb{L} \oplus \mathbb{L}^{\otimes 2} \oplus \cdots \oplus \mathbb{L}^{\otimes n}$. The object $\mathbb{L}$ is called the *Lefschetz motive* and it should be formally inverted in order to obtain the category of pure motives and morphisms should also be defined appropriately, but we gloss over these details here.

Restricting oneself to the subcategory of $\mathtt{Var}$ consisting of connected curves and applying the above-mentioned three steps one obtains the category of motives of curves. This category admits a better description when $\sim$ is chosen to be the rational equivalence relation and morphisms are tensored with $\mathbb{Q}$.

Proposition 1 ([48]) *The category of motives of curves is equivalent to the category of abelian varieties up to isogeny.*

Remark 8 *The functor associates to a curve its Jacobian variety. It turns out that the category of abelian varieties up to isogeny is abelian and semisimple.*

The category of motives is expected to be semisimple and Tannakian (Jannsen showed that the category of motives modulo numerical equivalence is semisimple [30]). The category $\mathtt{NCS}$ has some *motivic features*: it also has a tensor structure and an inner $\mathscr{H}om$ functor. But not all objects T are *rigid*, i.e., the canonical morphism $T \otimes T^{\vee} \to \mathrm{Hom}(T,T)$ is not an isomorphism for all $T \in \mathtt{NCS}$. However, the smooth and proper noncommutative spaces are rigid in the above sense.

3.2 Towards noncommutative motives

The first step of the construction of pure motives entails a linearisation of the category $\mathtt{Var}$ by including correspondences. We have argued that correspondences induce DG functors (indeed, the kernel of a Fourier–Mukai transform should be thought of as a correspondence). The following Theorem [64] says that all DG functors are described by a Fourier–Mukai *kernel*, and hence, more relevant to geometry than arbitrary exact functors between triangulated categories.

Theorem 3.2 (Toën) *Let k be any commutative ring and let X and Y be quasicompact and separated schemes over k such that X is flat over* $\operatorname{Spec} k$. *Then there is a canonical isomorphism in* NCS

$$\mathcal{C}_{dg}(X \times_k Y) \xrightarrow{\sim} \mathscr{H}om_c(\mathcal{C}_{dg}(X), \mathcal{C}_{dg}(Y)),$$

where $\mathscr{H}om_c$ denotes the full subcategory of $\mathscr{H}om$ formed by coproduct preserving quasifunctors, i.e., functors between the corresponding zeroth cohomology categories. Moreover, if X and Y are smooth and projective over $\operatorname{Spec} k$, *we have a canonical isomorphism in* NCS

$$\mathtt{Perf}_{dg}(X \times_k Y) \xrightarrow{\sim} \mathscr{H}om(\mathtt{Perf}_{dg}(X), \mathtt{Perf}_{dg}(Y)),$$

where $\mathtt{Perf}_{dg}$ denotes the full subcategory of $\mathcal{C}_{dg}$, whose objects are perfect complexes.

The above Theorem admits a natural generalisation to abstract DG categories (not necessarily of the form $\mathcal{C}_{dg}(X)$ for some scheme X), which can also be found in ibid. The above theorem asserts an equivalence of categories. It can be suitably *decategorified*, in order to have an understanding of the morphisms on the right hand side.

For DG categories $\mathcal{A}, \mathcal{B}$, let us define $\mathrm{rep}(\mathcal{A}, \mathcal{B})$ to be the full subcategory of the derived category $D(\mathcal{A}^{\mathrm{op}} \otimes \mathcal{B})$ of $\mathcal{A} - \mathcal{B}$-bimodules formed by those M, which (under $- \otimes_{\mathcal{A}} M : D(\mathcal{A}) \to D(\mathcal{B})$) send a representable $\mathcal{A}$-module to a $\mathcal{B}$-module, which, in $D(\mathcal{B})$, is isomorphic to a representable $\mathcal{B}$-module. The decategorified statement is that $\mathrm{Hom}(\mathcal{A}, \mathcal{B})$ in NCS is canonically in bijection with the isomorphism classes of objects in $\mathrm{rep}(\mathcal{A}, \mathcal{B})$ ibid.. If $\mathcal{B}$ is pretriangulated, the objects of $\mathrm{rep}(\mathcal{A}, \mathcal{B})$ are called *quasifunctors* as they induce honest functors $\mathrm{H}^0(\mathcal{A}) \to \mathrm{H}^0(\mathcal{B})$.

Generalising this intuition we conclude that the morphisms in NCS already contain all correspondences. However, NCS is not an additive category as there is no abelian group structure on the set of morphisms. However, there is a semiadditive structure on $\mathrm{Hom}(\mathcal{A}, \mathcal{B})$ given by the direct sum of the kernels of two DG functors or objects in $\mathrm{rep}(\mathcal{A}, \mathcal{B})$. We linearise them by passing on to the K_0-groups of the inner $\mathscr{H}om$ objects (see, for instance, [39],[58]).

It is also possible to talk about exact sequences in NCS. We provide one formulation of an exact sequence of DG categories (see, e.g., Theorem 4.11 of [37] for other equivalent definitions).

Definition 3.3 *A sequence of DG categories*

$$\mathcal{A} \xrightarrow{P} \mathcal{B} \xrightarrow{I} \mathcal{C}$$

such that $IP = 0$ is called exact if and only if P induces an equivalence of $\mathrm{per}(\mathcal{A})$ *onto a thick subcategory of* $\mathrm{per}(\mathcal{B})$ *and I induces an equivalence* *between the idempotent closure of the Verdier quotient* $\mathrm{per}(\mathcal{B})/\mathrm{per}(\mathcal{A})$ *and* $\mathrm{per}(\mathcal{C})$.

Remark 9 *In the classical setting, if X is a quasicompact quasiseparated scheme, $U \subset X$ a quasicompact open subscheme and $Z = X \setminus U$, then the following sequence*

$$\mathtt{Perf}_{dg}(X)_Z \longrightarrow \mathtt{Perf}_{dg}(X) \longrightarrow \mathtt{Perf}_{dg}(U)$$

is exact according to the definition, where $\mathtt{Perf}_{dg}(X)_Z$ denotes the full subcategory of $\mathtt{Perf}_{dg}(X)$ of perfect complexes supported on Z.

One knows that there is a well-defined K-theory functor on NCS, which agrees with Quillen's K-theory of an exact category $\mathcal{B}$, when applied to the Drinfeld quotient of $\mathcal{C}^b_{dg}(\mathcal{B})$ by its subcategory of acyclic complexes. Now we define a noncommutative analogue of the category of correspondences (a naïve version). A more sophisticated approach should treat the category enriched over *spectra*, a construction of which can be found in [61].

Definition 3.4 *The category of noncommutative correspondences* NCC *is the category defined as:*

- $\mathrm{Obj}(\mathtt{NCC}) = \mathrm{Obj}(\mathtt{NCS})$
- $\mathrm{Hom}_{\mathtt{NCC}}(\mathcal{C}, \mathcal{D}) = K_0(\mathrm{rep}(\mathcal{C}, \mathcal{D}))$

As a motivation we mention two Theorems: the first Theorem ensures *linearisation* of NCS, while the second one shows compatibility with localisation.

Theorem 3.5 *[23, 24, 58] A functor F from* NCS *to an additive category factors through* NCC *if and only if for every exact DG category $\mathcal{B}$ endowed with two full exact DG subcategories $\mathcal{A}, \mathcal{C}$ which give rise to a semiorthogonal decomposition $\mathrm{H}^0(\mathcal{B}) = (\mathrm{H}^0(\mathcal{A}), \mathrm{H}^0(\mathcal{C}))$ in the sense of [13], the inclusions induce an isomorphism $F(\mathcal{A}) \oplus F(\mathcal{C}) \xrightarrow{\sim} F(\mathcal{B})$.*

Such a functor is called an *additive invariant* of noncommutative spaces. The simplest example is $\mathcal{A} \longmapsto K_0(\mathrm{per}(\mathcal{A}))$.

Theorem 3.6 *[24] The functor $\mathcal{A} \longmapsto K(\mathcal{A})$ (Waldhausen K-theory) yields, for each short exact sequence $\mathcal{A} \to \mathcal{B} \to \mathcal{C}$ in* NCS, *a long exact sequence*

$$\cdots \longrightarrow K_i(\mathcal{A}) \longrightarrow K_i(\mathcal{B}) \longrightarrow K_i(\mathcal{C}) \longrightarrow \cdots \longrightarrow K_0(\mathcal{B}) \longrightarrow K_0(\mathcal{C}).$$

Remark 10 *The category* NCC *is additive and the composition is induced by that of* NCS. *Certain non-isomorphic objects of* NCS *become isomorphic in* NCC, *e.g., it is shown in [95] that each finite dimensional algebra of finite global dimension becomes isomorphic to a product of copies of k in* NCC, *whereas such a thing is true in* NCS *if and only if the algebra is semisimple.*

We should perform a formal idempotent completion (or pseudo-abelian completion) of NCC as discussed in Subsection 3.1 in order to obtain the category of noncommutative motives, which is denoted by NCM. It follows from Beĭlinson's description of the derived category of coherent sheaves on $\mathbb{P}^n$ [4] and the above remark that $\mathcal{C}_{dg}(\mathbb{P}^1) \simeq \mathcal{C}_{dg}(\mathbb{A}^1) \oplus \mathcal{C}_{dg}(\mathrm{pt})$ is also isomorphic in NCC to $\mathcal{C}_{dg}(\mathrm{pt}) \oplus \mathcal{C}_{dg}(\mathrm{pt})$, whence $\mathcal{C}_{dg}(\mathbb{A}^1) \simeq \mathcal{C}_{dg}(\mathrm{pt})$, i.e., the Lefschetz motive is isomorphic to the identity element.

A careful reader should have noticed that we have glossed over the issue of the choice of the equivalence relation, which was central to the construction of the category of pure motives in the classical setting. Manin mentioned in [48] (end of Section 3) that every cohomology theory should be a cohomological functor on the category of $\mathtt{Corr}_\sim$, i.e., every correspondence in $\mathtt{Corr}_\sim(X,Y)$ should induce a well-defined morphism $H^*(X) \to H^*(Y)$. Now we turn the argument around. Elements of $\mathrm{rep}(\mathcal{C},\mathcal{D})$ induce morphisms between $\mathcal{C}$ and $\mathcal{D}$. Our spaces are defined in terms of the (quasicoherent) cohomologies that they admit. Mostly cohomology theories appear as cohomology groups of a certain canonically defined cochain complex satisfying a bunch of axioms. We pretend that a morphism (a functor) in NCM is a morphism between the cohomology theories on the two spaces, as if given by some correspondence. If the question about universal cohomology theory is resolved, then probably one would like to argue that the elements of $\mathrm{rep}(\mathcal{C},\mathcal{D})$ are the ones which induce distinct morphisms between their universal cohomologies. If that turns out to be false then one can call an equivalence relation *universal* if it identifies two correspondences which induce isomorphic morphisms between the corresponding *universal cohomology theories* and then consider correspondences modulo this equivalence relation. Note that in NCM we set the Grothendieck group of $\mathrm{rep}(\mathcal{A},\mathcal{B})$ as morphisms between $\mathcal{A}$ and $\mathcal{B}$. Chow correspondences are obtained by taking the rational equivalence relation. The connection should be an analogue of the *Chern character* map which identifies the K-theory with the Chow group after tensoring with $\mathbb{Q}$. The readers are referred to [14] for a possibly relevant treatment of the Chern character.

3.3 Motivic measures and motivic zeta functions

We present a rather simplistic point of view on motivic measures. With respect to a motivic measure it is possible to develop a theory of *motivic integration* (see, e.g., [20]), which we shall not discuss here. This technology was invented by Kontsevich drawing inspiration from the works of Batyrev. A useful and instructive reference is, e.g., [44].

Let $\mathtt{Sch}_k$ be the category of reduced schemes of finite type (or reduced varieties) over k. Consider the Grothendieck ring of $\mathtt{Sch}_k$, denoted by $K_0(\mathtt{Sch}_k)$, which is defined as the free abelian group generated by isomorphism classes of objects in $\mathtt{Sch}_k$ modulo relations (often called scissor-congruence relations)

$$[X] = [Z] + [X \setminus Z], \tag{2}$$

where Z is a closed subscheme of X. The multiplication is given by the fibre product over k. There is a unit given by the class of $\operatorname{Spec} k$.

Every k-variety admits a finite stratification $X = X^0 \supset X^1 \supset \cdots \supset X^{d+1} = \emptyset$ such that $X^k \setminus X^{k+1}$ is smooth. Moreover, any two such stratifications admit a common refinement. Therefore $[X] = \sum_k [X^k \setminus X^{k+1}]$ is unambiguously defined and, in fact, it can be shown that $K_0(\mathtt{Sch}_k)$ is generated by complete and nonsingular varieties. The structure of $K_0(\mathtt{Sch}_k)$ as a ring is rather complicated. It is known that it is not an integral domain [54]. However, it admits interesting ring homomorphisms to some rings, which turn out to be quite useful in various cases.

Let A be any commutative ring. An A-valued *motivic measure* is a ring homomorphism $\mu_A : K_0(\mathtt{Sch}_k) \to A$. We write $\mu = \mu_A$ if there is no chance of confusion. If A has a unit the homomorphism is required to be unital.

Example 2 *Let $k = \mathbb{C}$, $A = \mathbb{Z}$ and $\mu(X) = \chi_c(X)$, i.e., the Euler characteristic with compact supports.*

Example 3 *Let $k = \mathbb{C}$, $A = K_0(\mathrm{HS})$, i.e., the Grothendieck ring of Hodge structures and $\mu(X) = \chi_h(X)$ such that*

$$\chi_h(X) = \sum_r (-1)^r [\mathrm{H}^r_c(X, \mathbb{Q})] \in K_0(\mathrm{HS}),$$

which is called the Hodge characteristic of X.

Example 4 *Let $k = \mathbb{F}_q$, $A = \mathbb{Z}$ and $\mu(X) = \#X(\mathbb{F}_q)$, i.e., the number of $\mathbb{F}_q$-points.*

Let us fix an A-valued motivic measure μ and, for a smooth $X \in \mathtt{Sch}_k$, let $X^{(n)}$ denote the n-fold symmetric product of X. Set $X^{(0)} := \operatorname{Spec} k$. Then associated to μ there is a *motivic zeta function* (possibly due to Kapranov [31]) of X defined by the formal series

$$\zeta_\mu(X, t) = \sum_{n=0}^{\infty} \mu\left(X^{(n)}\right) t^n \in A[[t]]. \tag{3}$$

Example 5 *If $k = \mathbb{F}_q$, $A = \mathbb{Z}$ and $\mu(X) = \#X(\mathbb{F}_q)$ as in Example 4 one recovers the usual Hasse–Weil zeta function of X. Indeed, the $\mathbb{F}_q$-valued points of $X^{(n)}$ correspond to the effective divisors of degree n in X.*

Let us denote $\mu(\mathbb{A}^1_k)$ by $\mathbb{L}$. Then we have the following rationality statement for curves (see Theorem 1.1.9 ibid.).

Theorem 3.7 *If X is any one-dimensional variety (not necessarily nonsingular) of genus g, then $\zeta_\mu(X, t)$ is rational. Furthermore, the rational function $\zeta_\mu(X, t)(1 - t)(1 - \mathbb{L}t)$ is actually a polynomial of degree $\leqslant 2g$ and satisfies the functional equation below.*

$$\zeta_\mu(X, 1/\mathbb{L}t) = \mathbb{L}^{1-g} t^{2-2g} \zeta_\mu(X, t) \tag{4}$$

Remark 11 *The rationality statement fails to be true in higher dimensions, e.g., if X is a complex projective non-singular surface of genus $\geqslant 2$ [41]. In fact, a complex surface X has a rational motivic zeta function if and only if it has Kodaira dimension $-\infty$ [42].*

3.4 Noncommutative Calabi–Yau spaces

This section attempts to introduce zeta functions of noncommutative curves *in a motivic framework* and possibly extract some arithmetic information out of them. That the zeta functions of varieties contain crucial arithmetic information is a gospel truth by now.

Before we move forward let us mention that such ideas are prevalent in noncommutative geometry, e.g., Connes' spectral realisation of the zeros of the Riemann zeta function [16, 17]. Some other important works in this direction are [18],[21, 22],[52] and [29], to mention only a few. Also the readers should take a look at [49] for a more holistic point of view.

Following Proposition 1 we argue that the category of noncommutative motives of noncommutative curves should be equivalent to the full subcategory of $\mathtt{NCM}$ generated by DG categories which resemble those of abelian varieties, *i.e.*, the inclusion of abelian varieties inside $\mathtt{NCM}$ (see Equation (2)). Given an abelian surface the cokernel of the multiplication by 2 map (isogeny) is a Kummer surface with 16 singular points, whose (minimal) resolution of singularities is a $K3$ surface. It is an example of a Calabi–Yau manifold of dimension 2. So even if we look at motives of curves Calabi–Yau varieties show up rather naturally. We propose to treat such varieties as they are, rather than working up to isogenies. Calabi–Yau varieties are interesting from the point of view of physics as well. We assume that a Calabi–Yau variety is just a variety, whose canonical class is trivial (no assumption on the fundamental group).

In a k-linear category $\mathcal{A}$ an additive autoequivalence S is called a Serre functor if there exists a bifunctorial isomorphism $\mathrm{Hom}(A, B) \xrightarrow{\sim} \mathrm{Hom}(B, SA)^*$ for any two $A, B \in \mathrm{Obj}(\mathcal{A})$. If it exists it is unique up to isomorphism. If X is a smooth projective variety of dimension n, the Serre functor is given by $(- \otimes \omega_X)[n]$, where ω_X is the canonical sheaf of X. The existence of a Serre functor corresponds to that of Grothendieck–Serre duality.

Definition 3.8 *A DG category $\mathcal{C}$ in $\mathtt{NCS}$ is called a noncommutative Calabi–Yau space of dimension n if $\mathrm{H}^0(\mathcal{C})$ is triangulated (i.e., $\mathcal{C}$ is pretriangulated as in Definition 2.2) with the finiteness condition $\sum_p \dim_k \mathrm{Hom}_{\mathrm{H}^0(\mathcal{C})}(X, Y[p]) < \infty$ for all $X, Y \in \mathrm{Obj}(\mathrm{H}^0(\mathcal{C}))$, and if there exists a natural isomorphism between the Serre functor and $[n]$. In other words, there exist bifunctorial isomorphisms $\mathrm{Hom}(A, B) \xrightarrow{\sim} \mathrm{Hom}(B, A[n])^*$ in $\mathrm{H}^0(\mathcal{C})$.*

Kontsevich originally defined a noncommutative Calabi–Yau space as a small triangulated category satisfying the strong finiteness condition mentioned above, with an isomorphism between the Serre functor and $[n]$. We have enhanced it to the DG level. It should be borne in mind that the homotopy category of a pretriangulated category is idempotent complete. It is clear that if X is a Calabi–Yau variety then $\mathcal{C}_{dg}(X)$ is a Calabi–Yau space in the above sense. Purely at the triangulated level there are other interesting examples of Calabi–Yau spaces of dimension 2 arising from quiver representations and commutative algebra cf. Section 4 of [38]. When such a triangulated Calabi–Yau category of dimension d is endowed with a *cluster tilting subcategory* it is possible to construct a Calabi–Yau DG category (in the above sense) of dimension $d+1$ [60].

Let us denote by $\mathtt{NCM}_{\mathrm{CY}}$ the full additive subcategory of $\mathtt{NCM}$ consisting of noncommutative Calabi–Yau spaces.

Example 6 *It is expected that via a noncommutative version of the construction of the Jacobian of a curve the category of motives of noncommutative curves can be seen as a full subcategory of* $\mathtt{NCM}_{\mathrm{CY}}$. *The way to view an abelian variety in this setting is not clear to the author yet. The category* $\mathtt{NCM}_{\mathrm{CY}}$ *contains honest elliptic curves (as they are their own Jacobians) as given by the inclusion of classical geometry in this setting (see Equation* (2)*). The noncommutative torus* $\mathbb{T}^{\tau}_{\theta}$ *is also included via its DG derived category of holomorphic bundles. It is isomorphic to* $\mathcal{C}_{dg}(X_\tau)$, *where* $X_\tau = \mathbb{C}/(\mathbb{Z}+\tau\mathbb{Z})$, *via a Fourier–Mukai type functor (see Proposition 3.1 [53]). J. Block [7, 8] suggests a more conceptual framework for such dualities to exist. The rough idea is to construct a differential graded algebra from a complex torus* X *associated to a deformation parameter in* $\mathrm{HH}^2(X)$ *and look at the DG category* $DG(X)$ *of twisted complexes, i.e., DG modules over that algebra equipped with a super connection compatible with the differential of the algebra. One can construct a (curved) differential graded algebra corresponding to the dual torus as well with a curvature contribution given by the deformation parameter, whose DG category of twisted complexes will be quasiequivalent to* $DG(X)$ *via some sort of a deformed Poincaré bundle (essentially a correspondence).*

3.5 The motivic ring of NCS

The author was kindly notified by B. Keller and the referee that the motivic ring constructed in [12], which is what we describe here, is actually isomorphic to the 0 ring [58]. However, with some appropriate finiteness conditions thrown in, this problem has been rectified, (e.g., [57]).

Let us recall from Section 3.3 that an A-valued motivic measure μ is a ring homomorphism from $K_0(\mathtt{Sch}_k) \to A$. We have replaced the category of k-schemes by the category of noncommutative spaces $\mathtt{NCS}$. We need an appropriate notion of the Grothendieck ring of NCS, which we call the motivic ring of NCS.

Since every object in NCS is quasiequivalent to a pretriangulated DG category we seek a Grothendieck ring of pretriangulated DG categories. In [12] the authors precisely construct a Grothendieck ring of pretriangulated DG categories, which is essentially the Grothendieck ring of Hqe. It was pointed out by the authors that it is crucial to work with DG categories (and not honest triangulated ones) as the tensor product of two triangulated categories does not have a natural triangulated structure in general. Let us briefly recall their construction.

The Grothendieck ring $\mathcal{G}$ is generated as a free abelian group by the isomorphism classes of pretriangulated DG categories in NCS (or quasiequivalence classes of objects in DGcat) modulo relations analogous to those of $K_0(\mathtt{Sch}_k)$. The authors reinterpret the excision relations as those coming from *semiorthogonal decompositions* (see [13] for the details of semiorthogonal decomposition). One writes $[\mathcal{B}] = [\mathcal{A}] + [\mathcal{C}]$ if and only if there exist representatives $\mathcal{A}'$, $\mathcal{B}'$, $\mathcal{C}'$ in $[\mathcal{A}]$, $[\mathcal{B}]$, $[\mathcal{C}]$ respectively such that

1. $\mathcal{A}'$, $\mathcal{C}'$ are DG subcategories of $\mathcal{B}'$,
2. $\mathrm{H}^0(\mathcal{A}')$, $\mathrm{H}^0(\mathcal{C}')$ are admissible subcategories of $\mathrm{H}^0(\mathcal{B}')$,
3. $(\mathrm{H}^0(\mathcal{A}'), \mathrm{H}^0(\mathcal{C}'))$ is a semiorthogonal decomposition of $\mathrm{H}^0(\mathcal{B}')$.

Remark 12 *Part* (3) *implies that* $\mathrm{H}^0(\mathcal{A}') = (\mathrm{H}^0(\mathcal{C}'))^{\perp}$, *which is Lemma 2.25 in [12]. An exact sequence* $\mathcal{A} \longrightarrow \mathcal{B} \longrightarrow \mathcal{C}$ *of pretriangulated DG categories (cf. Definition 3.3) induces an exact sequence of honest triangulated categories* $\mathrm{H}^0(\mathcal{A}) \longrightarrow \mathrm{H}^0(\mathcal{B}) \longrightarrow \mathrm{H}^0(\mathcal{C})$ *by definition. However, existence of a semiorthogonal decomposition is a stronger condition. It says that* $\mathrm{H}^0(\mathcal{C})$ *is a triangulated subcategory of* $\mathrm{H}^0(\mathcal{B})$ *and* $\mathrm{H}^0(\mathcal{A}) = (\mathrm{H}^0(\mathcal{C}))^{\perp}$, *i.e., the sequence is split (cf. Theorem 3.5). It is plausible that one obtains something sensible by allowing all possible exact sequences as relations.*

The product $\bullet$ is defined as follows:

$$\mathcal{A}_1 \bullet \mathcal{A}_2 := \mathrm{per}_{dg}(\mathcal{A}_1 \otimes \mathcal{A}_2),$$

where $\mathrm{per}_{dg}(\mathcal{A})$ is the pretriangulated DG category described in sub-section 2.4.

The product $\bullet$ preserves quasiequivalences of DG categories and hence descends to a product on $\mathcal{G}$. It is proven in [12] that the product is associative and commutative. There is a unit given by the class of $\mathcal{C}^b_{dg}(k)$, i.e., the DG category of finite dimensional chain complexes over k. That this product corroborates the fibre product of varieties is justified by Theorem 6.6 ibid.

Remark 13 *The motivic ring of* NCS *should be the above-mentioned ring with quasiequivalences replaced by Morita equivalence. There will be a canonical* *ring homomorphism corresponding to the localisation functor* $\mathtt{Hqe} \longrightarrow \mathtt{NCS}$. *Since a quasiequivalence is also a Morita equivalence the ring homomorphism will be surjective, identifying elements which are Morita equivalent but not quasiequivalent.*

It follows that the product of two noncommutative Calabi–Yau categories is again a noncommutative Calabi–Yau category, i.e., $\mathcal{A} \bullet \mathcal{B}$ is a noncommutative Calabi–Yau DG category of dimension $m+n$ for $\mathcal{A}, \mathcal{B} \in \mathtt{NCM}_{\mathrm{CY}}$ of dimensions m, n respectively. Indeed, the finiteness condition follows from Künneth formula. To check the existence of the Serre functor first observe that

$$\begin{aligned}
&\mathrm{Hom}_{\mathrm{H}^0(\mathcal{A}\otimes\mathcal{B})}(A \otimes B, A' \otimes B'[m+n]) \\
&= \mathrm{H}^0\mathrm{Hom}_{\mathcal{A}\otimes\mathcal{B}}(A, A') \otimes \mathrm{Hom}_{\mathcal{A}\otimes\mathcal{B}}(B, B')[m+n] \\
&= \mathrm{H}^{m+n}_{\mathcal{A}\otimes\mathcal{B}}(\mathrm{Hom}(A, A') \otimes \mathrm{Hom}(B, B')) \\
&= \mathrm{H}^{m+n}\left((\oplus_i \mathrm{Hom}^i_{\mathcal{A}\otimes\mathcal{B}}(A, A') \otimes \mathrm{Hom}^{k-i}_{\mathcal{A}\otimes\mathcal{B}}(B, B'))^{\bullet}\right) \\
&= \oplus_l \mathrm{Hom}^{l+m}_{\mathrm{H}^0(\mathcal{A}\otimes\mathcal{B})}(A, A') \otimes \mathrm{Hom}^{n-l}_{\mathrm{H}^0(\mathcal{A}\otimes\mathcal{B})}(B, B') \\
&= \oplus_l \mathrm{Hom}^{l}_{\mathrm{H}^0(\mathcal{A}\otimes\mathcal{B})}(A, A'[m]) \otimes \mathrm{Hom}^{-l}_{\mathrm{H}^0(\mathcal{A}\otimes\mathcal{B})}(B, B'[n]) \\
&= \oplus_l \mathrm{Hom}^{l}_{\mathrm{H}^0(\mathcal{A}\otimes\mathcal{B})}(A', A)^* \otimes \mathrm{Hom}^{-l}_{\mathrm{H}^0(\mathcal{A}\otimes\mathcal{B})}(B', B)^* \\
&= \left(\oplus_l \mathrm{Hom}^{l}_{\mathrm{H}^0(\mathcal{A}\otimes\mathcal{B})}(A', A) \otimes \mathrm{Hom}^{-l}_{\mathrm{H}^0(\mathcal{A}\otimes\mathcal{B})}(B', B)\right)^* \\
&= \mathrm{H}^0(\mathrm{Hom}^{\bullet}_{\mathcal{A}\otimes\mathcal{B}}(A' \otimes B', A \otimes B))^* \\
&= \mathrm{Hom}_{\mathrm{H}^0(\mathcal{A}\otimes\mathcal{B})}\mathrm{Hom}(A' \otimes B', A \otimes B)^*
\end{aligned}$$

This proves that $\mathrm{H}^0(\mathcal{A}\otimes\mathcal{B})$ has the right Serre functor $[n+m]$ on representable objects.

Now, it follows from [47] that the existence of the Serre functor $[n]$ is equivalent to the isomorphism $\mathrm{Hom}_k(\mathcal{A}(-,?), k) \simeq \mathcal{A}(?, -[n])$ in $D(\mathcal{A} \otimes \mathcal{A}^{op})$. This would imply that the Serre functor of $\mathcal{A} \otimes \mathcal{B}$ is $[n+m]$.

A $\mathcal{G}$-valued motivic measure (the ring homomorphism being the identity map) is a universal one. Let us denote the class of $X \in \mathtt{NCS}$ inside $\mathcal{G}$ by $[X]$. Given any variety Y we know that $\mathcal{C}_{dg}(Y \times Y)$ is quasiequivalent (in particular, Morita equivalent) to $\mathscr{H}om_c(\mathcal{C}_{dg}(Y), \mathcal{C}_{dg}(Y))$. Let us define inductively $X^n = \mathscr{H}om_c(X^{n-1}, X)$ and $X^1 = X$. Then the universal $\mathcal{G}$-valued motivic zeta function of $X \in \mathtt{NCS}$ is given by

$$\zeta_{\mu_{\mathcal{G}}}(X, t) = 1 + \sum_{n=1}^{\infty} [X^n] t^n \qquad \in \mathcal{G}[[t]].$$

It is shown in [12] that there is a canonical surjective ring homomorphism $K_0(\mathtt{Sch}_k) \to \mathcal{G}_{\mathrm{hon}}$ with $(\mathbb{L}-1)$ in the kernel, where $\mathcal{G}_{\mathrm{hon}}$ is the subring of

$\mathcal{G}$ generated by certain pretriangulated DG categories associated to honest smooth projective varieties over k.

Remark 14 *Since the DG category of holomorphic bundles on the noncommutative torus $\mathbb{T}_\theta^\tau$ is equivalent to $\mathcal{C}_{dg}(X_\tau)$, where $X_\tau = \mathbb{C}/(\mathbb{Z}+\tau\mathbb{Z})$ [53], the zeta functions of $\mathbb{T}_\theta^\tau$ and X_τ are the same. However, the equivalence is given by a non-trivial Fourier–Mukai type (correspondence) functor. The B-model of a conformal field theory associates to a complex torus its derived category of coherent sheaves. The equivalence perhaps indicates that deforming the complex torus to a noncommutative torus does not produce anything new for the B-model.*

One nagging point is that certain natural topological constructions do not allow us to define a category in which the composition of morphisms obeys associativity (it is associative only up to homotopy). Hence some mathematicians have resorted to working with A_∞ categories which encode such properties, e.g., [40]. The world of A_∞ categories subsumes that of DG categories. However, one knows that every A_∞ category is quasi-isomorphic to a DG category in a functorial manner (e.g., [19]).

References

1. M. Artin and J. J. Zhang. Noncommutative projective schemes. *Adv. Math.*, 109(2):228–287, 1994.
2. P. Balmer. Presheaves of triangulated categories and reconstruction of schemes. *Math. Ann.*, 324(3):557–580, 2002.
3. A. Beilinson and V. Vologodsky. A DG guide to Voevodsky's motives. *Geom. Funct. Anal.*, 17(6):1709–1787, 2008.
4. A. A. Beĭlinson. The derived category of coherent sheaves on $\mathbf{P}^n$. *Selecta Math. Soviet.*, 3(3):233–237, 1983/84. Selected translations.
5. R. Bezrukavnikov. Perverse coherent sheaves (after Deligne). *math/0005152.*
6. R. Bezrukavnikov, I. Mirković, and D. Rumynin. Singular localization and intertwining functors for reductive Lie algebras in prime characteristic. *Nagoya Math. J.*, 184:1–55, 2006.
7. J. Block. Duality and equivalence of module categories in noncommutative geometry I. *math.QA/0509284.*
8. J. Block. Duality and equivalence of module categories in noncommutative geometry II: Mukai duality for holomorphic noncommutative tori. *math.QA/0604296.*
9. A. Bondal and D. Orlov. Reconstruction of a variety from the derived category and groups of autoequivalences. *Compositio Math.*, 125(3):327–344, 2001.
10. A. Bondal and M. van den Bergh. Generators and representability of functors in commutative and noncommutative geometry. *Mosc. Math. J.*, 3(1):1–36, 258, 2003.
11. A. I. Bondal and M. M. Kapranov. Representable functors, Serre functors, and reconstructions. *Izv. Akad. Nauk SSSR Ser. Mat.*, 53(6):1183–1205, 1337, 1989.

12. A. I. Bondal, M. Larsen, and V. A. Lunts. Grothendieck ring of pretriangulated categories. *Int. Math. Res. Not.*, (29):1461–1495, 2004.
13. A. I. Bondal and D. Orlov. Semiorthogonal decomposition for algebraic varieties. *alg-geom/9506012.*
14. P. Bressler, A. Gorokhovsky, R. Nest, and B. Tsygan. Chern character for twisted complexes. In *Geometry and dynamics of groups and spaces*, volume 265 of *Progr. Math.*, pages 309–324. Birkhäuser, Basel, 2008.
15. A. Connes. *Noncommutative geometry.* Academic Press Inc., San Diego, CA, 1994.
16. A. Connes. Trace formula in noncommutative geometry and the zeros of the Riemann zeta function. *Selecta Math. (N.S.)*, 5(1):29–106, 1999.
17. A. Connes. Noncommutative geometry and the Riemann zeta function. In *Mathematics: frontiers and perspectives*, pages 35–54. Amer. Math. Soc., Providence, RI, 2000.
18. A. Connes, M. Marcolli, and N. Ramachandran. KMS states and complex multiplication. *Selecta Math. (N.S.)*, 11(3-4):325–347, 2005.
19. K. J. Costello. Topological conformal field theories and Calabi-Yau categories. *Adv. Math.*, 210(1):165–214, 2007.
20. J. Denef and F. Loeser. Motivic integration and the Grothendieck group of pseudo-finite fields. In *Proceedings of the International Congress of Mathematicians, Vol. II (Beijing, 2002)*, pages 13–23, Beijing, 2002. Higher Ed. Press.
21. C. Deninger. Number theory and dynamical systems on foliated spaces. *Jahresber. Deutsch. Math.-Verein.*, 103(3):79–100, 2001.
22. C. Deninger. Two-variable zeta functions and regularized products. *Doc. Math.*, (Extra Vol.):227–259 (electronic), 2003. Kazuya Kato's fiftieth birthday.
23. V. Drinfeld. DG quotients of DG categories. *J. Algebra*, 272(2):643–691, 2004.
24. D. Dugger and B. Shipley. *K*-theory and derived equivalences. *Duke Math. J.*, 124(3):587–617, 2004.
25. E. Frenkel. Lectures on the Langlands program and conformal field theory. In *Frontiers in number theory, physics, and geometry. II*, pages 387–533. Springer, Berlin, 2007.
26. E. M. Friedlander, A. Suslin, and V. Voevodsky. Introduction. In *Cycles, transfers, and motivic homology theories*, volume 143 of *Ann. of Math. Stud.*, pages 188–238. Princeton Univ. Press, Princeton, NJ, 2000.
27. P. Gabriel. Des catégories abéliennes. *Bull. Soc. Math. France 90, 323-448*, MR 38:1411, 1962.
28. V. Ginzburg. Lectures on Noncommutative Geometry. *math.AG/0506603.*
29. E. Ha and F. Paugam. Bost-Connes-Marcolli systems for Shimura varieties. I. Definitions and formal analytic properties. *IMRP Int. Math. Res. Pap.*, (5):237–286, 2005.
30. U. Jannsen. Motives, numerical equivalence, and semi-simplicity. *Invent. Math.*, 107(3):447–452, 1992.
31. M. Kapranov. The elliptic curve in the S-duality theory and Eisenstein series for Kac-Moody groups. *math.AG/0001005.*
32. M. Kapranov. Noncommutative geometry based on commutator expansions. *J. Reine Angew. Math.*, 505:73–118, 1998.
33. M. Kashiwara. The Riemann-Hilbert problem for holonomic systems. *Publ. Res. Inst. Math. Sci.*, 20(2):319–365, 1984.

34. B. Keller. Deriving DG categories. *Ann. Sci. École Norm. Sup. (4)*, 27(1):63–102, 1994.
35. B. Keller. Invariance and localization for cyclic homology of DG algebras. *J. Pure Appl. Algebra*, 123(1-3):223–273, 1998.
36. B. Keller. *A*-infinity algebras, modules and functor categories. In *Trends in representation theory of algebras and related topics*, volume 406 of *Contemp. Math.*, pages 67–93. Amer. Math. Soc., Providence, RI, 2006.
37. B. Keller. On differential graded categories. In *International Congress of Mathematicians. Vol. II*, pages 151–190. Eur. Math. Soc., Zürich, 2006.
38. B. Keller and I. Reiten. Cluster-tilted algebras are Gorenstein and stably Calabi-Yau. *math/0512471.*
39. M. Kontsevich. Notes on motives in finite characteristic. *math/0702206.*
40. M. Kontsevich and Y. Soibelman. Notes on A-infinity algebras, A-infinity categories and non-commutative geometry. I. *math.RA/0606241.*
41. M. Larsen and V. A. Lunts. Motivic measures and stable birational geometry. *Mosc. Math. J.*, 3(1):85–95, 259, 2003.
42. M. Larsen and V. A. Lunts. Rationality criteria for motivic zeta functions. *Compos. Math.*, 140(6):1537–1560, 2004.
43. O. A. Laudal. Noncommutative algebraic geometry. In *Proceedings of the International Conference on Algebraic Geometry and Singularities (Spanish) (Sevilla, 2001)*, volume 19, pages 509–580, 2003.
44. E. Looijenga. Motivic measures. *Astérisque*, (276):267–297, 2002. Séminaire Bourbaki, Vol. 1999/2000.
45. J. Lurie. Derived algebraic geometry II: Noncommutative algebra. *math/0702299.*
46. J. Lurie. Derived algebraic geometry III: Commutative algebra. *math/0703204.*
47. V. Lyubashenko and O. Manzyuk. A-infinity-bimodules and Serre A-infinity-functors. *math/0701165.*
48. Y. I. Manin. Correspondences, motifs and monoidal transformations. *Mat. Sb. (N.S.)*, 77 (119):475–507, 1968.
49. M. Marcolli. *Arithmetic noncommutative geometry*, volume 36 of *University Lecture Series*. American Mathematical Society, Providence, RI, 2005. With a foreword by Yu. I. Manin.
50. S. Mukai. Duality between $D(X)$ and $D(\hat{X})$ with its application to Picard sheaves. *Nagoya Math. J.*, 81:153–175, 1981.
51. D. Orlov. Derived categories of coherent sheaves on abelian varieties and equivalences between them. *Izv. Ross. Akad. Nauk Ser. Mat.*, 66(3):131–158, 2002.
52. J. Plazas. Arithmetic structures on noncommutative tori with real multiplication. *math.QA/0610127.*
53. A. Polishchuk and A. Schwarz. Categories of holomorphic vector bundles on noncommutative two-tori. *Comm. Math. Phys.*, 236(1):135–159, 2003.
54. B. Poonen. The Grothendieck ring of varieties is not a domain. *Math. Res. Lett.*, 9(4):493–497, 2002.
55. A. L. Rosenberg. The spectrum of abelian categories and reconstruction of schemes. In *Rings, Hopf algebras, and Brauer groups (Antwerp/Brussels, 1996)*, volume 197 of *Lecture Notes in Pure and Appl. Math.*, pages 257–274. Dekker, New York, 1998.
56. S. Schwede and B. Shipley. Stable model categories are categories of modules. *Topology*, 42(1):103–153, 2003.

57. G. Tabuada. Théorie homotopique des dg-categories. *arXiv:0710.4303.*
58. G. Tabuada. Invariants additifs de DG-catégories. *Int. Math. Res. Not.*, (53):3309–3339, 2005.
59. G. Tabuada. Une structure de catégorie de modèles de Quillen sur la catégorie des dg-catégories. *C. R. Math. Acad. Sci. Paris*, 340(1):15–19, 2005.
60. G. Tabuada. On the structure of Calabi-Yau categories with a cluster tilting subcategory. *Doc. Math.*, 12:193–213 (electronic), 2007.
61. G. Tabuada. Higher K-theory via universal invariants. *Duke Math. J.*, 145(1):121–206, 2008.
62. R. W. Thomason and T. Trobaugh. Higher algebraic K-theory of schemes and of derived categories. In *The Grothendieck Festschrift, Vol. III*, volume 88 of *Progr. Math.*, pages 247–435. Birkhäuser Boston, Boston, MA, 1990.
63. B. Toën. Lectures on dg-categories. *For the Sedano winter school on K-theory Sedano, January 2007.*
64. B. Toën. The homotopy theory of *dg*-categories and derived Morita theory. *Invent. Math.*, 167(3):615–667, 2007.
65. B. Toën and G. Vezzosi. From HAG to DAG: derived moduli stacks. In *Axiomatic, enriched and motivic homotopy theory*, volume 131 of *NATO Sci. Ser. II Math. Phys. Chem.*, pages 173–216. Kluwer Acad. Publ., Dordrecht, 2004.
66. B. Toën and G. Vezzosi. Homotopical algebraic geometry. I. Topos theory. *Adv. Math.*, 193(2):257–372, 2005.
67. M. van den Bergh. Blowing up of non-commutative smooth surfaces. *Mem. Amer. Math. Soc.*, 154(734):x+140, 2001.

Multiplicative Renormalization and Hopf Algebras

Walter D. van Suijlekom

Institute for Mathematics, Astrophysics and Particle Physics
Faculty of Science, Radboud Universiteit Nijmegen
Heyendaalseweg 135, 6525 ED Nijmegen, The Netherlands
waltervs@math.ru.nl

Summary. We derive the existence of Hopf subalgebras generated by Green's functions in the Hopf algebra of Feynman graphs of a quantum field theory. This means that the coproduct closes on these Green's functions. It allows us for example to derive Dyson's formulas in quantum electrodynamics relating the renormalized and bare proper functions via the renormalization constants and the analogous formulas for non-abelian gauge theories. In the latter case, we observe the crucial role played by Slavnov–Taylor identities.

AMS Classification Codes. 81T15,16W30, 81T18

1 Introduction

During the last decade much of the combinatorial structure of renormalization of perturbative quantum field theories has been understood in terms of Hopf algebras (starting with [10, 6]). Although this led to many insights in the process of renormalization one could argue that since the elements in the Hopf algebra are individual Feynman graphs, it is rather unphysical. Rather, one would like to describe the renormalization process on the level of the 1PI Green's functions. Especially for (non-abelian) gauge theories, the graph-by-graph approach of, for instance, the BPHZ-procedure is usually replaced by more powerful methods based on BRST-symmetry and the Zinn–Justin equation (and its far reaching generalization: the Batalin–Vilkovisky formalism). They all involve the 1PI Green's functions or even the full effective action that is generated by them.

The drawback of these latter methods is that they rely heavily on functional integrals and are therefore completely formal. The good thing about BPHZ-renormalization was that if one accepts the perturbative series of Green's function in terms of Feynman graphs as a starting point, the procedure

Ö. Ceyhan et al. (eds.), *Arithmetic and Geometry Around Quantization*,
Progress in Mathematics 279, DOI: 10.1007/978-0-8176-4831-2_10,

is completely rigorous. Of course, this allowed the procedure to be described by a mathematical structure such as a Hopf algebra.

In this chapter, we address the question whether we can prove some of the results on Green's functions starting with the Hopf algebra of Feynman graphs. We derive the existence of Hopf subalgebras generated by the 1PI Green's functions. We do this by showing that the coproduct takes a closed form on these Green's functions, thereby relying heavily on a formula that we have previously derived.

In [1] Hopf subalgebras were given for any connected graded Hopf algebra as solutions to Dyson-Schwinger equations. It turned out that there was a close relation with Hochschild cohomology. For quantum electrodynamics, certain Hopf subalgebras of planar binary tree expansions were considered in [2] (cf. also [3]). Via a noncommutative Hopf algebra of formal diffeomorphisms (see also [4]), the authors derived Dyson's formulas relating the renormalized and unrenormalized proper functions. The case of non-abelian gauge theories was discussed by Kreimer in [12, 11] where it was claimed that the existence of Hopf subalgebras follows from the validity of the Slavnov–Taylor identities *inside* the Hopf algebra of (QCD) Feynman graphs. We now fully prove this claim by applying a formula for the coproduct on Green's functions that we have derived before in [15]. In fact, that formula allowed us to prove compatibility of the Slavnov–Taylor identities with the Hopf algebra structure.

After recalling the preliminaries on Feynman graphs, Green's functions and some combinatorial factors, we state the key formula for the coproduct on Green's functions. We first consider a scalar field theory (ϕ^4) and derive the well-known relations between the renormalized and bare proper functions and the renormalization constants (Eq. (7) below).

Then, we consider quantum electrodynamics, for which we derive Dyson's formulas [7]:

$$\Gamma^\mu_{\text{ren}}(e) = Z_2 Z_3^{1/2} \Gamma^\mu(e_0), \qquad \Sigma_{\text{ren}}(e) = Z_2 \Sigma(e_0), \qquad \Pi_{\text{ren}}(e) = Z_3 \Pi(e_0),$$

with e_0 the bare electric charge and e the renormalized charge (this is Eq. (10) below).

Finally, we establish a Hopf subalgebra consisting of the Green's functions in a non-abelian gauge theory. Here the Slavnov–Taylor identities turn out to play a crucial role. Again, we have the well-known formulas for the renormalized and bare functions (Eq. (15) below).

Note that these formulas for the proper functions are not derived via the usual procedure of adding counterterms to the Lagrangian but follow from the Hopf algebraic structure in combination with the Feynman rules.

We have included an appendix with some basic notions on Hopf algebras.

2 Hopf algebra of Green's functions

We will prove the existence of a Hopf subalgebra in H generated by the three 1PI Green's functions relevant for renormalization of quantum field theories. In particular, we will consider Hopf subalgebras in the case of ϕ^4-theory, quantum electrodynamics (QED), and quantum chromodynamics (QCD). We start by briefly recalling the relevant definitions and results from [15] while referring the reader to that paper for more details.

2.1 Preliminaries

Our starting point is a renormalizable quantum field theory, given for instance by a Lagrangian $\mathcal{L}$. In perturbation theory, one usually associates to each term in the Lagrangian an edge or a vertex and starts to built Feynman diagrams from them. It is well-known that for the purpose of renormalization theory, it is enough to consider only one-particle irreducible (1PI) diagrams with external structure corresponding to each term in the Lagrangian. For example, in ϕ^4-theory, there is one vertex of valence 4 and one edge, and we consider only diagrams with 2 and 4 external edges. In general, we will consider sums over all 1PI diagrams with the same external structure and this defines the 1PI Green's functions

$$G^v = 1 + \sum_{\mathrm{res}(\Gamma)=v} \frac{\Gamma}{\mathrm{S}(\Gamma)}, \qquad G^e = 1 - \sum_{\mathrm{res}(\Gamma)=e} \frac{\Gamma}{\mathrm{S}(\Gamma)}.$$

with v a vertex and e, an edge. Here $\mathrm{res}(\Gamma)$ is the *residue* of Γ (i.e., the vertex/edge the graph Γ corresponds to after collapsing all internal points), and the symmetry factor $\mathrm{S}(\Gamma)$ is the order of the automorphism group of the graph. It is extended to disjoint unions of graphs by setting

$$\mathrm{S}(\Gamma \cup \Gamma') = (n(\Gamma, \Gamma') + 1)\, \mathrm{S}(\Gamma)\mathrm{S}(\Gamma'),$$

with $n(\Gamma, \Gamma')$ the number of connected components of Γ that are isomorphic to Γ'.

In [6], Connes and Kreimer defined a coproduct on Feynman diagrams. This encodes the procedure of renormalization in terms of a Hopf algebra (see the appendix for a quick overview of Hopf algebras). Let us briefly recall how the coproduct was defined. One considers the algebra H generated by Feynman diagrams (for some quantum field theory), on which a coproduct $\Delta : H \to H \otimes H$ is defined by

$$\Delta(\Gamma) = \Gamma \otimes 1 + 1 \otimes \Gamma + \sum_{\gamma \subsetneq \Gamma} \gamma \otimes \Gamma \tag{1}$$

where the sum is over all subdiagrams γ that are disjoint unions of 1PI diagrams. Moreover, a counit $\epsilon : H \to \mathbb{C}$ is defined as the algebra map that takes the value 1 on the identity and zero on any 1PI graph. The antipode $S : H \to H$ can be defined recursively by

$$S(\Gamma) = -\Gamma - \sum_{\gamma \subsetneq \Gamma} S(\gamma)\Gamma/\gamma.$$

It turned out that the BPHZ-procedure of recursively subtracting the divergent part of a Feynman amplitude $U(\Gamma)$ (for a given graph Γ) in order to give the renormalized amplitude $R(\Gamma)$ is given by a convolution product in the space of maps from H to some space of functions depending on the regularization parameter. Indeed, the Feynman amplitude U can be understood as such a map: $\Gamma \to U(\Gamma)$. The counterterms are given by the map C,

$$C(X) = \epsilon(X) - T\left[(C \otimes U)\big((\mathrm{id} \otimes (1-\epsilon))\Delta(X)\big)\right] \tag{2}$$

with T a map that projects onto the part of the amplitude that diverges when the regularization parameter goes to 0 (or infinity in the case of a cut-off). Crucial in proving that C is an algebra map is the following multiplicative property: $T(XY) = T(T(X)Y) + T(XT(Y)) - T(X)T(Y)$, which motivated the study of so-called Rota-Baxter algebras within the context of renormalization (see [8] and references therein). In the case of dimensional regularization, the regularizing parameter is the complex number z (working in $d - z$ dimensions) and T is the projection onto the pole part of the Laurent series in z which indeed satisfies the multiplicative property. The renormalized Feynman amplitude R is given as the convolution product:

$$R(X) = (C * U)(X) := (C \otimes U)\left(\Delta(X)\right).$$

Remark 1 *That this indeed encodes the BPHZ-procedure can be seen as follows. Let Γ be a 1PI graph. Then, with the coproduct given by* (1) *we obtain*

$$C(\Gamma) = -T\left[U(\Gamma) + \sum_{\gamma \subsetneq \Gamma} C(\gamma)U(\Gamma/\gamma)\right] = -T\left[\overline{R}(\Gamma)\right]$$
$$R(\Gamma) = \overline{R}(\Gamma) + C(\Gamma)$$

where $\overline{R}$ is the so-called prepared amplitude. See [5, Sect. 5.3.2] for more details on the BPHZ-procedure.

In the next sections, we would like to derive a closed form of the coproduct on Green's functions. We do this using a formula that we have derived in [15] and have shown to imply compatibility of the above coproduct with Ward identities in quantum electrodynamics and Slavnov–Taylor identities in non-abelian gauge theories. It turned out that the corresponding Hopf algebras can be consistently quotiented by these identities, giving Hopf algebras that have them 'built in.' From this, one can deduce that if the unrenormalized Feynman amplitudes satisfy the Ward or Slavnov–Taylor identities, then so do the renormalized ones as well as the counterterms.

Before stating the aforementioned formula, we introduce some notation. Let $L(\Gamma)$ denote the number of loops of Γ and $\Gamma \mid \gamma$ the number of ways to insert γ inside Γ. Explicitly, the latter is given by

$$\Gamma \mid \gamma = \prod_i n_{v_i}(\gamma)! \binom{V_i(\Gamma)}{n_{v_i}(\gamma)} \prod_j n_{e_j}(\gamma)! \binom{I_j(\Gamma) + n_{e_j}(\gamma) - 1}{n_{e_j}(\gamma)}. \tag{3}$$

Here $V_i(\Gamma)$ is the number of vertices in Γ of type i and $I_j(\Gamma)$ the number of internal edges in Γ of type j. Moreover, $n_r(\gamma)$ is the number of connected components of γ with residue r (r being a vertex or an edge). Indeed, then the binomial coefficients arise for each vertex v_i since we are choosing n_{v_i} out of V_i whereas for an edge e_j we choose n_{e_j} out of I_j *with repetition* because of multiple insertions of self-energy graphs on the same edge of Γ. See for more details [15], where we have also derived the key formula for the coproduct on the 1PI Green's functions:

$$\Delta(G^r) = \sum_\gamma \sum_{\mathrm{res}(\Gamma)=r} \frac{\Gamma \mid \gamma}{\mathrm{S}(\gamma)\mathrm{S}(\Gamma)} \gamma \otimes \Gamma. \tag{4}$$

The sum is over all γ which are disjoint unions of 1PI graphs, whereas Γ is 1PI with the indicated residue r. Note that this formula holds for the 1PI Green's functions for any quantum field theory, by simply allowing r to be vertices and edges of different types (photon, electron, gluon, etc...).

Example 2 *We illustrate the combinatorial factors introduced above with the following example. Consider the graph*

$\Gamma :=$ .

Then $\mathrm{res}(\Gamma) =$ *and* $\mathrm{S}(\Gamma) = 2$. *Moreover, for the number of insertion places, we have, for instance:*

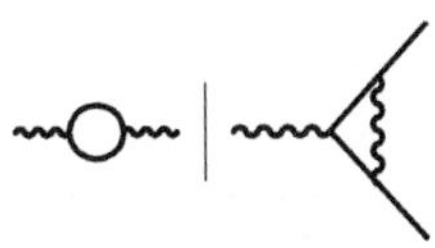 $= \binom{2}{1} = 2$

whereas

 $= 2!\binom{3}{2} = 6.$

2.2 Scalar field theory

As a warm-up for the next sections where we consider QED and QCD, we consider ϕ^4-theory. The Feynman diagrams are constructed from one type of vertex (of valence 4) and one type of edge. As mentioned above, we will consider only the 2- and 4-points Green's functions $G^{(2)} := G^{-}$ and $G^{(4)} := G^{\times}$.

We start by simplifying the expression $\Gamma \mid \gamma$ a bit. Recall that $n_v(\gamma)$ and $n_e(\gamma)$ denote the numbers that count the vertex and self-energy graphs in γ, respectively. From the presence of only one vertex in the theory, one easily obtains the two formulas $L = I - (V-1)$ and $4V = E + 2I$, relating the number of internal lines I, external lines E, and the number of vertices V of Γ to the loop number $L = L(\Gamma)$. Inserting the resulting expressions for I and V (in terms of L and E) into Eq. (3) yields

$$\Gamma \mid \gamma = \begin{cases} n_v(\gamma)!\binom{L}{n_v(\gamma)} n_e(\gamma)!\binom{2L+n_e(\gamma)-2}{n_e(\gamma)} & \text{if } \Gamma \text{ is a self-energy graph,} \\ n_v(\gamma)!\binom{L+1}{n_v(\gamma)} n_e(\gamma)!\binom{2L+n_e(\gamma)-1}{n_e(\gamma)} & \text{if } \Gamma \text{ is a vertex graph.} \end{cases}$$

Let us now consider the coproduct on the two Green's functions $G^{(2)}$ and $G^{(4)}$ by inserting these expressions in Eq. (4). Clearly, we can split the sum over γ into two parts: γ_V and γ_E containing only vertex and self-energy graphs respectively. This gives

$$\begin{aligned} \Delta(G^{(2)}) = \sum_{L=0}^{\infty} \sum_{\substack{E(\Gamma)=2 \\ L(\Gamma)=L}} & \left[\sum_{\gamma_V} n_V! \binom{L}{n_V} \frac{\gamma_V}{\mathrm{S}(\gamma_V)} \right] \\ & \times \left[\sum_{\gamma_E} n_E! \binom{2L+n_E-2}{n_E} \frac{\gamma_E}{\mathrm{S}(\gamma_E)} \right] \otimes \frac{\Gamma}{\mathrm{S}(\Gamma)}, \\ \Delta(G^{(4)}) = \sum_{L=0}^{\infty} \sum_{\substack{E(\Gamma)=4 \\ L(\Gamma)=L}} & \left[\sum_{\gamma_V} n_V! \binom{L+1}{n_V} \frac{\gamma_V}{\mathrm{S}(\gamma_V)} \right] \\ & \times \left[\sum_{\gamma_E} n_E! \binom{2L+n_E-1}{n_E} \frac{\gamma_E}{\mathrm{S}(\gamma_E)} \right] \otimes \frac{\Gamma}{\mathrm{S}(\Gamma)}, \end{aligned}$$

where we have used the shorthand notation $n_V := n_v(\gamma_V)$ and $n_E := n_e(\gamma_E)$. We will now evaluate each of the sums between square brackets. First, let us fix n_V and restrict the sum over γ_V to graphs consisting of n_V 1PI graphs. We use the notation $h^0(\gamma)$ for the number of connected components of a graph γ, in accordance with the usual notation for the Betti numbers. Then,

$$\sum_{h^0(\gamma_V)=n_V} n_V!\binom{L}{n_V}\frac{\gamma_V}{\mathrm{S}(\gamma_V)} =$$

$$\sum_{h^0(\gamma_V)=n_V}\left(\sum_{\substack{\gamma_v,\widetilde{\gamma}_V\\ \gamma_v\widetilde{\gamma}_V\simeq\gamma_V}}\frac{n(\widetilde{\gamma}_V,\gamma_v)+1}{n_V}\right)n_V!\binom{L}{n_V}\frac{\gamma_V}{\mathrm{S}(\gamma_V)},$$

where we have simply inserted 1. Indeed, for fixed γ_V we have

$$\sum_{\substack{\gamma_v,\widetilde{\gamma}_V\\ \gamma_v\widetilde{\gamma}_V\simeq\gamma_V}}\frac{n(\widetilde{\gamma}_V,\gamma_v)+1}{n_V}=\sum_{\gamma_v}\frac{n(\gamma_V,\gamma_v)}{n_V}=1.$$

A glance back at the definition of $\mathrm{S}(\gamma_v\widetilde{\gamma}_V)$ yields for the above sum

$$\sum_{\gamma_v}\frac{\gamma_v}{\mathrm{S}(\gamma_v)}\sum_{h^0(\widetilde{\gamma}_V)=n_V-1}(n_V-1)!\binom{L}{n_V}\frac{\widetilde{\gamma}_V}{\mathrm{S}(\widetilde{\gamma}_V)} =$$

$$(G^{(4)}-1)\sum_{h^0(\widetilde{\gamma}_V)=n_V-1}(n_V-1)!\binom{L}{n_V}\frac{\widetilde{\gamma}_V}{\mathrm{S}(\widetilde{\gamma}_V)}.$$

Iterating this argument n_V times and summing over n_V gives

$$\sum_{\gamma_V} n_V!\binom{L}{n_V}\frac{\gamma_V}{\mathrm{S}(\gamma_V)}=\sum_{n_V=0}^{\infty}\binom{L}{n_V}\left(G^{(4)}-1\right)^{n_V}=\left(G^{(4)}\right)^{L}$$

Similarly, we derive

$$\sum_{\gamma_E} n_E!\binom{2L+n_E-2}{n_E}\frac{\gamma_E}{\mathrm{S}(\gamma_E)}=\sum_{n_E=0}^{\infty}\binom{2L+n_E-2}{n_E}\left(1-G^{(2)}\right)^{n_E}$$
$$=\frac{1}{\left(G^{(2)}\right)^{2L-1}},$$

from which we obtain

$$\Delta\left(G^{(2)}\right)=\sum_{L=0}^{\infty}\frac{\left(G^{(4)}\right)^{L}}{\left(G^{(2)}\right)^{2L-1}}\otimes G_L^{(2)}. \tag{5}$$

In like manner, one can show

$$\Delta\left(G^{(4)}\right)=\sum_{L=0}^{\infty}\frac{\left(G^{(4)}\right)^{L+1}}{\left(G^{(2)}\right)^{2L}}\otimes G_L^{(4)}. \tag{6}$$

In other words, the coproduct closes on the proper 2- and 4-point functions in ϕ^4 and hence they form a Hopf subalgebra.

Renormalized amplitudes and counterterms

Via the Feynman rules one obtains the (unrenormalized) amplitudes — denoted $U(\Gamma)$ — for a graph Γ. By summing over all vertex and self-energy graphs and extending U by linearity, one defines the unrenormalized proper 2 and 4-point functions (in the presence of a regularization) by

$$\Gamma^{(n)}(\lambda) = U\left(G^{(n)}\right), \qquad n = 2, 4.$$

We have explicitly denoted the dependence on the coupling constant λ (as present in the original unrenormalized Lagrangian), but ignored for simplicity the momenta that are put on the external legs. The renormalized proper 2- and 4-point functions are given by

$$\Gamma^{(n)}_{\text{ren}}(\lambda) = R\left(G^{(n)}\right), \qquad n = 2, 4.$$

Finally, the renormalization constants Z_1 and Z_2 are defined by

$$Z_1 = C\left(G^{(4)}\right), \qquad Z_2 = C\left(G^{(2)}\right).$$

Recall that the maps R, C and U are related by the convolution product: $R = C * U$. In combination with Equation (5) this implies the following relation between the unrenormalized and renormalized proper functions and the counterterms:

$$\Gamma^{(2)}_{\text{ren}}(\lambda) = (C * U)\left(G^{(2)}\right) = \sum_{L=0}^{\infty} \frac{Z_1^L}{Z_2^{2L-1}} \Gamma^{(2)}_L(\lambda) = Z_2 \Gamma^{(2)}(\lambda_0),$$

where $\lambda_0 := \frac{Z_1 \lambda}{Z_2^2}$ is the bare coupling constant. Indeed, $\Gamma^{(2)}_L$ contains L powers of λ, one for each vertex. A similar computation can be done for the proper 4-point function, leading to the well-known relations (cf. for instance Equation (8-100) in [9])

$$\Gamma^{(n)}_{\text{ren}}(\lambda) = Z_2^{n/2} \Gamma^{(n)}(\lambda_0); \qquad (n = 2, 4). \tag{7}$$

2.3 Quantum electrodynamics

The three Green's functions that are of interest in renormalization of quantum electrodynamics are $G^{\sim\!<}, G^{-}$ and $G^{\sim}$ and correspond to the vertex, electron edge and photon edge. As in the previous subsection, we would like to establish that the coproduct has a closed form on these Green's functions. Let us start again by simplifying the expressions $\Gamma \mid \gamma$ that appear in Eq. (4). Since also in QED there is only one vertex, one can derive the following equalities for the number of vertices and electron and proton edges in a graph Γ at loop order L:

$$
\begin{aligned}
V &= 2L + E_e + E_p - 2;\\
I_e &= 2L + \tfrac{1}{2}E_e + E_p - 2,\\
I_p &= L + \tfrac{1}{2}E_e - 1.
\end{aligned}
\tag{8}
$$

If we insert these expressions in Eq. (3), then Eq. (4) implies

$$
\begin{aligned}
\Delta\left(G^{\sim\!<}\right) &= \sum_{L=0}^{\infty}\left[\sum_{\gamma_V} n_V!\binom{2L+1}{n_V}\frac{\gamma_V}{\mathrm{S}(\gamma_V)}\sum_{\gamma_E} n_E!\binom{2L+n_E-1}{n_E}\frac{\gamma_E}{\mathrm{S}(\gamma_E)}\right]\\
&\qquad\times\left[\sum_{\gamma_P} n_P!\binom{L+n_P-1}{n_P}\frac{\gamma_P}{\mathrm{S}(\gamma_P)}\right]\otimes G_L^{\sim\!<}\\
\Delta\left(G^{-}\right) &= \sum_{L=0}^{\infty}\left[\sum_{\gamma_V} n_V!\binom{2L}{n_V}\frac{\gamma_V}{\mathrm{S}(\gamma_V)}\sum_{\gamma_E} n_E!\binom{2L+n_E-2}{n_E}\frac{\gamma_E}{\mathrm{S}(\gamma_E)}\right]\\
&\qquad\times\left[\sum_{\gamma_P} n_P!\binom{L+n_P-1}{n_P}\frac{\gamma_P}{\mathrm{S}(\gamma_P)}\right]\otimes G_L^{-}\\
\Delta\left(G^{\sim}\right) &= \sum_{L=0}^{\infty}\left[\sum_{\gamma_V} n_V!\binom{2L}{n_V}\frac{\gamma_V}{\mathrm{S}(\gamma_V)}\sum_{\gamma_E} n_E!\binom{2L+n_E-1}{n_E}\frac{\gamma_E}{\mathrm{S}(\gamma_E)}\right]\\
&\qquad\times\left[\sum_{\gamma_P} n_P!\binom{L+n_P-2}{n_P}\frac{\gamma_P}{\mathrm{S}(\gamma_P)}\right]\otimes G_L^{\sim}.
\end{aligned}
$$

where $n_V := n_v(\gamma_V), n_E ;= n_e(\gamma_E)$ and $n_P := n_p(\gamma_P)$ for vertex, electron self-energy and vacuum polarization graphs.

A computation very similar to that of the previous section allows one to rewrite the terms in brackets as powers of $G^{\sim\!<}, G^{-}$ and $G^{\sim}$. Explicitly, we obtain for the three 1PI Green's functions:

$$
\begin{aligned}
\Delta\left(G^{\sim\!<}\right) &= \sum_{L=0}^{\infty}\frac{\left(G^{\sim\!<}\right)^{2L+1}}{\left(G^{-}\right)^{2L}\left(G^{\sim}\right)^{L}}\otimes G^{\sim\!<}\\
\Delta\left(G^{-}\right) &= \sum_{L=0}^{\infty}\frac{\left(G^{\sim\!<}\right)^{2L}}{\left(G^{-}\right)^{2L-1}\left(G^{\sim}\right)^{L}}\otimes G^{-}\\
\Delta\left(G^{\sim}\right) &= \sum_{L=0}^{\infty}\frac{\left(G^{\sim\!<}\right)^{2L}}{\left(G^{-}\right)^{2L}\left(G^{\sim}\right)^{L-1}}\otimes G^{\sim}
\end{aligned}
\tag{9}
$$

Hence, also in the case of quantum electrodynamics the coproduct closes on the 1PI Green's functions, thereby generating a Hopf subalgebra.

Remark 3 *From these formulas, the mentioned compatibility of the coproduct with the Ward identities* $G^{\sim\!<} = G^{-}$ *is now an easy consequence. Indeed,*

$$\Delta\left(G^{\prec} - G^{\text{—}}\right) = \sum_{L=0}^{\infty} \frac{\left(G^{\prec}\right)^{2L}}{\left(G^{\text{—}}\right)^{2L-1}\left(G^{\sim}\right)^{L}} \otimes \left[G_L^{\prec} - G_L^{\sim}\right]$$
$$+ \sum_{L=0}^{\infty} \left[G^{\prec} - G^{\sim}\right] \frac{\left(G^{\prec}\right)^{2L}}{\left(G^{\text{—}}\right)^{2L}\left(G^{\sim}\right)^{L}} \otimes G_L^{\prec},$$

from which it follows at once that the ideal I generated by $G_L^{\prec} - G_L^{\sim}$ ($L = 1, 2, \ldots$) is a Hopf ideal (see the appendix). Consequently, the Hopf algebra H can be quotiented by I to give again a Hopf algebra $\widetilde{H}$ which has the Ward identities built in.

Dyson's formula

In [7] Dyson derived formulas relating the unrenormalized and renormalized proper functions and counterterms for quantum electrodynamics; they are the analogues of Eq. (7) above. In this section, we will derive them using the above closed form of the coproduct on the 1PI Green's functions, while never referring to the Lagrangian.

As before, the Feynman rules give rise to amplitudes $U(\Gamma)$ for each QED Feynman diagram Γ. At the level of Green's functions, we define the unrenormalized proper vertex function, electron self-energy and vacuum polarization by the identities (adopting also the notation that is common in the physics literature):

$$\Gamma^{\mu}(e) = U\left(G^{\prec}\right), \qquad \Sigma(e) = U\left(G^{\text{—}}\right), \qquad \Pi^{\mu\nu}(e) = U\left(G^{\sim}\right).$$

We have explicitly indicated the dependence on the electric charge e, but ignored for simplicity the momenta that are put on the external legs. The renormalized proper functions $\Gamma^{\mu}_{\text{ren}}(e), \Sigma_{\text{ren}}(e)$ and $\Pi^{\mu\nu}_{\text{ren}}(e)$ are defined by replacing U by R in the above formulas. Finally, the three corresponding renormalization constants are defined by

$$Z_1 = C\left(G^{\prec}\right), \qquad Z_2 = C\left(G^{\text{—}}\right), \qquad Z_3 = C\left(G^{\sim}\right).$$

Dyson's formulas can now easily be derived by applying $R = C * U$ to $G^{\prec}, G^{\text{—}}$ and $G^{\sim}$ thereby using Eq. (9). Recall that the bare electric charge e_0 is related to e via the usual formula: $e_0 = \frac{Z_1 e}{Z_2 Z_3^{1/2}}$. A simple counting of the powers of e (i.e., the number of vertices) in the proper functions at loop order L then gives

$$\begin{aligned}
\Gamma^{\mu}_{\text{ren}}(e) &= \sum_{L=0}^{\infty} \frac{Z_1^{2L+1}}{Z_2^{2L} Z_3^{L}}\, \Gamma^{\mu}_{L}(e) = Z_2 Z_3^{1/2} \Gamma^{\mu}(e_0), \\
\Sigma_{\text{ren}}(e) &= \sum_{L=0}^{\infty} \frac{Z_1^{2L}}{Z_2^{2L-1} Z_3^{L}} \Sigma_L(e) = Z_2 \Sigma(e_0), \\
\Pi^{\mu\nu}_{\text{ren}}(e) &= \sum_{L=0}^{\infty} \frac{Z_1^{2L}}{Z_2^{2L} Z_3^{L-1}} \Pi^{\mu\nu}_{\mu\nu,L}(e) = Z_3 \Pi^{\mu\nu}(e_0).
\end{aligned} \tag{10}$$

Remark 4 *Let us come back once more to the Ward identities. Suppose we have chosen a regularization which respects the Ward identities, so that U satisfies them in the physical sense:*

$$\Gamma^{\mu} - \Sigma = 0.$$

Again we ignore the external momenta for the sake of simplicity; that this can be done was in fact shown in [14]. As a consequence, U vanishes on the ideal I (since it is generated by $G^{-\!<} - G^{-}$) and is thus defined on the quotient $\widetilde{H} = H/I$. Now, since I is a Hopf ideal (cf. Remark 3) it is shown in [6] that C is again a map of Hopf algebras so that C vanishes on I as well. Since $R = C * U$, we also have that $R(I) = 0$ so that both the renormalized proper functions as well as the counterterms satisfy the Ward identities, leading in particular to the well-known expression $Z_1 = Z_2$ [16].

2.4 Quantum chromodynamics

We next consider the case of a non-abelian gauge theory. In order to be as concrete as possible, we consider quantum chromodynamics. There are the quark, ghost and gluon propagators, denoted

respectively, and four vertices:

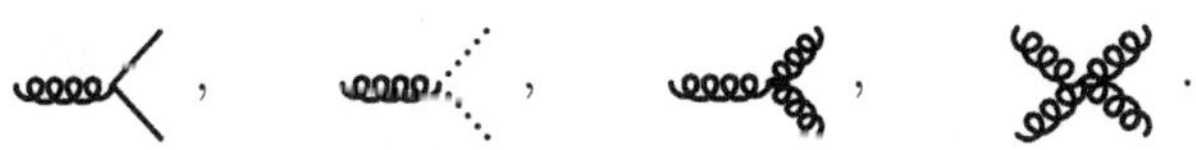

Corresponding to these edges and vertices, we define the following 7 1PI Green's functions:

$$G^{e} = 1 - \sum_{\Gamma^{e}} \frac{\Gamma}{S(\Gamma)}; \qquad G^{v} = 1 + \sum_{\Gamma^{v}} \frac{\Gamma}{S(\Gamma)}.$$

In [15] we have shown that the Slavnov–Taylor identities define an ideal in the Hopf algebra H of QCD Feynman graphs. More precisely, the coproduct is compatible with the following identities:

(11)

Hence, the quotient $\widetilde{H}$ of H by this ideal is still a Hopf algebra. In establishing a Hopf subalgebra for QCD, it is essential to work with $\widetilde{H}$ instead of H. Indeed, the Slavnov–Taylor identities are a crucial ingredient for a closed form of the coproduct on Green's functions.

Unfortunately, there is no simple expression for the number of insertion places $\Gamma \mid \gamma$ in QCD. This is due to the fact that there are many different vertices. Nevertheless, there are the following relations between the numbers of vertices, lines and loop number of a fixed graph Γ [15, Lemma 22]

$$I - V + 1 = L; \quad \text{(a)} \qquad V_{3F} = I_F + \tfrac{1}{2}E_F; \quad \text{(b)}$$
$$V_3 + 2V_4 - E + 2 = 2L; \quad \text{(c)} \qquad V_{3G} = I_G + \tfrac{1}{2}E_G. \quad \text{(d)}$$

The notation is as follows:

$$I = I_F + I_G + I_{YM} = \text{ number of internal quark, ghost and gluon lines}$$
$$E = E_F + E_G + E_{YM} = \text{ number of external quark, ghost and gluon lines}$$
$$V = V_3 + V_4$$
$$= V_{3F} + V_{3G} + V_{3YM} + V_4 = \text{ number of quark, ghost and gluon vertices}$$

We can use these expressions to simplify the coproduct on the Green's function corresponding to the vertex/edge r. Indeed, as in the previous sections, one can rewrite formula (4) as

$$\Delta(G^r) = \sum_{L=0}^{\infty} \sum_{\substack{\mathrm{res}(\Gamma)=r \\ L(\Gamma)=L}} \frac{\left(G^{-\!\!<}\right)^{V_{3F}} \left(G^{\cdot\!\!<}\right)^{V_{3G}} \left(G^{\sim\!\!<}\right)^{V_{3YM}} \left(G^{\times}\right)^{V_4}}{\left(G^{-}\right)^{I_F} \left(G^{\cdots}\right)^{I_G} \left(G^{\sim}\right)^{I_{YM}}} \otimes \frac{\Gamma}{S(\Gamma)}$$
$$= \left(G^{-}\right)^{\frac{1}{2}E_F} \left(G^{\cdots}\right)^{\frac{1}{2}E_G} \sum_{L=0}^{\infty} \sum_{\Gamma} \left[\frac{G^{-\!\!<}}{G^{-}}\right]^{V_{3F}} \left[\frac{G^{\cdot\!\!<}}{G^{\cdots}}\right]^{V_{3G}} \frac{\left(G^{\sim\!\!<}\right)^{V_{3YM}} \left(G^{\times}\right)^{V_4}}{\left(G^{\sim}\right)^{I_{YM}}} \otimes \frac{\Gamma}{S(\Gamma)} \tag{12}$$

where in going to the second line, we have applied the above equation (b) and (d). We have also understood the notation $E = E(\Gamma), E_F = E_F(\Gamma), \ldots$ We now insert the three Slavnov–Taylor identities in the following form:

$$\frac{G^{-\!\!<}}{G^{-}} = \frac{G^{\times}}{G^{\sim\!\!<}}; \qquad \frac{G^{\cdot\!\!<}}{G^{\cdots}} = \frac{G^{\times}}{G^{\sim\!\!<}}; \qquad \frac{\left(G^{\sim\!\!<}\right)^2}{G^{\sim}} = G^{\times},$$

and express everything in terms of the quartic gluon vertex function and gluon propagator. If we then apply the relations (a) and (c), we finally obtain

$$\Delta(G^r) = (G^{-})^{\frac{1}{2}E_F} (G^{\cdots})^{\frac{1}{2}E_G} (G^{\sim})^{\frac{1}{2}E_{YM}} \sum_{L=0}^{\infty} \left[\frac{\sqrt{G^{\times}}}{G^{\sim}}\right]^{2L+E-2} \otimes G^r_L. \tag{13}$$

Of course, the coefficients $E, E_F, \ldots$ are completely determined by the vertex/edge r; together with the factor $\frac{1}{2}$ they are precisely what one would expect from wave function renormalization. In the next subsection, we will see that the above equation allows us to derive the well-known relations between unrenormalized, renormalized amplitudes, and counterterms in QCD.

Remark 5 *The above argument also allows us to re-derive compatibility of the Slavnov–Taylor identities with the coproduct. In fact, the ideal I generated by the left hand sides of Eq.* (11) *defines a Hopf ideal. For this, observe that if we define X and Y by*

$$X = \frac{G^{\prec}}{G^{-}}; \qquad Y = \frac{G^{\times}}{G^{\prec}},$$

we can replace X^n (with $n = V_{3F}$ to lighten notation) in Eq. (12) *by Y^n after addition of $X^n - Y^n$. Now, by induction it follows that*

$$X^n - Y^n = (X - Y)\mathrm{Pol}(X, Y) \tag{14}$$

which is an element in I, and similar arguments apply to the other terms. Thus, at the cost of adding extra terms with elements in I on the first leg of the tensor product, one obtains the above formula (13). *When applied to the generators of I, one then easily obtains that $\Delta(I) \subset I \otimes H + H \otimes I$.*

Renormalized amplitudes and counterterms

Once again, the QCD Feynman rules induce a map U from H to the algebra of functions in the regularization parameter. We extend this map linearly and obtain the following self-energy functions:

$$\Sigma(g) = U\left(G^{-}\right), \quad \widetilde{\Pi}(g) = U\left(G^{\cdots}\right), \quad \Pi^{\mu\nu}(g) = U\left(G^{\sim}\right),$$

for the quark, ghost and gluon, respectively, as well as the four proper vertex functions:

$$\Gamma^{\mu}(g) = U\left(G^{\prec}\right), \quad G^{\mu}(g) = U\left(G^{\prec}\right),$$
$$\Gamma^{\mu\nu\sigma}(g) = U\left(G^{\prec}\right), \quad \Gamma^{\mu\nu\sigma\rho}(g) = U\left(G^{\times}\right).$$

We have adopted the notation of [13] and explicitly indicated the dependence on the strong coupling constant g. Again, due to the Slavnov–Taylor identities between the above self-energy and vertex functions, the map U vanishes on the ideal I and thus factorizes over I to give a map on $\widetilde{H}$. Moreover, the renormalized self-energy and proper vertex functions are obtained by adding a subscript 'ren' on the lhs and replacing U by R on the rhs of the above equations.

The renormalization constants are defined in terms of the counterterm map C of Eq. (2):

$Z_{1F} = C\big(G$  $\big)$, $\widetilde{Z}_1 = C\big(G$  $\big)$, $Z_{1YM} = C\big(G$  $\big)$,

$Z_{2F} = C\big(G$  $\big)$, $Z_{3YM} = C\big(G$  $\big)$, $\widetilde{Z}_3 = C\big(G$  $\big)$, $Z_5 = C\big(G$  $\big)$.

Since C is an algebra map from $\widetilde{H}$ to functions on the regularization parameter, it vanishes on the ideal I. Hence, we deduce the well-known *Slavnov–Taylor identities* between the renormalization constants (cf. for instance Eq. (III.59) in [13]):

$$\frac{Z_{3YM}}{Z_{1YM}} = \frac{\widetilde{Z}_3}{\widetilde{Z}_1} = \frac{Z_{2F}}{Z_{1F}} = \frac{Z_{1YM}}{Z_5}.$$

Let us now apply $R = C * U$ to Equation (13) to derive the well-known formula relating the renormalized and unrenormalized self-energy and vertex functions. First, recall the following formulas (cf. [13, Eq. (III.55)]) for the bare coupling constants

$$g_{0F} = Z_{1F} Z_{3YM}^{-1/2} Z_{2F}^{-1} g, \qquad \widetilde{g}_0 = \widetilde{Z}_1 \widetilde{Z}_5^{-1} Z_{3YM}^{-1/2} g,$$
$$g_{0YM} = Z_{1YM} Z_{3YM}^{-3/2} g, \qquad g_{05} = Z_5^{1/2} Z_{3YM}^{-1} g,$$

corresponding to the quark-gluon and ghost-gluon interaction and the cubic and quartic gluon self-interaction. We then obtain from Eq. (13)

$$\begin{aligned} \Sigma_{\text{ren}}(g) &= Z_{2F}\Sigma(g_0), \qquad \widetilde{\Pi}_{\text{ren}}(g) = \widetilde{Z}_3 \widetilde{\Pi}(g_0), \qquad \Pi^{\mu\nu}_{\text{ren}}(g) = Z_{3YM}\Pi^{\mu\nu}(g_0), \\ \Gamma^{\mu}_{\text{ren}}(g) &= Z_{2F} Z_{3YM}^{1/2} \Gamma^{\mu}(g_0), \qquad G^{\mu}_{\text{ren}}(g) = \widetilde{Z}_3 Z_{3YM}^{1/2} \Gamma^{\mu}(g_0), \\ \Gamma^{\mu\nu\sigma}_{\text{ren}}(g) &= Z_{3YM}^{3/2} \Gamma^{\mu\nu\sigma}(g_0), \qquad \Gamma^{\mu\nu\sigma\rho}_{\text{ren}}(g) = Z_{3ym}^{2} \Gamma^{\mu\nu\sigma\rho}(g_0). \end{aligned} \tag{15}$$

Here the argument g_0 on the rhs indicates that the regularized functions are computed using the Feynman rules involving the bare coupling constants $g_{0F}, \widetilde{g}_0, g_{0YM}$ and g_{05}. That the factors of $\sqrt{Z_5}/Z_{3YM}$ can indeed be absorbed in the bare coupling constants follows from the fact that due to the above Equation (c), the power of g that appear in the Green's function at loop order L is precisely $2L + E - 2$.

Acknowledgements

The author would like to thank ESI in Vienna for its hospitality during the Program 'Mathematical and Physical Aspects of Perturbative Approaches to Quantum Field Theory'. Alessandra Frabetti is thanked for pointing out the possibility of obtaining Dyson's formula from the closed formula of the coproduct on Green's functions.

Appendix: Hopf algebras

For convenience, let us briefly recall the definition of a (commutative) Hopf algebra. It is the dual object to a group and, in fact, there is a one-to-one correspondence between groups and commutative Hopf algebras.

Let G be a group with product, inverse, and identity element. We consider the algebra of representative functions $H = \mathcal{F}(G)$. This class of functions is such that $\mathcal{F}(G \times G) \simeq \mathcal{F}(G) \otimes \mathcal{F}(G)$. For instance, if G is a (complex) matrix group, then $\mathcal{F}(G)$ could be the algebra generated by the coordinate functions x_{ij} so that $x_{ij}(g) = g_{ij} \in \mathbb{C}$ are just the (i,j)'th entries of the matrix g.

Let us see what happens with the product, inverse, and identity of the group on the level of the algebra $H = \mathcal{F}(G)$. The multiplication of the group can be seen as a map $G \times G \to G$, given by $(g,h) \to gh$. Since dualization reverses arrows, this becomes a map $\Delta : H \to H \otimes H$ called the *coproduct* and given for $f \in H$ by

$$\Delta(f)(g,h) = f(gh).$$

The property of associativity on G becomes *coassociativity* on H:

$$(\Delta \otimes \mathrm{id}) \circ \Delta = (\mathrm{id} \otimes \Delta) \circ \Delta, \tag{A1}$$

stating simplfy that $f\big((gh)k\big) = f\big(g(hk)\big)$.

The unit $e \in G$ gives rise to a *counit*, as a map $\epsilon : H \to \mathbb{C}$, given by $\epsilon(f) = f(e)$ and the property $eg = ge = g$ becomes on the algebra level

$$(\mathrm{id} \otimes \epsilon) \circ \Delta = \mathrm{id} = (\epsilon \otimes \mathrm{id}) \circ \Delta, \tag{A2}$$

which reads explicitly $f(ge) = f(eg) = f(g)$.

The inverse map $g \mapsto g^{-1}$, becomes the *antipode* $S : H \to H$, defined by $S(f)(g) = f(g^{-1})$. The property $gg^{-1} = g^{-1}g = e$, becomes on the algebra level:

$$m(S \otimes \mathrm{id}) \circ \Delta = m(\mathrm{id} \otimes S) \circ \Delta = 1_H \epsilon, \tag{A3}$$

where $m : H \otimes H \to H$ denotes pointwise multiplication of functions in H.

From this example, we can now abstract the conditions that define a general Hopf algebra.

Definition 6 *A* Hopf algebra *H is an algebra H, together with two algebra maps $\Delta : H \otimes H \to H$ (coproduct), $\epsilon : H \to \mathbb{C}$ (counit), and a bijective $\mathbb{C}$-linear map $S : H \to H$ (antipode), such that equations* (A1)–(A3) *are satisfied.*

If the Hopf algebra H is commutative, we can conversely construct a (complex) group from it as follows. Consider the collection G of multiplicative linear maps from H to $\mathbb{C}$. We will show that G is a group. Indeed, we have the *convolution product* between two such maps ϕ, ψ defined as the dual of the coproduct: $(\phi * \psi)(X) = (\phi \otimes \psi)(\Delta(X))$ for $X \in H$. One can easily check that coassociativity of the coproduct (Eq. (A1)) implies associativity of the convolution product: $(\phi * \psi) * \chi = \phi * (\psi * \chi)$. Naturally, the counit defines the unit e by $e(X) = \epsilon(X)$. Clearly $e * \phi = \phi = \phi * e$ follows at once from Eq. (A2). Finally, the inverse is constructed from the antipode by setting $\phi^{-1}(X) = \phi(S(X))$ for which the relations $\phi^{-1} * \phi = \phi * \phi^{-1} = e$ follow directly from Equation (A3).

With the above explicit correspondence between groups and commutative Hopf algebras, one can translate practically all concepts in group theory to Hopf algebras. For instance, a subgroup $G' \subset G$ corresponds to a *Hopf ideal* $I \subset \mathcal{F}(G)$ in that $\mathcal{F}(G') \simeq \mathcal{F}(G)/I$ and viceversa. The conditions for being a subgroup can then be translated to give the following three conditions defining a Hopf ideal I in a commutative Hopf algebra H

$$\Delta(I) \subset I \otimes H + H \otimes I, \qquad \epsilon(I) = 0, \qquad S(I) \subset I.$$

References

1. C. Bergbauer and D. Kreimer. Hopf algebras in renormalization theory: Locality and Dyson-Schwinger equations from Hochschild cohomology. *IRMA Lect. Math. Theor. Phys.* 10 (2006) 133–164.
2. C. Brouder and A. Frabetti. Noncommutative renormalization for massless qed, hep-th/0011161.
3. C. Brouder and A. Frabetti. QED Hopf algebras on planar binary trees. *J. Algebra* 267 (2003) 298–322.
4. C. Brouder, A. Frabetti, and C. Krattenthaler. Non-commutative Hopf algebra of formal diffeomorphisms. *Adv. Math.* 200 (2006) 479–524.
5. J. Collins. *Renormalization.* Cambridge University Press, 1984.
6. A. Connes and D. Kreimer. Renormalization in quantum field theory and the Riemann–Hilbert problem. I: The Hopf algebra structure of graphs and the main theorem. *Comm. Math. Phys.* 210 (2000) 249–273.
7. F. J. Dyson. The S–matrix in quantum electrodynamics. *Phys. Rev.* 75 (1949) 1736–1755.
8. K. Ebrahimi-Fard and L. Guo. Rota-Baxter algebras in renormalization of perturbative quantum field theory. *Fields Inst. Commun.* 50 (2007) 47–105.
9. C. Itzykson and J. B. Zuber. *Quantum field theory.* McGraw-Hill International Book Co., New York, 1980. International Series in Pure and Applied Physics.
10. D. Kreimer. On the Hopf algebra structure of perturbative quantum field theories. *Adv. Theor. Math. Phys.* 2 (1998) 303–334.
11. D. Kreimer. Dyson–Schwinger equations: From Hopf algebras to number theory. *Fields Inst.Commun.* 50 (2007) 225–248.
12. D. Kreimer. Anatomy of a gauge theory. *Ann. Phys.* 321 (2006) 2757–2781.
13. P. Pascual and R. Tarrach. QCD: Renormalization for the practitioner. In *Lecture Notes in Physics*, volume 194. Springer-Verlag, Boston, 1984.
14. W. D. van Suijlekom. The Hopf algebra of Feynman graphs in QED. *Lett. Math. Phys.* 77 (2006) 265–281.
15. W. D. van Suijlekom. Renormalization of gauge fields: A Hopf algebra approach. *Commun. Math. Phys.* 276 (2007) 773–798.
16. J. C. Ward. An identity in quantum electrodynamics. *Phys. Rev.* 78 (1950) 182.

FSC
www.fsc.org
MIX
Papier | Fördert
gute Waldnutzung
FSC® C083411

Kolibri 360 GmbH
Ferdinand-Jühlke-Straße 7
99095 Erfurt, Deutschland
produktsicherheit@kolibri360.de